P9-CEX-865

PEOPLE OF THE WORLD

NATIONAL GEOGRAPHIC

PEOPLE OF THE WORLD

CULTURES AND TRADITIONS, ANCESTRY AND IDENTITY

CATHERINE HERBERT HOWELL
WITH K. DAVID HARRISON

NATIONAL GEOGRAPHIC

WASHINGTON, D.C.

CONTENTS

FOREWORD

BY SPENCER WELLS

As I write this, I'm lingering over a café au lait at the famous Café du Monde in New Orleans. Over the centuries, successive waves of humanity have descended on this city defined by a great river. Among the early peoples who lived here were Native Americans belonging to the southernmost extension of the Mississippian culture, with their massive earthwork mounds that can still be seen throughout the American South. French trappers, traders, and explorers arrived in the late 17th century, with the Spanish seizing control a century later. In 1803, Napoleon sold Louisiana to the United States in a deal engineered by Thomas Jefferson. New Orleans has since witnessed influxes of African slaves, Haitian refugees, and Irish, German, Italian, Lebanese, and other immigrants from Europe and the Middle East, as well as Latin Americans and Southeast Asians seeking a better life.

A human mélange such as the one in today's New Orleans is unique in this country to the past few centuries. We take for granted the cosmopolitan nature of a modern city such as this. But rewind the clock, and the world would have been a very different place. Five hundred years ago, cultural identity was more intimately linked with geographic location. Most people lived where their ancestors thousands of years back had lived, as far as they knew.

This book provides an overview of world cultures, from the indigenous and isolated to the cosmopolitan and syncretic. It is the culmination of more than 15 years of research, writing, and revision, having been published first in 2001, with a second edition in 2007. This period of time has coincided with the development of modern methods of genetic analysis that have helped scientists answer ancient questions about how humans are related to one another and how disparate cultures developed in such far-flung places around the globe. The National Geographic Genographic Project is perhaps the best known effort to map human history using genetics. The overarching theme that emerges from the genetic research is how similar we are—each culture connects to the others by the shared genetic threads of their ancestors. We can appreciate the diversity of the human world while understanding that we all share deeper similarities that tie us together in a human family.

Join us in the exploration of this extraordinary array of human cultural expression.

Opposite: Look-alike cousins wear traditional finery on the way to a wedding in Romania.
Pages 2–3: A sunset silhouettes youth in a cloud of dust kicked up by their dance in Kabrousse, Senegal, West Africa.
Page 4: A Buddhist monk poses in the doorway of the 12th-century Ta Phrom Temple in the Angkor region of Cambodia.

INTRODUCTION

BY CATHERINE HERBERT HOWELL

Currently, 7.3 billion of us live on Earth. We inhabit all seven continents. We have adapted to every kind of climate and terrain on the planet. We belong to more than 10,000 distinct ethnic groups and speak some 7,100 documented languages. We look different among ourselves, displaying a wide range of natural variation within a single species, and yet as a species, we all respond to the same needs to conduct and regulate our existence in ways that help us adapt to the exigencies of life.

We need to reproduce and to raise, educate, and protect our altricial young, which as primates have a long period of dependence. At the other end of our life span, we must look after our elders as they near their end. In between, we deal with the consequences to our daily lives of accident, disease, and aging. We cover our basic needs of food, water, and shelter. We make rules and laws that allow us to survive and—we hope—to thrive. The Ten Commandments cover basic topics that most cultures address in one way or another. We create traditions and conventions around coupling and reproduction. We afford special recognition to the dead, while struggling to comprehend the concepts of death and nonexistence. They and the other phenomena that confound us compel us to impose some kind of spiritual alignment on the unpredictability of life, giving rise to the cultural constellation of religion.

The Human Response

Through time and around the planet, humankind has covered life's challenges with an astonishing range and diversity of responses and adaptations. Small societies of hunters and gatherers, bands of nomadic herders, islands of intrepid fishers and seafarers, settlements of sedentary farmers, stakeholders in complex civilizations, and residents of modern, globally interconnected nations contribute to the shared traditions and understandings, polities and economies, languages and aesthetics that we recognize as human cultures.

Just as mindful travel widens our perception of the world and its people, *People of the World* affords the opportunity to learn about the many expressions of human life fashioned by the inhabitants of our planet. In this book, we celebrate the panoply of human cultures through time and space by a similarly diverse array of features. We introduce the emergence of *Homo sapiens* in Africa, the migrations of our species on globe-trekking journeys, and the milestones of human development. We present global responses to basic needs and aesthetic inclinations, and we illuminate a global sampling of cultures—in regional groupings—that allows readers to get a sense of the richness of human cultural experience. And because no culture exists in a vacuum, we describe the landscapes in which these ethnic groups live—the geography and ecology they call home.

This book also reminds us of what we have to learn from other cultures—that lifeways that seem simple and less advanced, or what anthropologists used to call "primitive," are anything but, and that humans as a species have been adapting and problem-solving from day one. Even in the 21st century, with technological advances surpassing and supplanting each other nonstop, we marvel at the accomplishments of traditional cultures: The Mongolian yak herder who sings to a mother yak and her nursing calf, simultaneously promoting lactation and enhancing the nursing-pair bond, a "scientific" outcome born from traditional wisdom. The Asian healer who uses the leaves and seeds of the neem tree *(Azadirachta indica)* to successfully treat cancerous skin lesions. The Polynesian navigator who consults the stars, planets, sea swells, and the extensive database of observations stored in his mind to travel with purpose across thousands of miles of ocean just as his ancestors did millennia ago.

Cultures in Peril

At this fraught juncture in world history, many cultures are facing annihilation. Some cultures change organically, just as cultures have changed and adapted since human life began. Some are steamrolled, swept up in changes not of their own making by powerful external forces. One powerful force is globalization, which creates artificially homogenized experiences at the expense of unique cultures. And some cultures change brutally, beaten down by disaster or violence.

As social anthropologist and former National Geographic Explorer-in-Residence (and contributor to this book) Wade Davis wrote in 2007:

> Anthropology suggests that when peoples are squeezed, extreme ideologies frequently emerge, inspired by strange and unexpected beliefs. The Shining Path in Peru, Pol Pot and homicidal mania of the Khmer Rouge in Cambodia, Osama bin Laden and al Qaeda are all examples of malevolent, atavistic movements that have sprung from conditions of chaos and disintegration, disenfranchisement and disaffection.
>
> Culture, in other words, matters. It provides the vital constraints of tradition and comfort that allow true civilization to exist.

To this I would add the unspeakable atrocities of ISIS/ISIL, staggering not only in terms of wanton human slaughter, but also of the wholesale destruction of the ancient cultural heritage of the Middle East. The barbarity that lurks somewhere in our nature is tempered for the most part by cultural stability. Cultures can be powerful and galvanizing forces. When directed toward noble goals, this is a good thing. But when cultural constructs are subverted and manipulated in a context of instability, the result can be inconceivably tragic.

Embracing the Ethnosphere

Each cultural loss is a diminution of the world's cumulative and intricately entwined web of human experience and expression we can usefully think of as the ethnosphere. Just as the biosphere represents the Earth's biological heritage, the ethnosphere embodies its human legacy. And just as the biosphere—and the planet's physical well-being and future—is under

Bedouin lead their camels across the Liwa Desert, part of the Rub al Khali, or Empty Quarter, of the Arabian Peninsula.

Young girls—and traditional bagpipers—await their star turns as participants in a wedding in Loch Achray, Scotland.

threat to an unprecedented degree, so too is the ethnosphere. If we consider just one component of human culture, language, we find that of the world's approximately 7,100 languages, about half seem destined to disappear by the end of this century. The scope of this potential loss dwarfs any anticipated losses we face in the biosphere.

While ethnic groups worldwide face uncertain futures, on the individual level people are more interested than ever in learning about their own ethnic identity and ancestry. Celebrity ancestry programs on TV also fuel the enthusiasm. Genealogy as a pastime has never been more popular or accessible. Old-school techniques for obtaining clues about forebears—corresponding with distant relations, poring through church registries, scrolling through rolls of brittle microfilm in dusty library basements—are still a productive pursuit. But for people yearning to capture their own identity and ancestry, much of what a genealogical newbie needs to get started is now available with a few clicks online. The Genealogy 101 feature of this book (see page 368) gives useful tips and a plan of action for how to begin your search.

The modern technology of DNA testing gives ancestry seekers additional tools that guide them further back toward their ancestral horizons than they ever imagined. When test results arrive, the individual becomes irrevocably linked to the ancestral past. The concept of "them" in the present suddenly becomes "us" in the past. And when these findings are put into perspective with the larger picture of human ancestry provided in the introduction to the Genographic Project that follows, we learn the ultimate ancestral truth: that we all are genetically linked. With that revelation, we need more than ever to truly understand the worldview of the millions and billions of other souls with whom we share the planet.

Human wisdom extends backward and forward in time and emanates from all quarters of the globe. No one culture or ethnic group, civilization or nation has a monopoly on understanding, intelligence, or innovation. As a species, we need every insight we can tap to address the challenges of the future.

ETHNIC GROUPS OF THE WORLD

How many ethnic groups are there in the world? A complete list of them all would have well over 10,000 entries, and perhaps many more. The exact number is unknown and ever changing. Ethnicity itself remains an elusive concept, and there is no solid, universally accepted definition of what constitutes an ethnic group.

In typical use, the term denotes a group of people who strongly identify themselves (or are identified by others, even against their will) as belonging together based on specific common traits they share. Such traits are largely involuntary: for example, skin color, clan or tribe membership, perceived or actual common ancestry, shared history, language, disability, or sexual orientation.

Other traits that group people into ethnicities may be chosen, abandoned, learned, or changed. These include culture, religion or sect, age, caste or social status, specific dialect, place or means of habitation, way of life, marriage into a group, immigration and naturalization—and on and on.

Language is perhaps the most salient and typical marker of ethnicity, but there are exceptions. The Baraguyu people of Kenya speak the Maasai language, but they consider themselves to be a separate ethnic group from the Maasai. Many speakers of Chinese cannot understand each other at all, yet they may feel that they all belong to the Han Chinese ethnic group.

The name people call themselves in their own language may be completely different from what outsiders or neighbors use: Albanians call themselves Shqiptarët, Germans say Deutsche. Naming preferences inside or outside a group may evolve over time. The label "Eskimo," which once applied to Inuit peoples of Canada and Greenland, is no longer accepted by them—but the Yupik people of Alaska, also Inuits, still prefer it. Groups may have multiple alternate names or alternate spellings of names depending on historical period or contemporary context.

As we move toward a deeper appreciation of human diversity, we embrace more flexible and less rigid notions of ethnicity. Many people have come to appreciate their mixed or multiethnic status as an asset, not a disadvantage, and resist being assigned to a single ethnic category. People all around the world can legitimately identify as a member of more than one ethnic group, depending on the context. For example, a Swahili-speaking Arab native to the island of Zanzibar may self-identify or be identified by others as being African, African American, Muslim, Ismaili, Swahili, Tanzanian, Arab, Zanzibari—or some combination of these affiliations.

Though we may be moving toward a more flexible notion of ethnicity, dark forces such as nationalism, ethnic chauvinism, ethnic cleansing, and terrorism work to reinforce rigid barriers that divide ethnicities. Ethnicity in itself is neither good nor bad. It can be a source of comfort, pride, and belonging. This book presents a revealing sampling of ethnic groups of the world. We encourage readers to use it as a jumping-off point for an exploration of the ethnosphere and its myriad expressions of human existence.

—*K. David Harrison*

Traffic swirls in downtown Ho Chi Minh City, home to nearly eight million Vietnamese.

THE GENOGRAPHIC PROJECT

BY SPENCER WELLS

Over the past 50 years, genetics research has changed our understanding of human origins. Early work on protein and blood group markers paved the way for discoveries such as "mitochondrial Eve," a name given to the most recent common ancestor of all living humans, and "Y-chromosome Adam," the most recent common ancestor of all men, firmly establishing the origin for all humans in Africa. As genetics defined its role in a field traditionally dominated by paleoanthropology and archaeology, initial skepticism within the anthropological community gave way to acceptance of its validity. Today, it would be difficult to discuss the origin and spread of our species without drawing on the genetic data.

By calculating patterns of genetic diversity, geneticists are beginning to understand both the age and the ancestry of groups living in different geographic regions. The data clearly show that African populations have a much greater diversity than non-African populations and that their mutations are ancestral to the genetic signals found in non-African groups. It is now widely accepted that humans originated in Africa, left that continent around 60,000 years ago, and proceeded to populate the rest of the planet. But many gaps remain in our knowledge, and we lack sufficient data to answer some questions fundamental to human evolutionary history, particularly events that have occurred over the past 15,000 years, during the period separating the end of the last ice age from recorded history.

Current efforts by geneticists to shed light on this period of human prehistory are facing increasing challenges. While the Internet, globalization, travel, and migration have brought the world closer together, that closeness and the pressures of rapid industrial development also threaten indigenous populations. Their numbers are dwindling, languages are dying out at alarming rates, and oral traditions and customs are dispersing as people worldwide increasingly enter the global melting pot. When indigenous peoples move to cities and lose touch with their traditional ways of life, the geographic and cultural context in which their genetic diversity arose is lost.

This process is happening with shocking speed. We may have only a single generation left to capture a genetic snapshot of the history of our species before much of it is lost forever. It may be that our best chance to work with indigenous communities to obtain valuable knowledge is now.

A New Archaeology

In 2005, the National Geographic Society and IBM launched the Genographic Project, an audacious research partnership. I served as the founding director of the project, whose stated mission is to retrace the earliest human migrations through the use of DNA donated from people around the world. Rather than gazing earthward in search of the stones and bones familiar to archaeologists and paleoanthropologists, the Genographic Project research team uses the DNA of people living today as a historical document to delve into our ancient past.

To trace recent migrations with scientific rigor, geneticists must pool data on the same set of markers from tens of thousands of individuals, representing a cross section of global genetic diversity. In the first phase of the project, completed in 2012 with funding from the Waitt Foundation, a consortium of 12 scientific centers around the world collaborated with indigenous populations in their respective regions. The work these centers has conducted has been guided by a core ethical framework that ensures the highest respect for the traditions of each participating indigenous community. Many in these communities consider the genetic information resulting from their participation in the Genographic Project as new pieces of the puzzle, potentially helpful in understanding their own histories. Participants around the world have overwhelmingly engaged in the project with great interest.

A Wider Net

Work with indigenous communities constitutes one aspect of the Genographic Project. Members of the general public can participate as well by purchasing a participation kit from the project's website *(genographic.nationalgeographic.com)*. Participants provide a cheek swab sample, learn about their own ancestors' migratory journeys, and track the overall progress of the research through the project's website. Proceeds from the sale of these kits are helping to fund both the field research and the Genographic Legacy Fund, which provides grants to indigenous communities for cultural conservation and revitalization efforts. The end goal is to leave a lasting legacy by empowering indigenous and traditional peoples on a local level while helping to raise global awareness of the challenges and pressures facing these communities.

In 2012, the project entered a new phase with the launch of the Geno 2.0 Public Participation kits and a completely new website. With the majority of the indigenous sampling complete, the focus of the project has now shifted to analyzing the data and expanding the scope for citizen science. The latter not only includes the active participation of members of the general public by swabbing their cheeks and sharing their data with scientific researchers but also extends further to encompass active data mining and analysis.

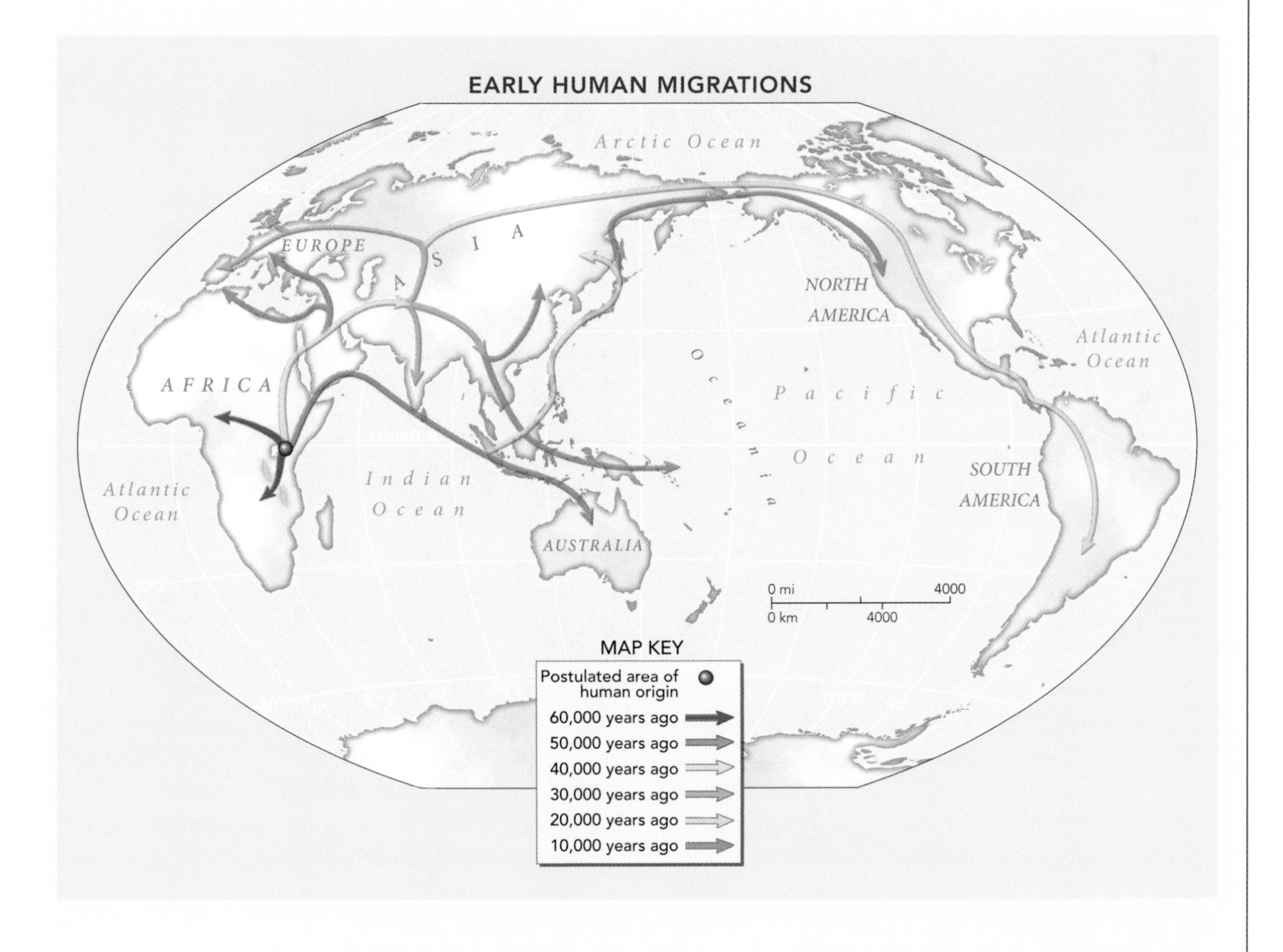

In Spain's ancient Basque region, a young soccer fan mourns a loss by his beloved Athletic Bilbao club.

These efforts are driven by the curiosity of the participants themselves, eager to discover more about their genetic results and place them in the context of the growing database, more than 720,000 strong as of mid-2015.

Genetic Breakthroughs

When the Genographic Project launched over a decade ago, the focus of the research was on the workhorses of genetic anthropology: mitochondrial DNA, tracing a purely maternal lineage, and its male equivalent, the Y-chromosome. Both have yielded important insights into our species' migratory history, but together they comprise less than 2 percent of the human genome. As the cost of examining genome-wide markers has declined and the technological power has increased, it has become clear that the Genographic Project needed to expand beyond mitochondria DNA and Y-chromosomes. Using genetic "chips" that scan hundreds of thousands of variable positions simultaneously, we can now detect subtle genetic patterns that are helping to fill in the gaps in our knowledge of human history.

Technology has also changed the way we collect data from ancient material. DNA, it turns out, is a remarkably stable molecule, often surviving for tens of thousands of years after the death of the individual carrying it. Recent technological advances have significantly improved our ability to obtain genetic data from ancient remains, allowing us to literally travel back in time to examine genetic patterns firsthand, at the time they were being created.

Among other things to emerge from this are new, previously unknown species of hominins (human ancestors), identified using only the genetic material. It is also revealing the genetic "turnover" associated with large-scale cultural shifts, as one group of people replaces another (and their DNA) in the archeological record. Ancient DNA will continue to be a critical component of our work in anthropological genetics in the future.

Ethnicity Versus Identity

The study of our origins raises the question of how we define ourselves. Our jobs, friends, family background, education, and hundreds of other factors go into creating this sense of self—what we can call an identity. One component of identity is one's ethnicity, the sum total of traditions, language, geography, and lineage that create a sense of a people, such as the Gujaratis or the Italians. Sometimes conflated with race, ethnicity is far broader, taking into account many cultural attributes that define geographic subregions. Genetic ancestry is simply one aspect of this, a reflection of a shared gene pool based on the migration patterns of our ancestors over the past 60,000 years. Since people living in a particular geographic location have historically tended to marry and reproduce with people in the same region, over time they have come to share more genetically with their neighbors than they do with people living elsewhere. These are subtle differences—humans around the globe are remarkably similar genetically—and they are detectable using the tools of modern genetics.

How have the past few centuries of global mass migration affected this? More than anything else, they have revealed how fluid categories such as ethnicity and identity truly are. In a melting-pot country like the United States, for instance, most people have ancestors from multiple ethnicities—German, Irish, Italian, Native American, Spanish, or perhaps West African—and genetic data can reveal how complex this tapestry truly is. But while genetics, and our knowledge of our ancestors' ethnicities, can show connections to populations around the world, our identities are forged in the here and now. Genetics and ethnicity affect who we are, but they alone don't define us.

A Wider Perspective

In this book on the ethnic identities of peoples around the world, the National Geographic Society's Genographic Project serves as a backdrop and an illuminating thread to help us understand the remarkable diversity described in its pages. Appearing throughout are essays, "Genographic Insights," that offer further evidence for the history of the human occupation of a wide range of geographic regions. They provide a perspective on the major migration routes by which humans departed from an African homeland and populated all the corners of our planet, in the process generating the dizzying patterns of cultural diversity we see today.

In addition, we have reached out to our public participants to share their stories, placing the genetic data into a more personal context in essays titled "Deep Ancestry." These personal stories of ancestry and identity show how our concepts of identity and ethnicity are still evolving, and the genetic data can reveal details that might be surprising.

These essays, and the work they reflect, show how other scientific disciplines—climatology, archaeology, linguistics, and paleoanthropology among them—combine with the genetic data, helping us construct a richer and more accurate understanding of our shared human past.

A Vietnamese farmer brings in the rice harvest with a centuries-old transport system.

ABOUT THIS BOOK

People of the World carries on a National Geographic Society mission reflected in two earlier editions to capture with respect and understanding the diversity of the world's myriad ethnic groups. This book expands that mission by presenting updated accounts and enhanced visual coverage of the included groups. We have also updated the findings of the Society's Genographic Project—now ten years into its path-breaking research—that connects the DNA of individual participants to the grand sweep of human migration and interaction over millennia.

This book describes in detail some 220 ethnic groups arranged and sequenced to reflect the migration of modern humans out of Africa, through time and into the world's regions. The cultures depicted here represent only a fraction of the world's more than 10,000 ethnic groups. But they form a well-considered sampling of humanity, ranging from little-known groups, some on the verge of extinction, to groups that might fairly be called nationalities—Egyptians, Japanese, Canadians, and Brazilians among them—with an understanding that for many, national identity is cultural identity.

Dozens more cultural groups are described in additional features and essays. "Across Cultures" essays deal with global expressions within a single component of culture, from language to music to shelter, presenting a variety of responses to addressing the same needs. "Genographic Insight" essays illuminate ancient migration and development in each world region, and "Deep Ancestry" profiles link contemporary individuals to their remote ancestral past.

Visually, *People of the World* offers a fresh array of photographs that depict each culture represented in the ethnic group entries. The photos capture moments large and small from daily life and ceremonial occasions.

New to this edition is an informative supplement that will help people interested in pursuing traditional genealogical inquiries. It also guides those who want to obtain DNA-based testing and helps them interpret their results.

It is hoped that the mindfully crafted content of *People of the World* will serve as an insightful vehicle for your continuing exploration of the human experience.

For each group featured in the book, we provide a statistical snapshot that includes current population statistics, the group's primary home location(s), and the language family to which the group's language belongs. Population figures represent homeland populations, not worldwide dispersals.

Within every chapter, we introduce each subregion with a summary of its geography and topography to make it easier to understand the physical settings in which human cultures operate. A complementary map helps readers chart the locations of the local ethnic groups.

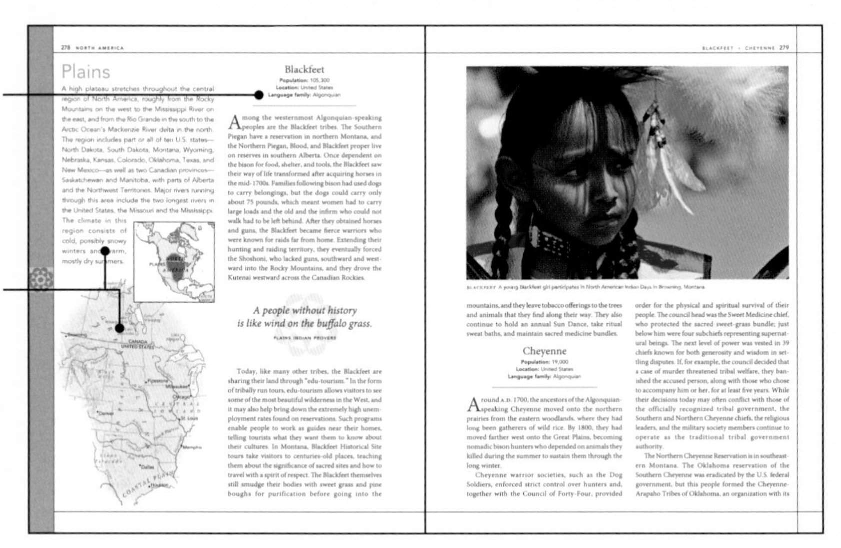

278 NORTH AMERICA

Plains

A high plateau stretches throughout the central region of North America, roughly from the Rocky Mountains on the west to the Mississippi River on the east, and from the Rio Grande in the south to the Arctic Ocean's Mackenzie River delta in the north. The region includes part or all of ten U.S. states—North Dakota, South Dakota, Montana, Wyoming, Nebraska, Kansas, Colorado, Oklahoma, Texas, and New Mexico—as well as two Canadian provinces—Saskatchewan and Manitoba, with parts of Alberta and the Northwest Territories. Major rivers running through this area include the two longest rivers in the United States, the Missouri and the Mississippi. The climate in this region consists of cold, possibly snowy winters and warm, mostly dry summers.

Blackfeet

Population: 105,300
Location: United States
Language family: Algonquian

Among the westernmost Algonquian-speaking peoples are the Blackfeet tribes. The Southern Piegan have a reservation in northern Montana, and the Northern Piegan, Blood, and Blackfeet proper live on reserves in southern Alberta. Once dependent on the bison for food, shelter, and tools, the Blackfeet saw their way of life transformed after acquiring horses in the mid-1700s. Families following bison had used dogs to carry belongings, but the dogs could carry only about 75 pounds, which meant women had to carry large loads and the old and the infirm who could not walk had to be left behind. After they obtained horses and guns, the Blackfeet became fierce warriors who were known for raids far from home. Extending their hunting and raiding territory, they eventually forced the Shoshoni, who lacked guns, southward and westward into the Rocky Mountains, and they drove the Kutenai westward across the Canadian Rockies.

A people without history is like wind on the buffalo grass.

PLAINS INDIAN PROVERB

Today, like many other tribes, the Blackfeet are sharing their land through "edu-tourism." In the form of tribally run tours, edu-tourism allows visitors to see some of the most beautiful wilderness in the West, and it may also help bring down the extremely high unemployment rates found on reservations. Such programs enable people to work as guides near their homes, telling tourists what they want them to know about their cultures. In Montana, Blackfeet Historical Site tours take visitors to centuries-old places, teaching them about the significance of sacred sites and how to travel with a spirit of respect. The Blackfeet themselves still smudge their bodies with sweet grass and pine boughs for purification before going into the

BLACKFEET · CHEYENNE 279

BLACKFEET A young Blackfeet girl participates in North American Indian Days in Browning, Montana.

mountains, and they leave tobacco offerings to the trees and animals that they find along their way. They also continue to hold an annual Sun Dance, take ritual sweat baths, and maintain sacred medicine bundles.

Cheyenne

Population: 19,000
Location: United States
Language family: Algonquian

Around A.D. 1700, the ancestors of the Algonquian-speaking Cheyenne moved onto the northern prairies from the eastern woodlands, where they had long been gatherers of wild rice. By 1800, they had moved farther west onto the Great Plains, becoming nomadic bison hunters who depended on animals they killed during the summer to sustain them through the long winter.

Cheyenne warrior societies, such as the Dog Soldiers, enforced strict control over hunters and, together with the Council of Forty-Four, provided order for the physical and spiritual survival of their people. The council head was the Sweet Medicine chief, who protected the sacred sweet-grass bundle; just below him were four subchiefs representing supernatural beings. The next level of power was vested in 39 chiefs known for both generosity and wisdom in settling disputes. If, for example, the council decided that a case of murder threatened tribal welfare, they banished the accused person, along with those who chose to accompany him or her, for at least five years. While their decisions today may often conflict with those of the officially recognized tribal government, the Southern and Northern Cheyenne chiefs, the religious leaders, and the military society members continue to operate as the traditional tribal government authority.

The Northern Cheyenne Reservation is in southeastern Montana. The Oklahoma reservation of the Southern Cheyenne was eradicated by the U.S. federal government, but this people formed the Cheyenne-Arapaho Tribes of Oklahoma, an organization with its

Unique to this edition of the book, each chapter contains one or more "Deep Ancestry" profiles. Each lays out the personal story of a participant in the National Geographic Society's Genographic Project and presents a surprising ancestral connection whose significance in the scheme of worldwide human migration is analyzed by Dr. Spencer Wells, the founding director of the Genographic Project.

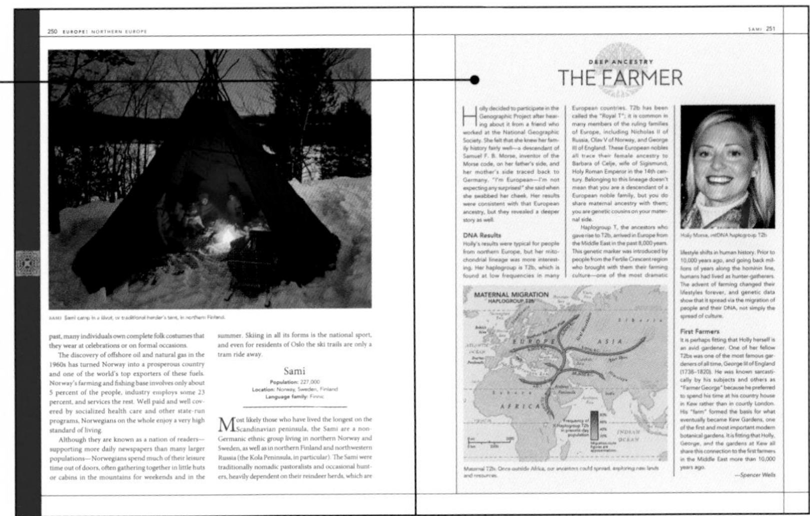

Every chapter also includes one or more "Genographic Insight." These discussions cover human migration into the region and the important milestones in early human development that took place there.

To help you hear the voices of the people of the world, we include translations of local proverbs. Through them, you can get a glimpse into the character of the people and their cultural idioms—their memes, if you will. We're certain you'll come to recognize that wisdom in one culture often serves as wisdom for all.

Each chapter of *People of the World* concludes with an "Across Cultures" essay. Written by experts in their respective fields, these compellingly illustrated pieces explore a worldwide range of solutions or responses to basic human needs, challenges—and aspirations.

150°
120°
90°
60°
30°
ARCTIC CIRCLE
60°
AMERICAN ARCTIC
NORTHWEST COAST
PLAINS
NORTHEAST
PLATEAU AND GREAT BASIN
30°
SOUTHWEST & FAR WEST
SOUTHEAST
TROPIC OF CANCER
NORTHER
WESTERN EUROPE
NOR
WEST
POLYNESIA
MESOAMERICA
CARIBBEAN
0°
EQUATOR
AMAZON AND ORINOCO BASINS
Pacific Ocean
Atlantic Ocean
ANDES
CENTRAL SOUTH AMERICA
TROPIC OF CAPRICORN
30°
0 mi
2000
0 km
2000
PATAGONIA
60°
ANTARCTIC CIRCLE
A

Arctic Ocean
ARCTIC CIRCLE
ASIAN NORTH
EUROPE
EASTERN EUROPE
CENTRAL EUROPE
CENTRAL ASIA
CAUCASUS
EAST ASIA
SOUTHERN EUROPE
MIDDLE EAST
SOUTH ASIA
Pacific Ocean
TROPIC OF CANCER
H AFRICA
SOUTHEAST ASIA
FRICA
CENTRAL AFRICA
EAST AFRICA
MICRONESIA
EQUATOR
MELANESIA
Indian Ocean
SOUTHERN AFRICA
AUSTRALIA
TROPIC OF CAPRICORN
Map Key
International boundary
Sub-Saharan Africa
North Africa and the Middle East
Asia
Oceania
Europe
North America
Central and South America
Note: Regional divisions in this book do not correspond to traditional geophysical continental boundaries.
ANTARCTIC CIRCLE
NTARCTICA
30° 60° 90° 120° 150°

SUB-SAHARAN AFRICA

SOUTHERN AFRICA
Afrikaners • San • Shona
Tswana • Zulu

WEST AFRICA
Asante • Bambara • Dogon • Fang • Fulani
Hausa • Igbo • Wolof • Yoruba

CENTRAL AFRICA
Efe • Kongo • Tutsi & Hutu

EAST AFRICA
Amhara • Baganda • Beta Israel • Dinka
Hadza • Kikuyu • Malagasy
Maasai • Somali • Swahili

Africa is the birthplace of humankind, an ancient land where our hominid, or humanlike, ancestors lived almost four million years ago. Extraordinarily diverse in its geography and cultures, this immense continent measures about 5,000 miles north to south and 4,600 miles east to west—from the Horn of Africa to the tip of Senegal—and over the millennia, it has given rise to a rich mix of peoples now inhabiting deserts and rain forests, mountains and valleys.

Even within similar geographic areas, one finds a wide range of social and political organizations. The Nuer of Sudan, for example, have no indigenous leadership roles, while the Zulu of southern Africa are known for their once heavily militarized kingdom. Societies such as the Asante and Yoruba in the west and the Baganda and Tutsi in the east maintained large states, while numerous others, like southern Africa's Shona and Tswana, composed smaller states or kingdoms. The Mbuti and Efe of central Africa and the San of the Kalahari Desert continue to live by hunting and gathering and have a fluid political organization without any recognized political authority.

LANGUAGE The languages spoken by Africa's many peoples can be grouped into a few general families, such as Bantu or Sudanic, but one group may include several different tongues. European languages are also spoken throughout Africa because of the long colonial period, which lasted until the 1960s. French is spoken in the former French and Belgian colonies of western and central Africa; English is heard in the west, east, and south; and Portuguese is spoken in the former Portuguese-occupied areas of Angola, Mozambique, and Guinea-Bissau.

Brightly clad women transport towering baskets up rock ledges in Mali.

Although African societies vary considerably in the ways they organize their families and social lives, some general patterns can be found. Marriage, for example, is usually considered a union of groups rather than individuals, and that union is symbolized and made legal by the transfer of bridewealth from the groom's family to the bride's. Many African groups permit polygamy, the marriage of a man to more than one woman. One area in Nigeria once practiced polyandry, the marriage of a woman to more than one man. Among the Lese of the Democratic Republic of the Congo and in some other societies of western and central Africa, an infertile wife may marry another woman. In a woman-to-woman marriage, the other wife reproduces with the husband, helping to make sure that he does not abandon the infertile wife for lack of children.

The vast majority of African societies reckon descent unilineally, through either the mother in matrilineal societies or the father in patrilineal societies. In both cases, however, the men have most of the authority. In matrilineal societies, descent traces through the mother's brother and is subject to his authority, not the mother's. A small number of African societies reckon descent in the same way that most Americans do: bilaterally, but with a bias toward the male side of the family.

Some African societies are well known for art that is both useful and aesthetic. Styles vary greatly, but the most common products are figurines used in sacrifice and worship, generally with heads oversized in relation to the body, and masks worn in ritual performances. Among the Dan of Liberia, masks hide the faces of secret-society members and protect the identity of judges who are settling disputes. The Akan of Ghana rely on wooden figures to facilitate childbirth. Africans also produce an enormous amount of "tourist art," much of it distributed in Europe and North America. Despite the name, this art is often of the same style and quality as art not created for export.

African people practice many religions, including Islam, Christianity, Judaism, and Hinduism. In western Africa, Islam now has numerous adherents. This is especially true among the Hausa of Nigeria and other societies that have engaged in trade with northern Africa, where Muslim Hamitic-speaking peoples have long been distinguished both culturally and religiously from the sub-Saharan societies.

The economies of Africa are also diverse, as disparate as hunting and gathering in the central African rain forest, diamond and gold mining in southern Africa, and large-scale farming in western Africa. Often the differences between the urban and rural areas are quite dramatic. Some families occupy high-rise apartments fitted with satellite dishes, and their members hold positions as engineers, teachers, doctors, lawyers, or software developers. But other Africans dwell in thatched-roof huts, with no plumbing, no electricity, and little direct access to cities. They may spend their lives as pastoralists, hunters, or farmers who grow crops of rice, yams, bananas, cassava, or coffee. Some Africans still follow nomadic lifestyles, taking only what they can carry with them and using shelters that they can quickly build and take apart.

Twins, like these Venda boys from South Africa, embody supernatural qualities in many African cultures.

In recent decades Africa has been besieged by many challenges, ranging from the health crises of the ebola virus and widespread HIV/AIDS to civil wars to terrorist operations. Great economic disparity and the corruption of many government officials add to the woes the continent has experienced. At the same time, however, people from all over Africa have begun to actively seek and promote better opportunities for themselves and future generations.

CULTURALLY INTERWOVEN Despite a tendency by scholars and others to study Africa as a collection of distinct cultures, the reality is that various groups have had long, extensive contacts with one another, with much cultural exchange. Neighboring societies may have borrowed elements of each other's rituals over the years, sometimes in a manner so subtle and inconspicuous that trying to determine the precise origins of a particular ritual becomes an impossible task. This rich cultural pluralism in Africa allows alternative beliefs and practices to coexist. The continent's so-called distinct cultures are, in fact, composed of many threads.

Southern Africa

Two rivers delineate this region in the southern quarter of the African continent. The Okavango originates in Angola and flows southeast, emptying not into a sea or ocean but into a broad delta in the center of the Kalahari Desert. The Zambezi, with headwaters in the Katanga Plateau, flows east to the Indian Ocean, swelling to a mile wide and plummeting more than 340 feet at Victoria Falls.

Namibia, Botswana, Zimbabwe, and Mozambique form the northerly stretch of this region; South Africa encompasses the tip of the continent at the Cape of Good Hope and is home to the region's larger cities.

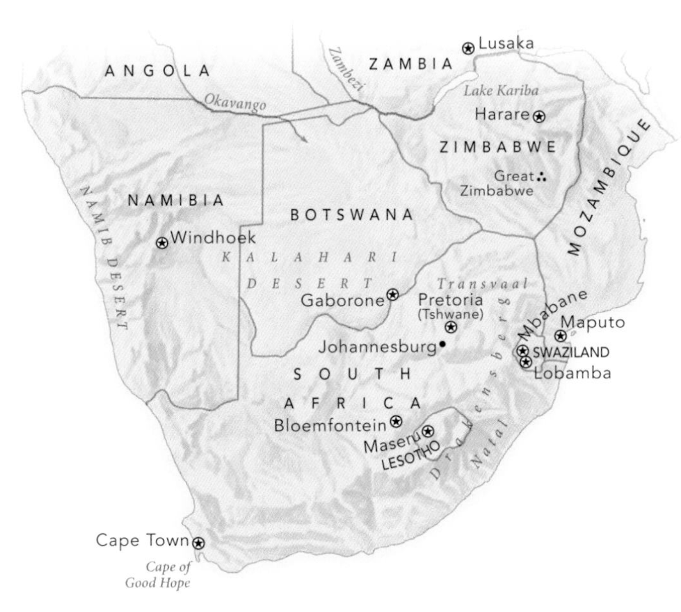

Afrikaners

Population: 3 million
Location: South Africa
Language family: Germanic

In 1652, Dutch farmers arrived at the Cape of Good Hope near the southern tip of Africa. They initially called themselves Boers—literally, "peasants" or "farmers" in Dutch—but in time they developed a distinctive South African identity and a new tongue related to Dutch with loanwords from English and local African languages. They began to see themselves as Africans, or Afrikaners, the term their Afrikaans-speaking descendants use today. Approximately 2.7 million Afrikaners now live in South Africa, which has a total population of more than 53 million.

The history of the Afrikaners is both heroic and tragic. On the one hand, their struggle for religious and economic freedom in a sometimes inhospitable land illustrates their endurance. On the other hand, their struggle entailed countless violent clashes with native Africans who had been dispossessed of their lands after the Boer arrival. The expansion of the Boers in South Africa was motivated in part by a deep-seated religious conviction that Africans, or Kaffirs—Afrikaans for "heathen"—were descended from the biblical character Ham and created by God to serve whites.

Working for the Dutch East India Company, the settlers of 1652 set up a trading post on the southwest coast of South Africa. But soon the company released many of the increasing number of workers to settle the land, grow food for its fleets, and expand eastward. In 1795, the Dutch East India Company ended its rule of the Cape colony, and the Boers found themselves in direct opposition to the British.

The British colonized all the Cape settlements, outlawed slavery, and gave native black Africans many civil rights previously denied them by the Boers. In 1835, feeling threatened by both Africans and the British, the Boers began their Great Trek into the Transvaal, where gold and diamonds would eventually be found. Neither the Boers (they called themselves Voortrekkers) nor any of the local African societies they encountered along the way would ever be the same. From that point on, the history of native South

Africans would be inextricably tied to the history of white expansion.

In 1899, President Paul Kruger of the South African Republic declared war on the British, who wanted to stop the rise of Afrikaner power and get control of the gold. Thus began the Boer War, a bloody conflict that the British won in 1902. The victors controlled South African politics until 1948, when Afrikaners won the legislature on a nationalist platform that advocated increased segregation between blacks and whites. During the Boer War, many Afrikaners had lost their land and become impoverished; they had also been forced to travel from their farms to the cities, where they competed with black South Africans for employment.

Once in command, the Afrikaner government designed a system of apartheid, or separateness, which divided the people racially, geographically, socially, economically, and politically. The people were grouped into four categories: White, Bantu (all-black Africans), Colored (people of mixed race), and Asian (Indians and Pakistanis). They were separate and unequal.

Beginning in the early 1990s, South Africans began to dismantle the apartheid system, but the legacy of differential access to economic and political resources remains. Many white South Africans, especially those who have descended from the Boers, have continued to farm and own land.

Today, the ethnic identity of Afrikaners is characterized by the Afrikaans language; their membership in the Dutch Reformed Church; their continued celebration of holidays remembering early historical struggles; and the realities of contemporary class and political relations within South Africa, particularly their relations with non-Afrikaner whites. Afrikaner identity continues to be defined by a long-standing, ambivalent relationship with South Africans of British descent.

AFRIKANERS Afrikaners attend a church service in South Africa. Most Afrikaners belong to the Dutch Reformed Church.

SAN A San hunter may track game for days across the red sands of the Kalahari in Botswana before achieving success.

San

Population: 100,000
Location: Namibia, Angola, Botswana, South Africa
Language family: Khoisan

After the Pygmies of central Africa, the San peoples of southern Africa are the best known hunter-gatherers in the world. Often referred to collectively and pejoratively as "Bushmen," they call themselves the *Zhun/twasi:* "the real people." They live in small camps composed of loosely organized kinship groups with no formal system of leadership.

The name *San* applies to a large group of people who have been living in remote desert regions of Botswana, Namibia, and Angola for perhaps thousands of years, but they are nonetheless ethnically and, to some extent, linguistically diverse. The San in the western Kalahari, for example, have a very different history from people in the desert's eastern part. Some San have lived closely with Bantu pastoralists for several hundred years and were much affected by the wars and population movements that followed the rise of the Zulu empire in the early 1800s; other San have remained relatively isolated from other ethnic groups.

In the European and American media, the San have been depicted as living in total isolation or, as they might have lived long before the advent of agriculture 10,000 years ago. Commercially successful films such as *The Gods Must Be Crazy* have perpetuated the false image of these people as living fossils. In fact, the San have interacted with neighboring societies for hundreds of years, and archaeological evidence strongly suggests that they were engaging in regional trade as far back as A.D. 500. Today, the San maintain relationships with cattle herders such as the Herero and the Tswana. They pay taxes, and they have even served in the South African and Namibian armies, despite the fact that the South African government has resettled San-speaking peoples on a number of occasions.

Compared with their neighbors, the San maintain a peaceful and egalitarian society in which sharing is highly valued but private property and individualism are disdained. They are known for their skills as hunters of small mammals, and they occasionally can bring down an elephant or a giraffe by using poisoned arrows. Once the arrow finds its mark, the hunter follows the wounded animal, sometimes for several days, until it collapses. This is one reason that the San have been enlisted by military forces as trackers. In a remarkably inhospitable environment, the San people can collect enough food and water to survive. But they must travel often to find resources, and sometimes the only available water that they can find is what they can squeeze out of deep roots.

> *Dance wins a wife but it does not keep the house.*
>
> DOGON PROVERB

The San also have a rich medical knowledge that they have incorporated into a series of rituals. During medicine trance rituals, for example, audiences dance and clap, and ill people who have come to be cured watch as healers go into a trance state. The healers may administer medicines derived from local plants or roots, and they may touch the patients to draw the sickness out of them and into their own bodies.

Shona

Population: 14 million
Location: Zimbabwe, Mozambique
Language family: Bantu

About 14 million Shona live in Zimbabwe and southern Mozambique, the majority of them in Zimbabwe. Like many other African ethnic groups, the Shona are actually a collection of peoples speaking mutually intelligible languages, such as Korekore, Ndau, and Karanga. The Shona language is among the most widely spoken in southern Africa.

Many contemporary archaeologists believe that Great Zimbabwe, the ruins of which make up one of Africa's most formidable archaeological sites, was built by ancestors of today's Shona. The rich artistic tradition of these people has continued into the 21st century, with intricate wooden headrests perhaps their best known creations.

The Shona people are primarily an agricultural group. Maize is their main crop, but they grow millet, sorghum, rice, peanuts, and a variety of tubers too. They also raise cattle, sheep, and chickens; the cattle work is taboo for women and therefore done exclusively by men. Women may supplement their income by selling utilitarian pottery and baskets.

In their traditional political organization, Shona people still live mostly in villages that typically accommodate one or more interrelated families and consist of clustered mud-and-wattle huts, granaries, and common cattle kraals, or pens. Although villages have often been quite distant from each other, they were once part of central states with hereditary kings who made large-scale decisions for the territory, such as waging war. These policies were financed by substantial tributes from

SHONA Using a long paddle for stirring, a woman prepares *sadza*, a Zimbabwean staple made from cornmeal.

GENOGRAPHIC INSIGHT

OUR HUMAN BIRTHPLACE

We are intrigued with our own humanity. That this book is open before you is itself testament to the fact that we are fascinated with the sea of faces around us. Celebrating our differences is a wonderful part of being human. Learning new languages and traveling to exotic places are practically universal curiosities, uniting people in a way that underscores the fact that our own DNA is 99.9 percent identical to that of everybody else on this planet. But simply recognizing diversity falls short of answering the fundamental questions motivating this global phenomenon. In short, what of that other 0.1 percent? After all, who hasn't asked at some point, "Where do we all come from, really? And if we're all so closely related, why do we look so different?"

In South Africa, a Zulu woman harvests sugarcane with a machete.

Looking for Home

Getting to the root of these differences boils down to questions of origin that have fueled centuries of scientific research in search of the putative homeland of our species. Archaeology and paleoanthropology have focused on the evolutionary emergence of modern humans from a number of ancestral species spread geographically and temporally throughout the fossil record. With each new finding, theories became more refined, and over time a clearer picture emerged, placing the origin of our species squarely in sub-Saharan Africa, a region now widely referred to as the "cradle of humanity."

In fact, southern Africa itself is home to a number of sites containing some of the earliest evidence of modern humans found anywhere in Africa. Border Cave, a rock shelter in the Lembombo Mountains between South Africa and Swaziland, has produced both anatomically modern human skeletal remains and advanced tool artifacts dated to 75,000 years ago. East of the Klasies River, a series of caves at the southernmost tip of South Africa show human habitation over the past 125,000 years and provide important clues into what life might have been like during the Middle Stone Age. Plant and animal remains show that these people knew how to seasonally hunt and gather, cook, and fish. Each new finding seems to produce a new theory of human evolution, and only recently have these remains been put to the test.

What Genes Begin to Tell Us

Geneticists help by studying DNA for signals of the past. Most of our genome is passed down in equal parts from our parents, with each half representing a shuffled combination of genetic material from their ancestors. The shuffling process, called recombination, is vital to maintaining the genetic diversity that enables species to adapt to changing environments and survive new diseases. But it makes life difficult for anthropological geneticists, because it creates a genetic mix of everyone who has come before. Fortunately, some segments of our genome—such as the maternally inherited mitochondrial DNA (mtDNA) and the paternally inherited Y chromosome—get passed down unshuffled from parent to child and vary only through occasional mutations. When inherited by the next generation, these mutations become markers of descent that geneticists can use to decipher human prehistory.

Studies of mtDNA have shown that the Khoi and the San, groups indigenous to southern Africa, are among the oldest on the continent. Matching the genetic patterns with the archaeological findings suggests that in times of drastic climate fluctuation, the ancestors of these groups did the smartest thing possible and headed to the coast. Offering abundant shellfish, the sea acted as a buffer against changing conditions; picking up camp and moving with the shoreline kept life relatively stable. Meanwhile, distant inland populations around the same time were probably struggling in East and Central Africa to survive on the changing savannas. Even in the evidence of the earliest humans, we see signs of the same ingenuity and adaptability that set in motion the technological innovations of today.

the surrounding settlements. Local decision-making was left to elders residing within each village, a policy that has changed somewhat because of the practice of centralized decision-making in the nation-states where today's Shona live.

In Shona traditional religion, the *vadzimu,* or ancestor spirits, play major roles. They represent the essence of the Shona way of life and embody the ideals and values of greatest importance, such as preserving and maintaining harmonious social networks. Vadzimu may be recently departed ancestors or heroes from the past. They offer protection if it is merited, but they will disappear if Shona values are not upheld.

Tswana

Population: 5.1 million
Location: South Africa, Botswana, Namibia, Zimbabwe
Language family: Bantu

The Tswana people constitute most of the population in Botswana—"land of the Tswana." Twice as many, however, live in northeastern South Africa, where they inhabit what was known as Bophuthatswana—"the place of gathering of the Tswana"—during the time of apartheid. Significant numbers of Tswana also live in neighboring Namibia to the west and Zimbabwe to the east.

TSWANA A Tswana mother and child take in the sunshine outside the conical thatched huts of their village.

A southern Bantu people, the Tswana are closely related to the Sotho of Lesotho and South Africa. Both groups speak mutually intelligible languages, have agricultural economies, share certain customs, and trace their descent from a common ancestor known as Mogale. The meaning of *Tswana,* essentially an invention of European colonists who encountered the group, is not known, but the name is believed to derive from a Bantu word. The people themselves have no common name; instead, they use the various names of their eight constituent subgroups.

During the second half of the 19th century, the Ngwato subgroup's Chief Khama III converted to Christianity while still a boy. He attended a missionary school and recognized the growing European interest in the area. By the 1880s, most Tswana were at least nominal Christians.

The human heart is not a bag into which one can plunge one's hand.

KONGO PROVERB

Traditionally, the Tswana have included ancestors along with the living as members of their society; they call the ancestor spirits dwelling among them and possessing metaphysical powers the *Badimo.* But European settlers did not understand the powerful role of social networks among the Tswana, including their ongoing relationships with ancestors.

With their Victorian notions of privacy and individualism, the Europeans did not comprehend the subtlety of Tswana life, where prestige and power are gained by obtaining followers and allies. In this way of life, a person who is not part of a larger group is nothing. Through a highly developed system of clientage, men gained rights over others in Tswana society, a connection that was once known as "eating" one another.

Since gaining independence from Britain in 1966, the Tswana of Botswana have maintained a stable, democratic government, one of the few African peoples to do so.

Zulu

Population: 11.9 million
Location: South Africa
Language family: Bantu

Perhaps best known for having one of the most powerful state and military organizations in 19th-century Africa, the Zulu remain a vital political and cultural force in South Africa.

Under Shaka (ca 1787–1828), its founder, the Zulu empire dominated sub-Saharan Africa. Shaka's military actions led to political consolidation elsewhere, with new kingdoms rising and stimulating large population movements into frontier territories. But behind the florescence of the Zulu and other African kingdoms, such as the Swazi and Sotho, were oppression and the ejection of many people from their homelands. And the political intrigue sometimes rivaled the best fiction: Shaka was assassinated by his half brother Dingane, who was deposed by Mpande, another brother.

Though dominant over other African societies, the Zulu eventually found themselves in conflict with the Boers, the Dutch colonists who escaped British oppression by trekking east to occupy Zululand. Dingane agreed to let the Boers into Zululand in early 1838 but then changed his mind and sent his men to murder dozens of men, women, and children. On December 16, 1838, the Boers retaliated against Dingane in a fight that became known as the Battle of Blood River. Thousands of Zulu were killed; not a single Boer soldier died.

The Zulu nation has long been defined as several hundred clans held together by their loyalty to the king, who owned all land. Men demonstrated their allegiance by fighting for the king's regiments and chiefs; in the time of Shaka, a man was called *isihlangu senkosi,* meaning "the war shield of the king." Today, the Zulu remain a stratified society that values political power. The Zulu chief plays an important role in South African national politics and civil rights. Now numbering about 11 million, the Zulu generally live in the Natal region of eastern South Africa and speak a language that has three distinctive tones and fifteen varieties of clicks. Many contemporary Zulu live in cities or near the gold mines where they work, but a large number live in the countryside, herding cattle and growing

vegetables. They are known for their beadwork, basketry, and circular houses.

Zulu men and women are easily recognized by their distinctive traditional dress. Men tie a beaded belt over a goatskin apron and wear certain clothing for rituals and other special occasions. The Zulu identify themselves as a warrior people, and their shield and spear designs distinguish different regiments. Zulu women make elaborate and multicolored beaded necklaces in symbolic colors. Blue, for example, means fidelity, and white means purity.

The cock with the very fine comb is the one we kill for a sacrifice.

WEST AFRICAN PROVERB

Zulu society is patrilineal. Descent is thus reckoned through the male line, and after marriage, women almost always move to the homes of their husbands. Marriages continue to be conducted with the exchange of *lobola*, or bridewealth—goods given by the groom's kin to the bride's kin. Although cattle are preferred, money and other goods are increasingly given. On the surface, the practice appears to be a sort of bride "purchase," but like other societies that have with bridewealth customs, the Zulu consider it more of an entitlement to domestic and sexual rights. Bridewealth also makes the marriage legal, and so women are able to hold their husbands responsible for their actions.

Believing in a single creator called Nkulunkulu, the Zulu look to a spiritual world inhabited by ancestors who intervene in everyday life. These ancestors are known as shades and appear as dreams and omens. They are not considered distant from the living, and they may even be called the living dead.

Many Zulu are Christian, but many follow the faith first revealed by Isaiah Shembe, a Zulu prophet who brought Christian and Zulu beliefs and practices together in a religion he called Zionism.

ZULU Zulu men don traditional 19th-century warrior garb for rituals, a stark contrast to the Western clothing they may wear daily.

West Africa

The Gold Coast, the Ivory Coast, the Grain Coast, the Slave Coast—the many names of commodity and abundance that have been attached to this region indicate its history of importance in world trade and economics. These lands once saw great empires built over centuries: the Mali Empire from the 13th century, the Songhay from the 15th. Great rivers, including the Gambia and the Niger, flow from mountains to shore, through forests, savannas, and arid plains to an often swampy coastline. Numerous small countries cluster in West Africa today, including Senegal, Gambia, Guinea-Bissau, Guinea, Sierra Leone, Liberia, Côte d'Ivoire, Ghana, Togo, Benin, as well as the larger surrounding countries of Mauritania, Mali, Niger, and Nigeria. The complex patchwork of national boundaries signals the region's long and complicated tribal, colonial, and political history.

Asante

Population: 2.8 million
Location: Ghana
Language family: Kwa

Nearly 2.8 million Asante dwell in southern Ghana. They speak the Asante language, which is part of the Akan cluster of Kwa languages, and they are perhaps best known for their colorful, geometrically designed kente cloth.

The mighty Asante kingdom emerged in the forests of central Ghana in the late 17th century after the union of several smaller states. At that time, the Golden Stool, symbol of Asante statehood, appeared. This sacred relic is said to have descended from the heavens in answer to the prayers of Okomfe Anokye, a legendary priest and adviser. The stool came to rest on the knees of Osei Tutu, who ascended to power as the first Ashantehene, or Asante, ruler. Treated with the reverence given to a king, the Golden Stool travels with a retinue and occupies its own palace. It has never been sat on.

In the early Asante economy, the people depended on the gold-and-slave trade that developed between them and the Manding and Hausa, other peoples of West Africa, as well between them and the Europeans who came to their shores. By acting as middlemen in the slave trade, the Asante received firearms, which increased their already dominant power in the region. Other luxury goods became symbols of status and political office. The Asante turned to the surrounding forest for another economically important item: kola nuts, sought after as gifts and used as a mild stimulant among the Muslim peoples to the north.

Today, the Asante are primarily farmers, growing cocoa for export and planting yams, plantains, and other produce for local consumption. They also include master stool makers and accomplished carvers, jewelers, metalsmiths, and weavers. Originally worn by rulers, the famous kente cloth is now donned by commoners and has become an important cultural symbol for many African Americans.

ASANTE Asante pottery is one of the many items for sale in the famous Kejetia Market in Kumasi, Ghana.

Most of the Asante reside in south-central Ghana's Asante region, which is centered around Kumasi, the country's second largest city, after Accra. Villages include the extended families of clans that descend from common female ancestors along the maternal lines. The Asante maintain customs and beliefs in the spirit world that all share, while some practice Christianity, Islam, and animism as well.

Bambara

Population: 4 million
Location: Mali, Guinea, Burkina Faso, Senegal
Language family: Mande

Mali's largest ethnic group, the Bambara live primarily in the western part of the country and along the Niger River. They speak Bambara, a language widely used throughout Mali, particularly in business and trade.

During the 1700s, there were two Bambara kingdoms: Segu and Karta. Muslim groups overthrew these kingdoms in the 1800s, and many Bambara remained wary of Islam until the arrival of the French as a colonial power. As resistance to French colonialism grew, so did acceptance of Islam, and today nearly 80 percent of Bambara peoples are Muslim.

The Bambara are farmers who produce large quantities of sorghum and groundnuts; their main crop is millet. They also grow maize, cassava, tobacco, and numerous other plants in private gardens. Environmental hardships such as drought often make farming difficult, so people may keep livestock to supplement their diet. Often they trust their neighbors, the Fulani herdsmen, to look after their domestic animals. This arrangement lets the Bambara focus on farming during the short rainy season from June to September. Both men and women share farming duties, although the division of labor allows women to leave the fields before the men to prepare meals for their families.

Bambara villages are collections of different family units, usually all from one extended family or lineage,

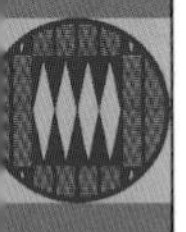

BAMBARA An elaborately turbaned young Bambara man pauses along the road in Mali.

and each *gwa,* or household, provides for all of its members and helps them with their farming duties. The members of each gwa work together every day except Monday, usually considered market day. Generally larger than the homes of other West African groups, Bambara dwellings may house 60 or more people.

For the Bambara, marriage carries heavy expenses, but most men are prepared to sacrifice a great deal because they view it as an investment. The return on that investment is the arrival of children, who then help with farming and contribute to the growth of a family's fortunes. Many Bambara women give birth to eight or more children. All adults marry, and it is not unusual for widows in their 70s or 80s to have suitors, for marriage enhances a person's reputation.

Although most Bambara are Muslims, old beliefs and rituals persist. In many areas, traditional male religious societies perform ceremonies to honor ancestors and promote the fertility of both the land and the people. Each society has its own distinctive masks, carved in wood and often depicting animals.

Dogon

Population: 600,000
Location: Mali, Burkina Faso
Language family: Volta-Congo

Now inhabiting the Bandiagara cliffs region of south-central Mali, as well as parts of Burkina Faso, the Dogon are thought to be descended from the first inhabitants of the Niger River Valley. From there, they probably migrated to Burkina Faso between the 13th and 15th centuries. Then, around 1490, they fled an onslaught of their territory by Mossi people and found a haven among the cliffs of southern Mali.

Contemporary populations speak a language also known as Dogon. They live in a stratified society, with each individual's status determined by his or her position within important subgroups, such as clan, village, male lineage, and age cohort. Each subgroup has a hierarchy based on age and rules of descent, meaning that individuals who enjoy high status in their families are the people most likely to occupy important positions in society.

The Dogon build their dwellings into and atop the dramatic cliffs of their homeland, as well as on the land below. Family homes are made of stone or adobe and are arranged around a courtyard open to domestic animals. Often set above the corners of this courtyard are thick-walled, fortress-like towers where millet and wheat are stored; such towers have also been used in defensive efforts against attackers.

In the woman's womb,
a child takes shape through the
ancient stories of the world.

DOGON PROVERB

Because the Dogon are an agricultural people, the land is of great importance in their lives and in their religious views and practices. For example, the Lebe cult is dedicated to agricultural renewal; its altars contain clumps of earth to encourage fertility. Perhaps the most significant agricultural rite is the *bulu,* performed before the rains come and the planting begins.

Dogon religion involves the worship of ancestors, as well as spirits the people encountered as they moved across the landscape of West Africa. An important institution is the Awa society, whose rituals enable the deceased to leave the world of the living for the world of the dead. Many ceremonies are public, and these include funerary rites known as *bago bundo*, with masks and dancing, and the *dama* ceremony to mark the end of mourning. The society's most important responsibility is the planning of the *sigui* ceremonies, held every 60 years when the star Sirius appears between two mountain peaks.

Rites and ceremonies are performed by mask-wearing men who personify supernatural beings and speak a special ritual language, *sigi so*. The Dogon's carved masks and wooden figurative art—usually in red, black, and white, with spirals and checkerboard patterns—draw admiration around the world.

Fang

Population: 3.3 million
Location: Gabon, Equatorial Guinea, Cameroon
Language family: Bantu

The Fang people have not occupied their rain forest territories for very long, having migrated from the northeast a couple of centuries ago. Although they were feared at first by Europeans who met them on the coast, they later traded ivory and forest products with them. They did not take part in the widespread slave trade.

The Fang grow crops such as manioc and plantains using swidden cultivation, an agricultural technique

DOGON In the highly stratified Dogon society, a revered elder displays the confidence of his status.

that creates temporary growing plots. They rotate crops to help preserve fertility and prevent erosion. Many Fang cultivate cash crops of cocoa and coffee, while others make a living by working in cities.

Outside of Africa, the Fang are perhaps best known for their wooden reliquary figures, stylized carvings meant to embody aesthetic and social values of opposition and vitality. Features in Fang art reflect the idea that through opposition will come strength, growth, and energy. From antagonism or difference comes new life; the simplest example of this concept is the joining of men and women to produce children.

The principle of dichotomies also extends to architecture. Entering a Fang village, one sees straight rows of houses flanking a central courtyard or plaza, creating a square of houses. Those who build opposite one another often say that "it is better to shout insults to your brother across the court than to whisper them to his ear as your neighbor." They believe conflict festers in secret and is best resolved in the open.

FANG A young Fang girl tends her fishing net in the rain forest of Equatorial Guinea.

Fulani

Population: 22 million
Location: Nigeria, Guinea, Senegal
Language family: Atlantic

One of the most wide-ranging African tribes, the Fulani are found from Senegal and Gambia on the western coast of Africa as far east as the Nile River. They speak many dialects of the Fulfulde language. Though some animist practices continue, the majority of Fulani are Muslim, and many Fulani men have become respected religious leaders and teachers.

Traditionally, the Fulani are pastoralists who practice transhumance: They move their herds seasonally from one area to another in search of pasturage and available water, away from diseases that affect cattle, and to the markets where they can transact business and learn important news. The pastoral Fulani exist for their cattle, doing everything they can to manage their herds successfully. In return, the cattle give them the dairy products on which they base their nutrition. Fulani migrate as households of one man with from one to four wives and numerous children. They travel light, with minimal household possessions—with one exception: the decorated calabashes, or gourds, that Fulani women tend to accumulate in numbers far greater than any domestic need.

Today's Fulani pursue economic strategies on the spectrum between nomadism and a totally sedentary lifestyle. Those who remain pastoralists are known as Cattle Fulani, while those who are assimilated into urban life and follow a profession such as religious leader, teacher, trader, or government worker are known as Town Fulani. Variations combine a connection to cattle—sometimes owning a herd managed by others—with settled life as a farmer or in a town. Some Town Fulani have allied themselves with the Hausa of Nigeria and now are totally sedentary (and nearly indistinguishable from the Hausa).

Fulani people have a reputation of being aloof and inscrutable, traits that have been linked to their values of reserve, patience, and gentleness. They view

FULANI The Fulani who still practice nomadic pastoralism migrate as households with all their possessions.

themselves as proper Muslims and look askance at those who are lax. This attitude drew the Fulani into jihads, or holy wars, in the 18th and 19th centuries and allowed them to achieve ethnic dominance in parts of West Africa. Today, clashes with farmers outside their group are common.

If the music changes, so does the dance.

HAUSA PROVERB

Fulani carry personal decoration to the height of elaboration for both men and women. As women do in many nomadic societies, Fulani women make and wear large amounts of jewelry as a way to transport and safeguard wealth. Their treasures may include beads of glass and bronze, silver, gold, and coins. In times of need, they can sell a piece of jewelry to raise cash.

Fulani clothing differs from group to group but includes embroidered and brightly colored and patterned tunics, skirts, pants, turbans, and head wraps. Many Fulani people practice body decoration in the form of multiple tattoos, colorful face powders and paints, and intricately braided hairstyles.

Fulani have strongly developed notions of male and female physical beauty, or *boodal,* that perhaps is carried out to the extreme among the Wodaabe, a nomadic Fulani group found mainly in Niger and northern Nigeria. At the annual *geerewole* ceremony, the men dress and make themselves up painstakingly to appear as a group in front of eligible young women. They jump, dance, and make facial contortions that they hope will catch the eye of a potential second wife, or perhaps just start a dalliance, as first wives usually are chosen in a more circumspect way among first cousins by the young man's family.

Hausa

Population: 25 million
Location: Nigeria, Niger
Language family: Chadic

The history of the Hausa, who inhabit northern Nigeria and southern Niger, is marked by the rise of powerful kingdoms built on lucrative trade markets. Living along caravan routes that reached to the Middle East, the Hausa became expert traders, even though the majority of the people were, and remain, settled farmers.

Hausa origin myths hold that the hero Bayajidda rescued the queen of Daura and her people from a powerful serpent. As a reward, the queen married Bayajidda and gave birth to seven sons, each of whom founded a Hausa kingdom. Allied politically, all of the kingdoms specialized in economic and military activities that were based on their locations and natural resources. These substates were known for indigo, slaves, markets, government, war, and so forth. Constant military vigilance was required to fend off such aggressive neighbors as the Akan and Songhay.

From at least the 11th century, the history of the region has been intimately tied to Islam and the Fulani ethnic group, which seized power from the Hausa in the early 1800s through a series of jihads. In many areas, intermarriage has since blurred the distinctions between Hausa and Fulani, and they are considered one group in contemporary Nigeria. Together, they form the largest ethnic group in the country and the most numerous Muslim group in sub-Saharan Africa.

Political leadership in the early Hausa states was based on ancestry, and individuals who could trace their relations back to Bayajidda were considered royal. When Islam arrived, many Hausa rulers embraced the religion and combined it with traditional ways. Rulers also welcomed praise singers, known as *maroka,* who were historians of lineage and were granted access to important personages.

As the British colonized the region in the early 20th century, the Hausa became part of Nigeria. Today, they

HAUSA Hausa tribesmen pick up their nets from Lake Chad on the border of Nigeria and Niger. Many Hausa also farm, and others excel in trade.

are still traders and farmers, well known for their production of indigo-dyed cloth. Hausa, a language related to Arabic, Berber, and Hebrew, is spoken throughout Africa as a language of trade.

Igbo

Population: 32 million
Location: Nigeria
Language family: Benue-Congo

In the corner of southeastern Nigeria divided by the Niger River and encompassing its delta, the Igbo (EE-boh) people have been a dominant presence for at least a thousand years, the age of their earliest surviving art. They may have come from an area a hundred miles to the north of the confluence of the Niger and Benue Rivers, from which they fanned out to the south. Today, some 18 million Igbo inhabit a region referred to as Igboland. Most contemporary Igbo people speak the Igbo language, and they also often speak English, Nigeria's official language, which has gained them access to educational opportunities since the British colonial period.

IGBO An Igbo man dons traditional garb—plus sunglasses—for a ceremonial occasion.

Unlike neighbors to the north such as the Yoruba and the Hausa, the Igbo did not establish kingdoms or hereditary chiefdoms. They traditionally live in villages composed of households formed mostly on the basis of patrilineal descent. Village size ranges anywhere from a few dozen members to several thousand. Igbo men often have several wives who share the same homestead but keep their own rooms, kitchens, and storage areas for their provisions.

Villages are governed by a council of elders, who include the village's oldest members and titled men. Igbo women form their own political organizations under the leadership of respected matriarchs, which sometimes rally together, as they did against the colonial British in the 1929 Women's War.

An Igbo man accrues status by obtaining a number of titles that indicate particular achievements. For example, the "killer of cows" has acquired enough cows to slaughter and then hold a large feast. Igbo men work hard to prove they are worthy of the titles, and some may spend quite a bit of money in this quest. Women can acquire titles independently and can also receive them by association from their titled husbands. Other villagers morally judge titled and important Igbo by whether they use their high positions to help their less successful kin.

Traditionally, the Igbo are farmers. They follow a division of labor by gender, with the men growing yams—culturally, their most important crop—and the women growing cassava, cocoyams, pumpkins, peppers, and their own varieties of yams. Igbo also cultivate and process the fruit of the palm to obtain products to use and also sell, such as palm oil, which is even exported to Europe.

All Igbo, particularly the women, participate actively in a local market economy. Women sell surplus crops along with mats that they weave and pottery that they make, sometimes with considerable economic success. Women also decorate their homes with painted geometric patterns and participate in the construction and decoration of *mbari,* elaborately decorated houses built in honor of gods that often incorporate life-size pottery

WOLOF Wolof women with their babies attend an adult literacy class in a Senegal village.

figures. Igbo men are renowned blacksmiths and wood-carvers, specializing in stylized painted masks, their designs inspired by the deities whose images artisans create to be worn in ritual dances.

During the British colonial period, most Igbo converted to Christianity. Today, they combine the Christian religion with their traditional beliefs, which revolve around a creator god, Chukwu, and a pantheon of lesser gods, goddesses, and spirits. Divination priests and priestesses discern the deities' wishes, so that appropriate actions can be taken to thwart misfortune and promote good consequences.

The period leading up to and following Nigerian independence was tumultuous for the Igbo. In 1967, the largely Igbo region of Nigeria's southeastern coast seceded from the rest of the country and became known as Biafra. During the ensuing civil war, many Igbo were killed, and mass starvation mobilized international relief efforts. Biafra became reincorporated into Nigeria in 1970, but regional instability led to a diaspora that took many Igbo to overseas destinations, including the United Kingdom and the United States.

Wolof

Population: 6 million
Location: Senegal
Language family: Atlantic

Neighbors to the Mandinka and the Fulani, the Wolof are the dominant ethnic group of coastal Senegal. Their history dates from about the 12th or 13th century and is preserved in the art of the griots, or oral praise singers.

After the fall of the Ghana Empire in the 11th century, the ancestors of today's Wolof no longer felt protected in their old territory and migrated westward from Mali to the Atlantic coast. By the 15th century, they had established a powerful presence in what is now the Wolof homeland. The Portuguese established a presence on the African continent around that time as well, building a fort on Gorée Island, off the coast of Dakar; that fort would be used in the Atlantic slave trade. Islam arrived in the 15th century, brought to Senegal by Mauritanian imams. Today, the majority of Wolof people are Muslims, and many of the men belong to Muslim brotherhoods, each with its own unique features.

The traditional Wolof political system involved the rule of powerful headmen from elite lineages. These headmen acted as a sort of parliament, electing a supreme ruler from a pool of candidates. Slavery was a part of Wolof society, and the people were themselves involved in the slave trade. In West Africa, there were two types of enslaved people: those born into a household as slaves and those who were either captured in war or purchased.

In today's world, the Wolof people are well known for their rich musical tradition. They have developed an enormously sophisticated body of percussion instruments and harps, the most famous of which may be the *tama,* or talking drum, so called because its resonance can be altered by pressing strings that attach the drumhead to the drum, resulting in changing pitches that sound like human speech. Also important is the *xalam,* a five-stringed guitarlike instrument. Some contemporary Wolof musicians—Youssou N'Dour and Baaba Maal, to name two of them—have established international reputations and made the sounds of Wolof music better known around the world.

The Wolof are known for their spicy stews too. Sorghum and millet are the staple crops, but tomatoes, peanuts, beans, and peppers are important fare as well. Fish and rice are also staples, and all of these ingredients find their way into the stews and other dishes that are characteristic of Wolof cooking.

Yoruba

Population: 38 million
Location: Nigeria, Benin, Togo
Language family: Benue-Congo

One of the largest groups in Africa, the Yoruba people live primarily in southwestern Nigeria and the neighboring countries of Benin and Togo. While some of the Yoruba make their homes in grasslands and forests, members of this group have also been city dwellers for centuries. They are among the most urbanized people of all those living on the African continent.

In the past, Yoruba city-states were governed by an *oba,* or king, and a supreme council. Ancient city-states frequently went to war, and in the mid-18th century, the Oyo kingdom became dominant, providing a cohesive identity among the Yoruba that continues to this day.

YORUBA A Yoruba woman of Nigeria carries a basket of food on her head, perhaps to market.

Religion has always been important in Yoruba society, with traditional beliefs still exercising a major influence on daily life, even though both Christianity and Islam are now widespread. Olorun is the high god, but he has no specific shrines. People consider him relatively inactive in human affairs and instead seek help from traditional deities known as *orisha,* which are said to number 401. The Yoruba believe that when they die, they enter the ancestors' realm, from which they continue to influence earthly affairs. In rare instances, they may even become orisha like Shango, the thunder god who was once a celebrated king. Yoruba spiritual beliefs have had an enormous influence on New World syncretic religions such as Cuba's Santeria, Trinidad's Orisha, and Brazil's Candomblé.

Along with their skills in wood carving, weaving, pottery making, and metalwork, the Yoruba are famous for something else: their high incidence of twinning. With 45 births in 1,000 producing twins, the rate is four times that of the United States or Great Britain. This biological phenomenon has yielded a special status for twins in the Yoruba cosmology; they are known as thunder children, because the first twins were believed to be the children of Shango. Twins are considered both a burden and a blessing: Twice as much trouble to raise, they also bring twice as much good fortune to parents who give them proper care.

DEEP ANCESTRY

THE ELDER

Charles Dorsey was unsure of what he would find when he tested his DNA through the Genographic Project. "My father died in an accident when I was eight years old, and I never learned his family history," he said. Although the family lore related that they were of mixed African and Native American ancestry, he knew very little for sure.

DNA Results

His results intrigued him. As with most other African Americans, his DNA showed a mix of sub-Saharan African and European ancestry. "I was surprised because I thought that my maternal ancestry had American Indian mixture—not European." But his biggest surprise came from his Y-chromosome result—his paternal line. He belonged to haplogroup A0, one of the deepest branches in the Y-chromosome tree and the one that helps to place the ancestors of all modern humans in Africa more than 140,000 years ago.

The haplogroups specifying the Y-chromosome and mitochondrial DNA trees are defined by markers that have accumulated as DNA is copied and passed down through the generations. Random changes, called mutations, arise spontaneously during the copying process, and over time, they have accumulated in particular combinations that define the haplogroups we see today. The A0 branch of the human family tree is one of the most divergent, occupying one side of the DNA tree, with almost every other haplogroup falling on the other side of the tree.

Ancestral Branch

In fact, until 2013, A0 was the most divergent branch known on the Y-chromosome tree, defining its root.

Charles Dorsey, Y-chromosome haplogroup A0

Interestingly, an African-American man who tested himself through the Genographic Project had a Y-chromosome that seemed to be ancestral to all of the other branches. Further analyses revealed a new, more ancient branch than A0. It was named A00 and was estimated to be around 270,000 years old—older than the earliest known modern humans. It was later found in the Mbo people of Cameroon as well, and some geneticists have suggested that it might be the remnant of an admixture event with another species of hominid—in much the same way that we see Neanderthal admixture outside Africa. Apart from this rare lineage, though, A0 is the oldest branch on the male family tree—the branch that shows us all the oldest part of our own ancestry.

Charles Dorsey summed up his test result: "Before taking part in the Genographic Project, there was no connection to the past. Now I can pass along information we did not know."

—*Spencer Wells*

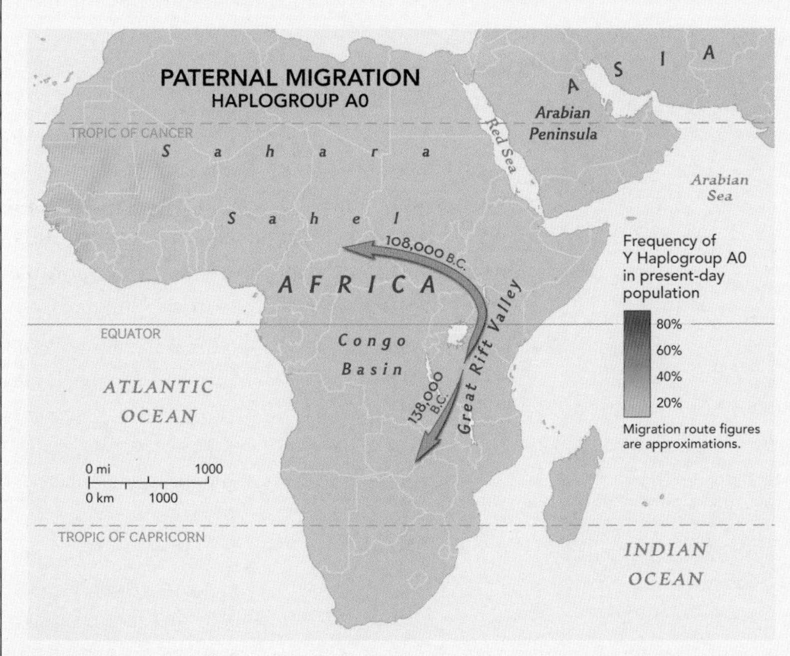

Paternal A0: Our earliest migrations were likely limited by the large deserts and mountains of Africa.

Central Africa

South of the Sahara, west of the Great Rift Valley, and intersected by the Equator, the region of Central Africa (also called Middle Africa) contains some of the densest tropical rain forest of the world. Through it flow the Congo River and its tributaries, a complex river system that drains about 1,425,000 square miles into the Atlantic. The nations that make up Central Africa are Chad, Sudan, South Sudan, Gabon, Cameroon, Angola, the Central African Republic, the Democratic Republic of the Congo, and the Republic of the Congo, often called simply the Congo.

Efe

Population: 20,000
Location: Democratic Republic of the Congo
Language family: Central Sudanic

The central African rain forest is home to the Efe, one of several hunter-gatherer groups known as Pygmies, a term increasingly viewed in the region as a pejorative. With a total population of nearly 200,000, these groups also include the Mbuti, the Sua, and the Aka.

Today, the Efe—like their immediate neighbors, the Mbuti—are almost exclusively hunter-gatherers. Some Efe men may reach a height of five feet, but the vast majority of men and women never exceed four feet, ten inches. Researchers believe that their shortness is an adaptation: Smaller individuals have more surface area in proportion to volume and can therefore dissipate body heat better in the hot rain forest.

The Efe live in camps of between 4 and 20 people, occupying quickly built huts for only a few months before moving on to new hunting territory. Unlike the Mbuti, who use nets, the Efe hunt exclusively with bows and arrows, usually for monkeys and small African antelopes known as duikers. But the Efe cannot survive by eating only what they find in the rain forest. For carbohydrates they depend on cassava, rice, peanuts, potatoes, and yams collected from their farmer neighbors, with whom they engage in long-term, hereditary exchange partnerships, conversing in mutually understandable dialects of the same language.

Ideally, a Lese farmer inherits as his partner the son of his father's Efe trading partner. The Efe give their partners meat, honey, and other forest foods in return for cultivated foods and metal. Farmers sometimes marry Efe women, but it is taboo for Efe men to marry farmer women. Hence, the farmer group has lost height over time while the Efe have remained the same.

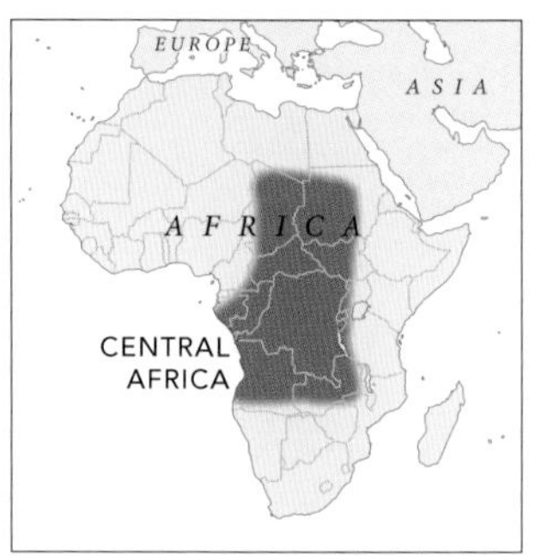

With a somewhat nomadic lifestyle, the Efe do not possess property or wealth that they cannot move.

EFE A hand mirror allows an Efe woman to paint her face precisely with vegetable dyes.

Artistically, they are perhaps best known for their bark cloth paintings with simple black geometric designs that are common to nearly all Pygmy groups.

Kongo

Population: 5 million
Location: Democratic Republic of the Congo, Congo
Language family: Bantu

More than five million Kongo live in the southwest part of the Democratic Republic of the Congo, in northwestern Angola, and in southern Congo. These people reckon descent through the maternal line, believing they descend from 12 female ancestors who assisted the founders of the ancient kingdom of Kongo.

In 1482, Portuguese explorers first encountered this kingdom. They saw a vibrant and complex political and economic organization of chiefs, governors, and judicial officers, with a strong military and a system for collecting taxes. Though the Portuguese had limited success in converting the people to Christianity, they still retained respect for them and later agreed to help when the Kongo were plundered by nearby Jaga warriors in the 16th century. Even so, the Kongo never fully recovered, and in 1885, they became colonial subjects of Belgium. Over the years, the Kongo have occasionally gained power in national politics, especially in 1960, when Joseph Kasavubu, a Kongo, became the first head of state of the newly independent Republic of the Congo.

The economy of the ancient kingdom depended on taxes, tribute, and proceeds from the slave trade, but now the Kongo subsist primarily through agriculture and fishing. They maintain a conventional division of labor: Women harvest fields of cassava, beans, and squash, and men tend orchards of palm oil trees.

Kongo live in cities and in the countryside, but in the Democratic Republic of the Congo, many live in independent religious communities, some arising from the religion of the prophet Simon Kimbangu (1889–1951), a Zulu Zionist who developed a syncretic religion based on Christianity and native beliefs. Kimbangu became so powerful that the Belgian colonial government imprisoned him for life, and he died a martyr. Kimbangu's church was the first in Africa to be admitted to the World Council of Churches.

KONGO Young Kongo dandies, known as *sapeurs,* pursue elegance. Here, they exhibit their flashy dress in a cemetery.

Tutsi & Hutu

Population: 23 million
Location: Rwanda, Burundi
Language family: Bantu

The Hutu and the Tutsi, along with the Twa, a Pygmy people, occupy the nations of Rwanda and Burundi. While they have lived side by side for several centuries, their relationship has been marked by hostility and occasional violence.

For a long time the Tutsi, the minority group, ruled the Hutu. According to tradition, the cattle-herding Tutsi came into the area in the 14th or 15th century and began to incorporate the Twa hunter-gatherers and the Hutu farmers into their society as subordinates. They established a hierarchical political system headed by a king, or *mwami,* considered divine. The German and Belgian colonists who took control of the area in the 19th and early 20th centuries reinforced the distinctions, calling the Tutsi "lords" and the Hutu "vassals."

The integration of these two peoples socially and economically was achieved primarily through a patron-client relationship. Since cattle were the key to social power, the dominant Tutsi needed to retain control over livestock, and they achieved this through a process called *ubuhake.* A Tutsi cattle owner or patron permitted his less-well-off client—a Hutu or perhaps another Tutsi—to use his cattle in return for labor and goods. Thus, most Hutu and Tutsi families have long been linked in some way, including by intermarriage. The Hutu and Tutsi cultures also became integrated through language, with Tutsis adopting mutually intelligible Bantu languages and religion: Tutsi and Hutu animism has largely given way to Christianity.

The colonial powers continued to perpetuate these socioeconomic distinctions, and the powerful Tutsi received greater access to European resources and education, fueling resentment among the Hutu. Following independence in 1962, two countries emerged from what had been Ruanda-Urundi, or Belgian East Africa. The Hutu seized power, and some of the worst violence of the 20th century ensued. Military campaigns on both sides resulted in the deaths of thousands.

In 1994, civil war erupted after Rwanda's and Burundi's Hutu presidents were killed when their helicopter was shot down. The retaliation left half a million Tutsis dead and created one of the largest groups of refugees in African history. The Tutsis took back power in Rwanda, and the country became enmeshed in complex regional politics. Many Hutu and Tutsi have stood trial before the United Nations for crimes of genocide, and many others have been judged in communal grassroots courts called *gacaca.*

TUTSI A Tutsi man performs a ritual dance in Rwanda, scene of civil warfare in recent decades.

GENOGRAPHIC INSIGHT

THE GENES OF OUR ANCESTORS

The National Museums of Kenya reopened their flagship Nairobi Museum in 2008 after two years of renovations. This East African country's permanent exhibit showcases arguably the strongest evidence of an evolutionary origin for our species. Displayed throughout the museum are the clearest links to our distant ancestors yet uncovered, a temporal continuum of hominid species stretching back four million years to the time when apes first stood upright. Headlining the collection is the 1.6-million-year-old Turkana Boy, the most complete hominid skeleton ever unearthed, a juvenile *Homo erectus* excavated less than 250 miles north of the capital city, in the heart of Africa's Rift Valley. These are Kenya's national treasures, and nowhere else on Earth can one peer through display glass and look directly into the hollow eyes of four million years of evolution.

Adorned with traditional beaded accessories, Maasai women gather for a ritual in Kenya.

Meet Mitochondrial Eve

It is quite a jump from the ancient bones of Turkana Boy to those of modern humans—and that jump represents a gap that paleoanthropologists have been hard-pressed to bridge. The remains of anatomically modern humans have been discovered across Africa and the Levant. They date from as early as 195,000 years ago and raise important questions—without offering answers—about both the location and timing of the dawn of our species.

In 1987, geneticist Rebecca Cann and colleagues took up the cause in a revolutionary study on the mitochondrial DNA of people from populations both African and non-African, looking at the diversity of genetic lineages around the world. What they found rang in a new era of molecular anthropology, showing that all humans today share a common ancestor, now famously known as "mitochondrial Eve," a hypothetical human being who lived in sub-Saharan Africa. Perhaps more important, because DNA changes at a relatively constant rate over time, Cann and her colleagues were able to use the diversity of genetic material as a sort of molecular clock, leading them to conclude that this woman lived around 200,000 years ago—a date that matches nicely with the earliest fossil evidence of our species.

Out of Africa

With an origin firmly planted in the sub-Sahara, geneticists turned their focus to the rest of the world and found the age of human lineages outside Africa to be far younger than those within it. If the entire world's genetic diversity was baked into a single pie, they discovered, non-African DNA would account for no more than a sliver. Dating the age of non-African lineages (again using that molecular clock), the findings indicate that from the dawn of our species, it took a very long time—more than 100,000 years—before our *Homo sapiens* ancestors finally left the African cradle in their successive waves of migrations.

But it if took so long, why did they finally succeed?

First Cultural Revolution

Just before the time of those first successful waves out of Africa, one each to Australia and the Middle East between 50,000 and 60,000 years ago, something peculiar started happening in East Africa.

Evidence for complex art and refined tools—projectile points, knife blades, bone artifacts—pops up all over sites dating to this period. The Late Stone Age, as it is known, signified a cultural revolution and a great leap forward for our species. Findings suggest that by this time, people were able to transmit complex ideas and work in large groups, newfound advantages that kicked off the exponential population boom that has continued ever since.

The exact location of our earliest African ancestors is still hotly debated, but it is clear from archaeological and genetic evidence that the first hunter-gathers to leave this sub-Saharan homeland likely did so from the same East African savannas that produced the earliest signs of human evolution.

East Africa

East Africa encompasses the lands sometimes called the Horn of Africa, bordering the Gulf of Aden; the great island of Madagascar and companion islands in the Indian Ocean; and the lands defined by the Great Rift Valley, a 3,000-mile-long north-south fault line caused by massive tectonic plate shifts some hundreds of thousands, perhaps a million, years ago. Some of the most complete remains of hominid ancestors have been found in this terrain. The geographical highlights of East Africa include Mounts Kilimanjaro and Kenya, Africa's highest peaks, and Lake Victoria, Africa's largest lake (and the world's second largest freshwater lake) and the source for the White Nile, the largest branch of the Nile River, which flows northward through this region and eventually empties into the Mediterranean Sea.

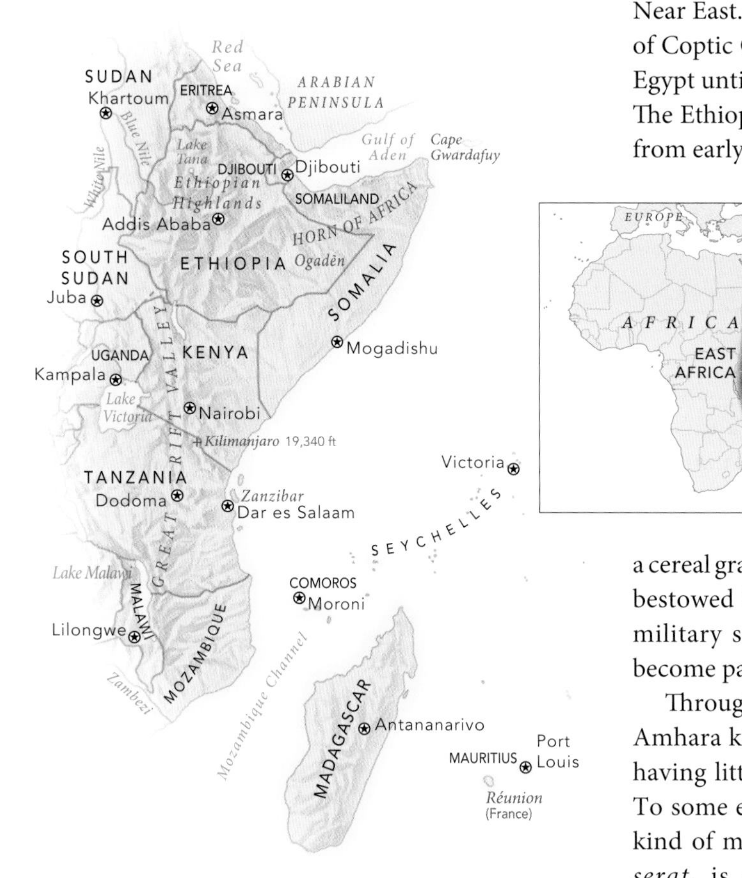

Amhara

Population: 21 million
Location: Ethiopia
Language family: Semitic

The Amhara people live in the highlands of Ethiopia. For most of their history, they have been farmers and part-time pastoralists, but they have also been the most powerful ethnic group in the region. Amhara kings fought for power in Ethiopia throughout the 18th and 19th centuries, and in 1889, Menelik II, an Amhara, declared himself emperor of an area nearly equal in size to today's country. Except for its occupation by Italy from 1935 to 1941, when Emperor Haile Selassie went into exile, Ethiopia escaped European domination in the 20th century—the only African nation to do so.

Like many other residents of the highlands, the Amhara maintain religious traditions that derive from extensive and early contacts between Ethiopia and the Near East. Some groups practice Islam, others a form of Coptic Christianity, which was widely practiced in Egypt until the Arab conquest in the seventh century. The Ethiopian Orthodox Church eventually emerged from early Greek and Coptic influences.

According to linguists, Amharic belongs to the Afro-Asiatic family of about 300 languages spoken mostly in the Middle East, the Horn of Africa, and North Africa. Within the family, Amharic is most closely related to Hebrew and Arabic.

The Amhara economy is based primarily on cultivation of maize, wheat, sorghum, and a cereal grass called teff. In the past, landownership was bestowed by the emperor in thanks for loyalty and military service, and new landowners would then become patrons to client farmers.

Throughout history, the social organization of the Amhara kingdom has been patriarchal, with women having little authority or potential to own property. To some extent, the status of a woman is tied to the kind of marriage she has. The most common form, *serat,* is a civil union not bound by religious

AMHARA Guiding yoked oxen, an Amhara man plows his field on a hillside in Ethiopia.

restrictions. In such a union, a woman is more likely to be able to seek divorce or to air a dispute with the help of a mediator; she may also claim part of her husband's wealth if they divorce. A smaller number of people marry in the Ethiopian Church, a process called *qurban,* and this kind of marriage cannot be dissolved even if a partner is an Ethiopian priest. The form of marriage giving a woman the lowest status, *damoz,* occurs often, despite being illegal in Ethiopia. In this arrangement, women are paid to become short-term wives or sexual partners, sometimes for just a few weeks.

In the 20th century, the Amhara expanded militarily and became the political elite of Ethiopia. People who did not speak Amharic, the language of the elite, were soon at a disadvantage in everyday life and before the law. By the middle of the century, after Amharic had become the national language, neighboring groups declared they would no longer be dominated by the Amhara. One of the central forces behind the rebellion was the growth of Oromo nationalism; although the Oromo have far greater numbers, they were long subordinate to the Amhara. Ethiopia continues to struggle for the national integration of its different ethnic groups, and in this respect, Ethiopia shares much with the rest of sub-Saharan Africa.

Baganda

Population: 6.2 million
Location: Uganda
Language family: Bantu

Uganda's largest ethnic group, the Baganda live on the northern and western shores of Lake Victoria. They speak Luganda, or Ganda, which also is an alternate name for the people.

BAGANDA Women preside over the sale of attractively displayed vegetables in a Ugandan market.

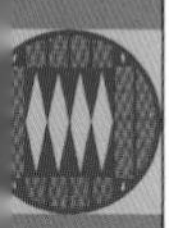

The Baganda kingship remained intact during the British colonial period, and when Uganda gained its independence in 1962, the king of Buganda became the nation's first president. Within a few years, however, all kingships in Uganda were abolished, not to be reinstated until 1993.

The majority of Baganda are Christian, about equally divided between Catholic and Protestant, and about 15 percent are Muslim. Some Baganda believe in sorcery and witchcraft as a cause of misfortune and death; some also practice ancestor worship.

Baganda villages occupy flat-topped hills on an elevated plateau and contain a number of households, each representing a patrilineage. Family compounds are surrounded by their own banana groves, which provide the Baganda's staple food, along with yams. Baganda grow a number of other food crops—sweet potatoes, taro, millet, and peanuts, along with cash crops such as cotton, coffee, and tea—and they also hunt and fish. Even Baganda who reside in towns often cultivate a kitchen garden.

From cleverness of hand, you get serfdom. From cleverness of mouth, you become boss.

AMHARA PROVERB

British explorers coming to the area in the mid-1800s encountered a kingdom called Buganda, ruled by a powerful *kabaka,* or hereditary king. The king presided over a network of patrilineal clans headed by chiefs who administered his policies, distributed land, and collected tribute in the form of food, livestock, and services. The kabaka took wives from many different clans, inspiring loyalty by keeping open the possibility that a future king could come from any clan. Behavior in the presence of the kabaka was highly ritualized and required that all commoners lie prone on the ground before him. The Baganda trace their kinship to the first kabaka, Kintu, who ruled about seven centuries ago. Myths attribute a heavenly origin to Kintu, regarded by some as the progenitor of all Baganda.

From early childhood, Baganda children are encouraged to be loquacious and to manipulate the Luganda language for fun. With peers, they engage in games of *ludikya,* or talking backward, often to befuddle adults. All ages participate in riddle contests, a favorite pastime in the evening. These games, along with a rich folklore tradition, have helped make modern Baganda some of Africa's most prolific novelists, playwrights, poets, and songwriters.

Beginning in the 1980s, many Baganda succumbed to the effects of HIV/AIDS, prompting the Ugandan government to undertake one of Africa's most far-reaching and effective programs for dealing with the disease. Within a relatively short time it managed to reduce rates of infection significantly, although in recent years they have begun to rise.

Beta Israel

Population: 150,000
Location: Ethiopia, Israel
Language family: Semitic

Outsiders have often referred to the African Jews of Ethiopia as Falasha, meaning "moved" or "gone into exile" in the ancient Ge'ez language. Today, the community considers the word derogatory and prefers *Beta Israel,* a Hebrew term for "House of Israel." For millennia, this ethnic group from northwestern Ethiopia has practiced a form of Judaism. The thousands of its people who remain in Africa are concentrated in Addis Ababa, Ethiopia's capital city. Since the 1980s, most Beta Israel have migrated to Israel—many in a series of airlifts dubbed Operation Moses—and today, about 130,000 Beta Israel reside in Israel, most of them born there.

Although the origins of Judaism in Ethiopia remain a mystery, it is likely that the community's roots go back 2,500 years. Until relatively recent times, the Beta Israel were cut off from the rest of the Jewish world. This isolation left them unaware of the presence of other Jews and without access to important Jewish scripture such as the Talmud. Furthermore, most Ethiopian Jews had no idea that most Jews were light-skinned and of Middle Eastern and European heritage. Nevertheless, the Beta Israel have always considered themselves to be exiles from the land of Israel, followers of Moses, and one of the lost tribes. Some trace the group's history to the biblical parting of the Red Sea, claiming that the Jews of Ethiopia were those who did not cross before the waters returned and so had to escape from Egypt by heading south.

Some scholars have suggested that the Beta Israel descend from a people who spoke Agaw, a Cushitic language, and were converted roughly 2,000 years ago by Jews from southern Arabia (today's Yemen). In the second century A.D., the Ethiopian kingdom of Aksum arose. It adopted Christianity as its religion in the fourth century, and the Beta Israel relocated to the mountains around Lake Tana. During the 13th

BETA ISRAEL Kessim, or religious elders of the Beta Israel, jointly consult a prayer book.

century, the people numbered in the hundreds of thousands and ruled a powerful state in what is now Ethiopia. But beginning around 1400, the Solomonic dynasty of Ethiopia gradually subdued the Beta Israel. In the 17th century, the Ethiopian emperor finally defeated the Jewish state and seized their lands. Most Beta Israel eventually gave up their Agaw language and adopted the Tigrinya or Amharic language of their neighbors.

The religious life of the Beta Israel is based on the Torah, oral interpretations passed down through generations, and the community's holy writings. Beta Israel strictly adhere to the kosher diet, observe the Sabbath, and celebrate all the festivals mentioned in the Torah, as well as those of their own tradition. The timing of holidays is determined by their own religious lunar calendar.

For centuries, the Ethiopian Jews have grouped their Sabbaths into seven-week cycles. The seventh Sabbath, called Legata Sanbat, is considered particularly holy and includes special prayers, festivities, and a sanctification service; at this time, priests may have their sins absolved completely. A traditional priesthood, the Kohanim, leads the community's religious life, and its importance as the repository of the group's traditions has increased during the upheavals of the past century.

Dinka

Population: 4.3 million
Location: Sudan, South Sudan
Language family: Nilotic

Decades of civil strife in southern Sudan have dramatically altered the way of life of the tall, stalwart Dinka, who revere and rely on the herds of long-horned cattle that have formed the basis of their economy for centuries.

Traditionally, the tribe moves between dry-season camps near the Nile's banks, where Dinka men pasture and manage their herds, and rainy-season settlements away from Nile flooding that turns those grasslands into swamps. Spears at their shoulders, they guard the

DINKA A Dinka man attends his valuable cattle at a camp along the Nile River.

cattle vigilantly from animal predators and possible marauders from rival Nuer and other groups. They also hunt and fish to supplement their diet, sometimes taking down hippopotamuses. At their settlements, women work in gardens, cultivating crops such as millet, pumpkins, okra, and cassava.

God is grinding fine flour;
what remains is the sifting.

DINKA PROVERB

Dinka live in patrilineal groups that are linked by descent from a common ancestor, but they observe no overall tribal leadership. Traditionally, men often supported more than one wife, and some wealthy men could have upward of 50. Each wife contributed to the compound's chores but raised her own children. While newborn daughters are welcome for the cattle they bring at marriage, a newborn son will carry on a man's name and lineage.

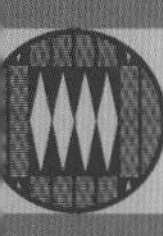

As children, Dinka boys milk the cows, tie them up, and collect their dung. As youth, they are initiated with others of their age into warriorhood, never again to milk the cows but to assume adult responsibility for their protection. During initiation, the boys' foreheads are incised with tribal marks, and the ritual leaves distinctive permanent scars.

Every part and product of the cow has a use, and often several. While alive, the cow provides milk and butter and also blood, which is sometimes mixed with milk and drunk. Cow urine is used for washing hair, as an astringent for insect stings, and for tanning hides. Cow dung provides fuel for fires, and the ash becomes a tooth powder and an insect repellent for cows and people. Cattle are not slaughtered, except as a sacrifice on ceremonial occasions, such as a birth, a wedding, or a death. When a cow dies, either naturally or ritually, it provides meat and hides that are tanned. Cow horns and bones are used as musical instruments.

As the traditional form of wealth among the Dinka, cattle are offered as bride-price to secure spouses for young Dinka men. It can take as many as 40, 50, or 100 cows to seal a deal, depending on the bride's status. The recent widespread disruption of Dinka life has affected many aspects of their culture. Many have lost their cattle, and so they can offer only the promise of cows in the future to a prospective bride's family.

In the 1980s, civil war erupted once again in Sudan. Forced Islamization policies of the Arab Muslim government in the north enraged the largely black animist south, and oil discovered in the southern Sudan in 1979 added to the conflict. The government also played to established rivalries between Dinka and Nuer.

Large populations of refugees have fled to northern Sudan, neighboring countries, and places such as the United States, Europe, and Australia. Among the refugees were the so-called Lost Boys of Sudan, young Dinka men who were off guarding cattle when their families were wiped out. Banding together, they set out on a long, perilous journey. Those who survived have begun new lives vastly removed from their Dinka past.

Southern Sudan achieved independence in July 2011; A 2005 peace accord had effectively ended the north-south conflict. The terrible conflict in Darfur in western Sudan, however, still awaits resolution.

Hadza

Population: Fewer than 1,000
Location: Tanzania
Language family: Language isolate

The Hadza, also called Hadzabe, live near Lake Eyasi in northwest Tanzania, just to the south of Ngorongoro Crater. Numbering fewer than 1,000, the Hadza are one of the last tribes of hunter-gatherers on Earth, with a way of life largely unchanged by recent developments in human history. They have no cultivated plants such as wheat or corn, no livestock, and they make their tools and weapons from materials collected in the bush, with some trading for iron used to make their arrowheads.

The Hadza speak a language isolate—a language unrelated to any other—that includes clicks, like the Khoisan languages spoken in southern Africa. While the clicks are a component of the language, its structure is very different from the Khoisan languages, and some linguists think that the Hadza might have adopted clicks into an originally unrelated tongue. Their

DEEP ANCESTRY

THE CONQUEROR

Xenia Fuller learned about the Genographic Project through a college professor. She knew her ancestry was mixed, but she felt confident that her deep roots were in Africa. "My mother is from the Caribbean (Barbados) and my father is African American from New York City. Since Barbados is known to have been active in the slave trade, I believed the test would tell me my ancestry traced directly back to Africa," explained Xenia. "I was correct, but found it interesting that as a reference group, the Genographic site provided Yoruba."

DNA Results

Xenia's results revealed that she was primarily of sub-Saharan African ancestry, but her results also showed several small yet significant ancestral components (between 2 and 6 percent) from other regions of the world. She carried a small percentage of South African ancestry, a component associated with the earliest hunter-gatherer settlers of the deserts and savannas of Botswana, Namibia, and South Africa. This segment of her ancestry came from prehistoric mixing among various African groups. Yet Xenia's ancestry also showed components from northern Europe, the Mediterranean, and southwestern Asia. These may have originated from either side of the family, since it's not uncommon to see European mixture in African Americans, as well as Afro Caribbean populations, such as those from Barbados. This mixing commonly occurred during colonial times in the United States and certainly throughout the Caribbean.

Her mitochondrial DNA lineage, in contrast, showed a much deeper and ancient story of migration and conquest. Her lineage was L1c1a, a common Bantu lineage from West Africa. *Bantu* is the name for a language group associated with a prehistoric culture that at one time conquered and dominated most of western and central Africa. By studying the relationship among living languages across the region, anthropologists have postulated that the homeland of the Bantu was somewhere in modern Nigeria and Cameroon. From there some of the earliest expansions into central and eastern Africa began some 3,000 years ago, with the conquering Bantu culture spreading its language and genes across thousands of miles. Today, Bantu languages are spoken in more than two-thirds of the continent, from Senegal east to Kenya and from Nigeria south to South Africa.

Xenia Fuller, mtDNA haplogroup L1c1a

Interestingly, the Yoruba ethnic group she referenced is from Nigeria, a likely birthplace of the Bantu. This interesting fact helped her in understanding her personal results. "My father travels to Africa often," commented Xenia, "and knowing where our ancestry traces back to could be of use to him when interacting with his friends and colleagues there."

—*Spencer Wells*

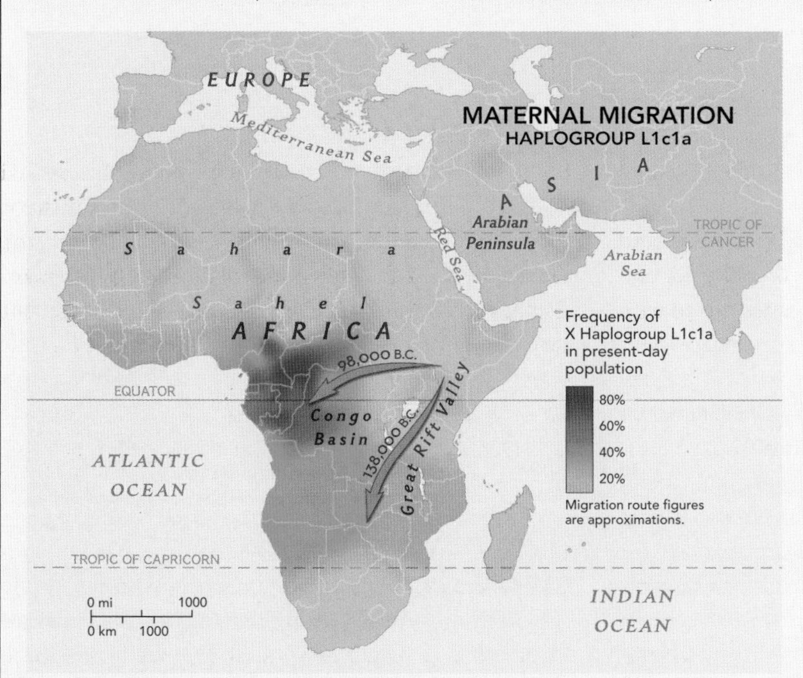

Maternal L1c1a: Some ancestors thrived in Africa's equatorial climates, while others expanded south.

HADZA A Hadza man relaxes in the doorway of his hut made of bent tree branches and dried grasses.

distribution likely once extended over a much greater area, but they were displaced by the migration of Nilotic peoples such as the Maasai over the past several hundred years. Now all that remains is the small group living near Lake Eyasi.

The Hadza live in bands of 20 to 30, moving camps with the seasons and the availability of resources. Like most other hunter-gatherers, they have a gender-based division of labor in which men hunt and women forage for tubers, berries, and other plant-derived components of the diet, although men also forage, particularly for honey and baobab fruit. The men hunt a variety of game using poison-tipped arrows, often focusing on birds because they are more plentiful than larger game. It is an ancient way of life, one that our ancestors followed for more than 99 percent of our history on this planet, being replaced by agriculture only in the past 10,000 years. Attempts by governments over the past century to settle the Hadza and turn them into farmers have failed, with the Hadza returning to their hunter-gatherer way of life.

The Hadza have no formal system of government, in common with other hunter-gatherer groups, and disputes are typically settled by group fission, with a small subgroup moving elsewhere to settle the dispute. Their communal social structure involves a great deal of storytelling and social interaction, and the relatively small amount of time they devote to hunting and gathering (a few hours a day) leaves them with a great deal of free time for socializing and child care.

Kikuyu

Population: 10.1 million
Location: Kenya, Tanzania, Uganda
Language family: Bantu

The largest ethnic group in Kenya, the Kikuyu live throughout modern Kenya and in parts of Tanzania and Uganda as well. The earliest Kikuyu arrived in what is now Kenya during the massive Bantu migrations of A.D. 1200 to 1600. Over the years, this ethnic group has played an important and highly visible role in the country's politics.

Kikuyu legend holds that a man named Gikuyu was the group's founder. He and his wife had nine daughters, who eventually married, had their own families, and retained a dominant role in Kikuyu society. The story helps explain the Kikuyu's nine family groupings. In fact, there are ten clans, but because the number ten is taboo, the usual practice is to count nine clans (plus one).

The Kikuyu were mainly farmers during the period of English colonization, but as the city of Nairobi grew under British rule, Kikuyu migrants made their way there. Many of the migrants had been educated at mission schools and had held municipal jobs that gave them unique exposure to colonial abuses. As a consequence, the city's Kikuyu began to organize politically in the 1920s; by the 1950s, the Kikuyu protest had solidified into a nationalist movement involving all ethnic groups. The government declared a state of emergency during the Mau Mau rebellion (1952–1960), arresting approximately 15,000 Africans in Nairobi, many of them Kikuyu, and sending them to detention camps. With Kenyan independence in December 1963, Nairobi became the national capital, and Jomo Kenyatta, a Kikuyu, soon became the first president of the Republic of Kenya.

KIKUYU Stepping out of his everyday life, a Kikuyu man of Kenya wears ceremonial dress and face paint.

The British at one time banned the Kikuyu language, often called Gikuyu, but today, many schools teach it to children, following a policy of teaching the local language first, then Swahili and English.

Service rendered means service is owed.

EAST AFRICAN PROVERB

Many Kikuyu name their children after important family members, alternating between families of the father and the mother. Kikuyu recognize a supreme deity called Ngai, who habitually dwells invisibly in the sky but periodically manifests himself within an enormous cloud on Mount Kenya. When the Kikuyu people see clouds on the mountaintop, they know that Ngai is present.

Malagasy

Population: 17.6 million
Location: Madagascar
Language family: Malayo-Polynesian

The term *Malagasy* denotes the language and identity of a complex and diverse group of people who live on Madagascar, the fourth largest island in the world. As many as 20 different ethnic groups—including the Merina, Betsimisaraka, Betsileo, Sakalava, and Antandroy—occupy the island. But while the millions of members of these ethnic groups claim particular cultural affiliations, these people also tend to say they are Malagasy and to speak Malagasy, the national language. Despite their many differences, most Malagasy communities have similar religious beliefs, bilateral kinship systems, and irrigated rice farming economies. Perhaps the most salient ethnic division is not between particular groups but between the central highlands people, mainly of Indonesian descent, and the coastal people, of black African descent. Today, these two regions are engaged in a fierce political, social, and economic competition.

Although Madagascar's native inhabitants are known to be descendants of people who came from islands off the coast of eastern Africa or from the Indonesian archipelago, the specific migrations that led to today's population have not yet been precisely identified by scholars. The Malagasy language shares many linguistic features with Indonesian tongues. French, the language of colonists who occupied Madagascar and its close neighbor Mauritius, is a secondary national language spoken by about a quarter of the population, mainly city and town dwellers.

Like other Malagasy, the large Merina ethnic group divides its society into a hierarchy of elites, commoners, and descendants of slaves. This social hierarchy is evident not only in politics, where elites hold the most power, but also in small-scale activities such as the evening meal: The youngest family members and more distant relatives from commoner or slave kinship groups always eat last.

Although the French abolished slavery on the southeastern islands of Africa in 1897, the Malagasy people retain extensive genealogical memories of slave ancestry. There is still some freedom in kinship affiliation, since people can reckon descent through either the father or the mother's brother. The results can become confusing, with people tracing their descent from different Malagasy ethnic groups as well as from Indians, Chinese, Arabs, and non-Malagasy Africans.

Perhaps the culture's most distinctive aspect—the one that has stimulated the most anthropological research—is the burial tomb. The Malagasy take great care to protect and honor the dead because they, like many other African peoples, believe that their ancestors influence their everyday lives. Often they build elaborate wood, stone, or concrete tombs that remain partly above the ground. Relatives can thus enter a tomb to adorn it or to offer sacrifices and prayers. They may also practice what anthropologists sometimes call secondary funerals: Certain Malagasy groups in central Madagascar will place a deceased relative in a tomb, then remove the body a short time later for reburial in a larger tomb or to rewrap and reposition the body. Failure to properly care for the dead during and after burial, believe the Malagasy, can bring harm to the living.

MALAGASY Their faces smeared with turmeric-paste sunblock, Malagasy women play with an endangered lemur.

Maasai

Population: 1.2 million
Location: Kenya, Tanzania
Language family: Nilotic

The Maasai pastoralists living near the Rift Valley of Kenya and Tanzania are among the most easily recognizable people in East Africa. Tall, with dark skin, Maasai men wear jewelry and colorful red or blue togalike garments, and they often carry spears as symbols of their manhood. Very few of them escape the attention of the annual swarms of tourists in East Africa. Some marketing efforts include photographs of Maasai in native costumes and even offer visits to their households. The Maasai appear to have profited little from the safari business that goes on all around them, though they have been recruited to assist in lion conservation efforts and as occasional guides.

Unlike most other African pastoralists, who rely on crops as well as animals, the Maasai do not practice agriculture. They depend on cattle for subsistence, consuming the milk, blood, and meat of their animals. The Maasai are one of the few African societies who struggle, usually with success, to practice pastoralism exclusively, even in the context of East Africa's extensive tourist industry. Often moving their cattle to available grasslands and water, they see themselves ideally as nomadic; in the 1970s, however, they began privatizing herds into ranches, and that process has increased inequality among the people.

The neck of a cow cannot overtake the head.

MAASAI PROVERB

Each Maasai settlement contains a cattle enclosure and homesteads consisting of houses made of mud and wood, built by women. A man may have more than one wife, and each wife and her children have their separate house within the family compound that is ringed by an acacia thorn fence built by the men. Among the more distinctive aspects of Maasai society is the age-set kinship system, which defines relations among men and also between men and women. Age sets are groups of people born into the same generation. Members share the same life events, such as circumcision and marriage, and remain in the same groups for life. The age sets group men into three main categories—uncircumcised, circumcised, and elders—and these are divided into smaller categories having to do with property ownership, marital status, parentage, and so on. Women, by comparison, are grouped into only two categories: circumcised and uncircumcised.

MAASAI A young Maasai man of Kenya checks his messages. Traditional cow herders, Maasai now pursue other types of work, often in cities.

In Maasai society, the elders among the men have great power, for they make most of the important decisions. They are responsible for deciding on the physical layout of villages; the production, sale, and distribution of livestock; and marriage partnerships within the community.

The Maasai of different territories employ numerous rituals to initiate men and women into age sets, and all of the rites involve recognition of changing status. There are various sacrifices of oxen and the eating of certain foods. In some areas, rituals incorporating a man into the warrior or elder class may take up to 15 years to complete. Once initiated into the warrior set, young Maasai men assume new duties: protection of elders, killing or fending off a lion if necessary, delivering messages long distances, and migrating the herds.

A ceremony called *e unoto* is held years after the last member of an age set has been circumcised; it signals passage into true adulthood. Preceding it is a ritual in which adult men and their mothers accuse each other of incest, then slaughter oxen that the men feed to their mothers to show they now can care for them. At this time, men still wear their hair in braids, but at the close of e unoto, they cut their hair to signal their new status.

Somali

Population: 14.7 million
Location: Somalia, Ethiopia
Language family: Cushitic

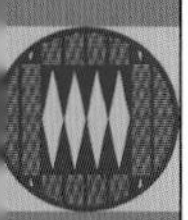

East Africa's entire Horn area is home to the Somali people, the majority of whom live in the country of Somalia and in the breakaway nation known as Somaliland. In 1991, after escalating civil conflict, northwestern Somalia declared independence and became Somaliland, but it still awaits diplomatic recognition from the international community. Somali are the principal inhabitants of the Ogadēn region of southeastern Ethiopia; they can also be found in Djibouti and in northeastern Kenya.

According to archaeological and historical evidence, Somali have been living in the Horn of Africa for nearly 3,000 years. In some local genealogical accounts, certain Somali clans have been traced to the Arabian Peninsula and associated with the Sharifs, the family of the Prophet Muhammad. But linguistic, cultural, and historical evidence indicates that the Somali came originally from the southern highlands of what is now Ethiopia and are a Cushite people with ties to a broad range of Cushite ethnic groups. Part of an ancient region known as Nubia, Cush gave rise to a group of languages that eventually defined numerous peoples in northern and northeastern Africa; the Somali tongue is a member of the Eastern Cushitic language family.

Claims to an Arab origin may arise not only from trade relationships and marriage alliances with Arab colonies on the Somali coast but also from a desire for stronger ties to the heart of Islam. The Somali people had accepted Islam by the 1400s, and according to some historians, they may have done so as early as the

SOMALI In transition to adulthood, a Somali girl covers her head.

1100s. In recent years, the Ayaana religion, which emphasizes fertility and the intersection of the spirit world in the human, has been growing among the Somali. At the same time, Islamic fundamentalism has been gaining ground over the traditional Sufi mystical practice.

As with many other African groups, the Somali people were never a unified polity before the arrival of various European powers. Instead, they relied on clan federations that provided a broad, loose identity. Traditionally a nomadic people, they did maintain boundaries for the herding area of each clan and subclan. Somali clans resisted invasions into recognized settlements and herding areas; actual borders were contested, however, and military clashes were common among the clans themselves.

During the colonial era, the British, French, and Italians established territories, or Somalilands, roughly following the geographical areas of the clan federations. In 1960, Britain and Italy combined their territories into a unified, independent nation known as the Somali Democratic Republic. The French territory remained separate, and in 1977, it gained independence under the name Djibouti.

As a nomadic, pastoral people, the Somali center their economic culture around camels, raising a few cattle, goats, and sheep in the more fertile areas. Women and young children care for sheep and goats, while young men and boys herd the highly esteemed camels. In a land where rainfall averages less than four inches a year, Somali lives are consumed with finding water and grazing land for their livestock. Their diet once consisted almost entirely of cow's milk and milk products, but for most people, it now includes maize meal and rice.

Swahili

Population: 500,000
Location: Tanzania, Kenya
Language family: Bantu

Although the word *Swahili* usually refers to an East African lingua franca, it is also the name for a large mercantile civilization on the East African coast and on islands such as Zanzibar and Lamu. The word is widely thought to have come from the Arabic *sawahil,* meaning "coast," thus making the Swahili the "people of the coast."

Descendants of Arab and Persian traders who intermarried with East Africans over the years, the Swahili are merchants with a long history as brokers between East Africa and Middle Eastern, European, and central African countries. They also frequently served as middlemen between Africa and Asia during the slave trade. Today, the extensive interaction between Africans and Arabs is reflected in art and architecture. Carved wooden doors, for example, are a renowned Swahili art form. Because Swahili art is Islamic art, the human form is never depicted. Instead, the people create elaborate carvings or paint intricate designs, and these can be seen decorating doors as well as chests and other furniture, house walls, and even boats known as *dhows.*

Swahili, the national language of Tanzania and Kenya, is a mixture of Bantu grammar with Arabic vocabulary. As evidence of the Swahili's success as long-distance traders, variants of the language are spoken today throughout Uganda, Rwanda, Burundi, and the eastern portion of the Democratic Republic of the Congo. The extensive literature in Swahili dates back to the mid-17th century.

About 500,000 Swahili live on East Africa's coast and nearby islands, but some traders live on islands as far south as Mozambique and as far north as Somalia. Spread out over such a large area, they have significant cultural differences, with many groups claiming different origins. The Swahili are almost always integrated politically and socially into other communities, yet even so, those who consider themselves to be Swahili share many beliefs, traditions, and attributes. Most of them are Muslim. They are often wealthier than their neighbors, residing in stone houses with the distinctive, carved wooden doors considered signs of wealth.

Swahili men and women place a high value on a good physical appearance, which they believe can symbolize inner purity. Swahili men wear amulets that contain Koranic verses. Swahili women use reddish henna dye to create intricate designs on their hands, and they tend to adorn themselves with jewelry, cosmetics, interesting hairstyles, and aromatics.

SWAHILI A smiling Swahili woman leans on an elaborately carved doorway on Zanzibar.

ACROSS CULTURES

THE EMERGENCE OF CULTURE

When in the human lineage did an aesthetic sense develop? While science tries to pinpoint origins and trajectory, we all can observe—and recognize the sublime.

Culture is born of the imagination—and the human imagination as we know it arose in the distant past as our species, *Homo sapiens,* evolved from its direct ancestor, *Homo erectus,* and, infused with consciousness, embarked on a journey that would carry it to every habitable domain on the planet. It is an extraordinary story—and one we continually revise with each new discovery.

Asking when the first sparks of culture appeared, we acknowledge that the answer keeps getting nudged further and further into the past. Currently, attention lights on a zigzag engraved onto a mussel shell that has been hiding in plain sight in a trove of fossils collected in Indonesia by a Dutch paleoanthropologist in 1891. With a recent dating of the shell to 430,000 years ago, the attribution of the engraving lies squarely in the realm of *Homo erectus,* a mind-bending reckoning. *Homo erectus* often has been thought of as a kind of beta-version ancestor, someone to get past to get to the updated model. Now we're not so sure, although in the absence of any associated content, it is difficult to ascribe artistic intent to the zigzag design. Was it symbolic, magical, decorative—a deliberate aesthetic expression? We don't yet know, but we must be open to any possibility.

The first evidence of our existence as *Homo sapiens* is found in Africa, dating back some 195,000 years ago. But long before our species embarked on its migrations out of the continent, it was mixing ochre in South African caves some 70,000 to 100,000 years ago. The assemblage of items and their residues suggest a knowledge of chemistry and the ability to plan for future needs.

For most of our history we shared the world with another branch on the hominid tree, our remote cousins the Neanderthals, who were descendants of the same progenitor. Neanderthals clearly had more than just a spark of awareness. They used tools, and there is evidence of deliberate burial as early as 70,000 years ago. They also crafted what can reasonably be thought of as jewelry. Incised eagle talons bored with holes (as for stringing) first came to light in a cave site in present-day Croatia dating to 130,000 years ago. Like the engraved mussel, the talons didn't draw much attention at first. But then their modifications were noticed, along with polished surfaces suggesting that they had rubbed together while worn. Suddenly, the Neanderthal emerged from the beetle-browed cave man stereotype to demand reevaluation.

The Turning Point

Beginning about 60,000 years ago, two waves of human migration, *Homo sapiens* this time, moved out of Africa. One followed the southern shores of Asia, reaching Australia 45,000 years ago. The second went north through the Middle East into Central Asia, with one pulse heading west into Europe, another going east and ultimately reaching the Americas. We took with us some competitive advantages—whether it was an increase in the size of the brain, the development of language, or some other evolutionary catalyst—that were to launch our destiny in an astonishing manner.

The last vestiges of Neanderthal life slipped away in Europe some 39,000 years ago. Within the next 10,000 years, our direct ancestors began to fill the caves of Europe with stunning works of art. Reaching deep into the Earth, through narrow passages that opened into chambers illuminated only by the flicker of tallow lamps, men and women drew with stark realism the

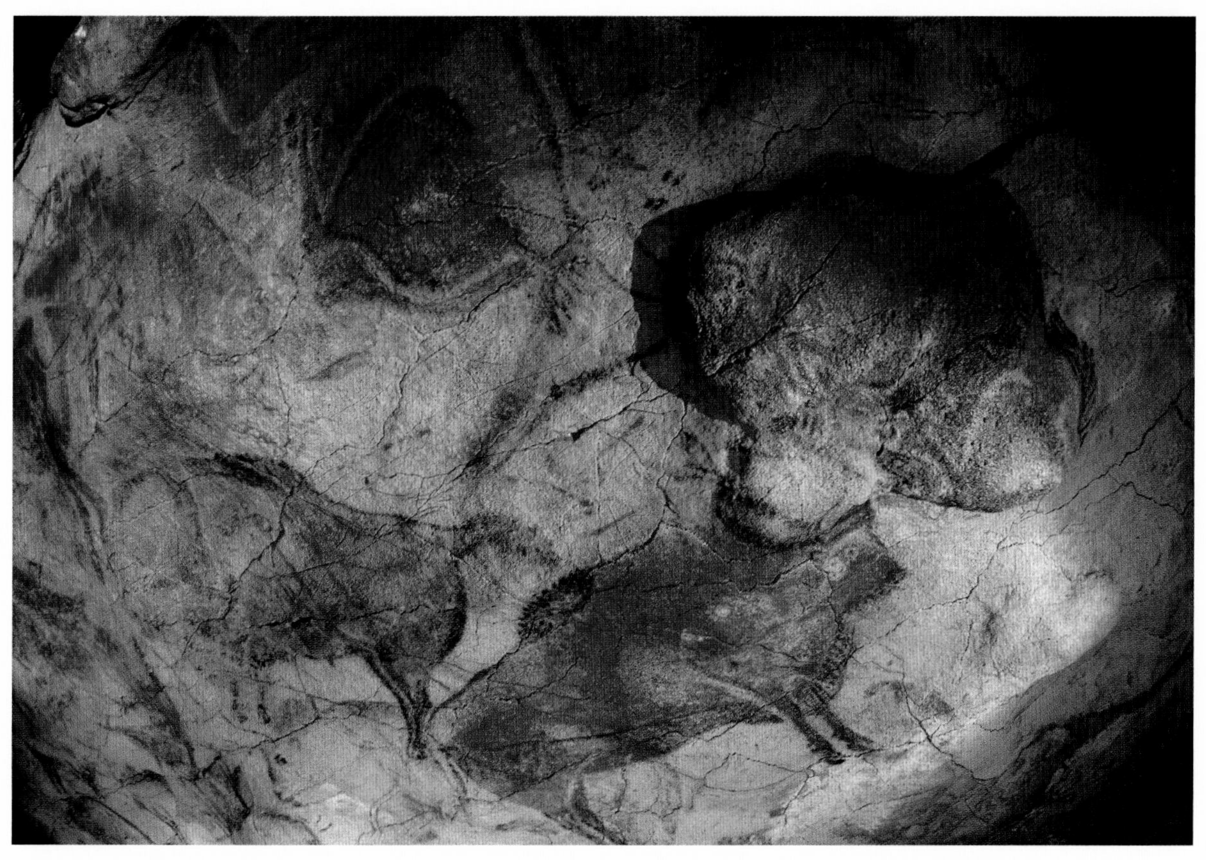

Vibrant animal figures reveal the depth of early human expression in this Upper Paleolithic cave art in Altamira, Spain.

animals they revered, singly or in herds, using the contour of the stone to animate the forms so dramatic that entire caverns come alive even today with creatures long since lost to extinction.

The sophistication of the figurative art found at in caves near Chauvet, France, and at later sites such as Lascaux and Altamira, is astonishing not only for its transcendent beauty but also for what it tells us about the florescence of the human potential once brought into being by culture. The technical skills—the exploitation of ochre and manganese, iron oxide, and charcoal to yield a full palette of colors; the use of scaffolding; the diverse techniques to apply the pigments—are themselves remarkable. All suggest relatively high levels of social organization and specialization, which are echoed in the genius of the Upper Paleolithic tool kit, the elegant scrapers and blades pounded from flint. The use of negative space and shadow, the sense of composition and perspective, the superimposition of animal forms through time—all indicate a highly evolved aesthetic that itself implies the expression of some deeper yearning.

The cave art does much more than invoke the magic of the hunt. It pays homage to that moment when human beings through consciousness separated themselves from the animal realm, emerging as the unique entity that, whether we like it or not, we know ourselves to be. Viewed in this light, the art may be seen almost as postcards of nostalgia, laments for a lost time when animals and people were as one. Proto-shamanism, the first great spiritual impulse, grew as an attempt to reconcile and even reestablish through ritual a separation that was irrevocable. What is perhaps most remarkable is the fact that the fundamentals of Paleolithic art remained essentially unchanged for more than 25,000 years, six times the chronological span that separates us today from the builders of the Great Pyramid at Giza. Ours was a long farewell.

—Wade Davis

Chapter 2

NORTH AFRICA & MIDDLE EAST

NORTH AFRICA

Arabs • Berbers

Egyptians • Nubians • Tuareg

THE MIDDLE EAST

Jews • Kurds • Lebanese • Palestinians

Persians • Qashqai • Turks

The history and cultures of North Africa have long been influenced by those of the Middle East, a region rich in ethnic groups and ancient traditions. Encompassing the Near Eastern civilizations of antiquity, the Middle East was also the birthplace of Judaism, Christianity, and Islam, all of which trace their roots to the biblical Abraham. Today, religion remains a key defining factor for ethnicity, more so even than language, region of origin, or class. Still, religious differences among groups are somewhat diminished by a shared past and similar traditions.

Around 8,000 to 10,000 years ago, the Tigris and Euphrates Rivers in Mesopotamia and the Nile in Egypt gave rise to the world's earliest agriculture. Cultivation of crops in turn helped lead to the first civilizations, in Sumeria and Egypt. But despite the presence of these major rivers and a few other fertile areas, much of the Middle East and North Africa is arid, with some of the most hostile deserts in the world. Nomadic herding has been the means of subsistence for generations, and human life has centered around oases scattered through the deserts. On the Arabian Peninsula, domestication of the camel between 3000 and 1000 B.C. offered a way to travel in these otherwise inhospitable lands, opening new routes for trade and expansion across the peninsula and the Sahara.

CULTURAL INTERSECTION Four major cultural groups with ancient roots still predominate in the Middle East: Arabs, Turks, Persians, and Jews. While Turks, Persians, and Jews are dominant in Turkey, Iran, and Israel, respectively, Arabs are spread throughout the region and across North Africa.

The word *Arab* comes from the Semitic-speaking A'raab people, who by 900 to 800 B.C. had established trade and pastoral networks between Syria and the Arabian Peninsula. They are related to speakers of such other Semitic languages as Hebrew and Aramaic. Over time, some Arab groups became sedentary, while others continued as nomads. The settled Arabs called the nomadic Arabs *Bedu,* from which the word *Bedouin* comes. Today Arabs are quite diverse, with their own cultural differences, but they continue to be unified by a common language—Arabic—spoken in many dialects from Morocco into East Africa, throughout the eastern Mediterranean, and into the Persian Gulf region.

Distinctive hats protect Yemeni women from the sun as they gather clover for their cattle in Wadi Hadramaut.

Because the Bedouin were the ancestors of modern Arabs, these people are regarded by other Arabs as true and original representatives of their culture and language. A few Bedouin continue to travel the deserts of the Arabian Peninsula, Egypt, and Libya by camel, living in tents and tending their herds as they have for centuries. But modern pressures are redefining the Bedouin way of life, and increasingly fewer of them continue to live as nomads.

Arabic and Hebrew are Semitic languages spoken by Arabs and Jews. Although these two groups share common religious roots, relations between them have been strained to the point of war on a number of occasions since the mid-20th century, when the state of Israel was created as a Jewish nation by the partitioning of Palestine. Israel's population is now roughly 80 percent Jewish, while Islam is the major religion of most other Middle Eastern and North African nations. Islam itself is split into two sects, the Sunnis and the Shiites, each of which holds different beliefs concerning the successor to Muhammad, the Prophet of Islam. Most Shiites reside in Iran and southern Iraq, while Sunnis predominate in the rest of the region.

Ancient traditions straining against modern trends introduced from the West have created friction between cultures in recent decades, resulting in the pan-Islamic movement that has swept the region. A manifestation of Arab unity occurred during the so-called Arab Spring of 2011, in which Arabs of North Africa and the Middle Eastern countries protested against their ruling elites. Today, a self-proclaimed Islamic State (also known as ISIS or ISIL) wages war and commits terrorist atrocities in Syria and Iraq in the quest to establish a region-dominating Sunni caliphate, destroying cultural heritage and creating streams of refugees in the process.

Age, wisdom, and a lifetime of experience etch the face of this elderly Muslim man in northern Egypt.

BREAD IS LIFE Despite entrenched conflicts, common customs unite the various peoples of North Africa and the Middle East in their day-to-day lives, down to the foods they eat. Central Asian and Mediterranean influences flavor the cuisine, and bread—flat or in loaves—is a staple so important that in Egypt it is called *aish,* "life," in Arabic. Rice dishes with herbs, beans, fruits, or nuts abound, except in North Africa, where couscous takes the place of rice. Yogurt salads and kebabs made of lamb, beef, or chicken are widely eaten, as are lamb or chicken stews with vegetables or fruits. Meze, a plate of appetizers featuring eggplant, beans, olives, and pickles, often starts a meal in the eastern Mediterranean. Rice pudding, halvah (a sesame-based confection), or honey-soaked pastries frequently end it.

LIFE PASSAGES Shared customs include rites of passage. Most groups bury the deceased within specified periods of time; they also observe prescribed periods of mourning. Marriage customs also show similarities. Though "love matches" are becoming more prevalent, arranged marriages occur among all religious communities, with a first-cousin marriage traditionally deemed an ideal match. Whole families are involved in the negotiations concerning bride-price (the amount of money the groom should give the bride's family), the marriage contract, the time of the marriage, and the future residence of the couple—traditionally the family home of the groom. Marriages tend to be multiday celebrations. Among Arab Muslims and Coptic Christians in Egypt, the bride and women from both families gather to decorate their hands and feet with geometric and floral henna designs and to become better acquainted. Regionally, women from different communities tend to dress modestly, though often to quite different degrees.

In an area of the world where conflict and political uncertainty have been consistent realities, long-cherished traditions and the ethnic identities that engender them offer constancy and comfort. Commonly held customs help to keep communities intact from one generation to the next.

North Africa

Containing the vast Sahara and the diverse Mediterranean coastline, the region of North Africa spans some 3,500 miles west to east. It contains the Atlas Mountains of Morocco and Algeria; the ancient cities of Tunis, Alexandria, Cairo, and Giza; and the mighty Nile River, longest in the world. The Sahara itself extends southward more than a thousand miles, terminating in the Sahel, a steppe that borders the region of West Africa. Morocco, Algeria, Tunisia, Libya, and Egypt all share a Mediterranean coastline; the Sahara stretches into Mauritania, Mali, Niger, Chad, and Sudan.

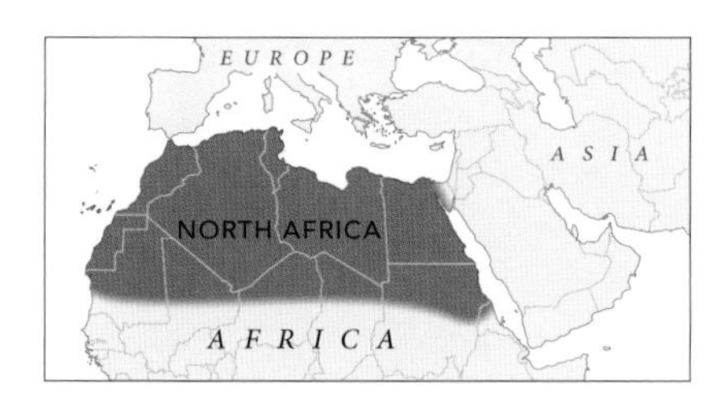

Arabs

Population: 377 million
Location: North Africa and Middle East
Language family: Semitic

Until the advent of Islam, Arabs and their culture remained almost entirely on the Arabian Peninsula. But the revelations of God to the Prophet Muhammad, himself an Arab, resulted in a cultural and economic flourishing in seventh-century Arabia. This in turn led to a spectacular spreading and mixing of cultures, as Arabs traveled throughout the Middle East, North Africa, and Spain and into central and southern Asia. The result today is an incredibly diverse culture, where both nomadic Bedouin and longtime urban dwellers call themselves Arabs.

Despite their diversity, Arabs throughout the world share cultural ideas and practices, including a common faith (more than 90 percent are Muslims) and language (Arabic), traditional gender roles, male-oriented inheritance laws, and ritualized ideas of honor, hospitality, and generosity.

Classical Arabic is the sacred language of Muslims because it is believed to be the language in which God revealed the Koran to Muhammad. Well-educated Arabs and religious scholars speak and read classical Arabic, and its use in radio and television has increased its comprehension among less educated listeners. Every Arab,

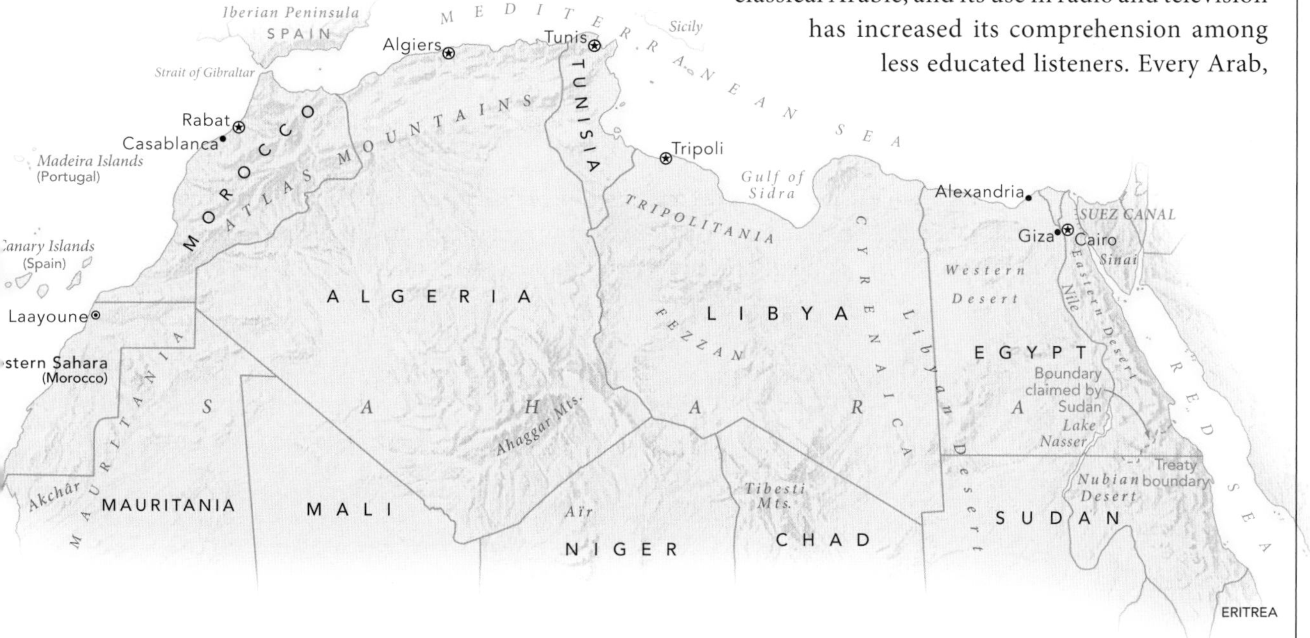

however, speaks at least one of the many colloquial forms of Arabic, which may be quite different from one another. Moroccan Arabic, for example, is incomprehensible to an Arab of the eastern Mediterranean or the Persian Gulf area. In recent times, movies and television soap operas, produced in Cairo but widely distributed, have familiarized most Arabs with the vernacular Egyptian dialect.

As with all other Muslims, Arab Muslims are enjoined to make the hajj, the sacred pilgrimage to Mecca, before they die. The most important rite of passage next to birth, marriage, and death, the hajj is a spiritual and religious event, a reenactment of the journey of Muhammad. Muslims believe that if they die on the pilgrimage, they will immediately go to heaven.

The Bedouin, believed to be ancestral to other Arab groups, are considered the culture's purest representatives, although they now make up less than 10 percent of the modern Arab population. Living in Libya, Egypt, Lebanon, Jordan, Syria, Israel, Iraq, and throughout the Arabian Peninsula, they are distinct from other Arabs because of their nomadic, pastoral lifestyle and their more extensive kinship networks, which provide them with community support and the basic necessities for survival. Such networks traditionally served to ensure the safety of families and protect their property.

As contemporary Middle Eastern governments have encouraged the Bedouin to settle, large family networks and the nomadic lifestyle have been diminishing. Nevertheless, the Bedouin continue to be revered by other Arabs, especially because of their rich oral poetic tradition, their herding lifestyle, and their traditional code of honor.

While poetry and literature are very important for all Arabs, the Bedouin are masters at oral verse, an

ARABS Camel farmers, such as this Arab man, have had ranch lands reclaimed to create the Dubai Desert Conservation Reserve.

emotional and evocative form of poetry that is recited from memory in classical Arabic. These people believe displays of unchecked emotions weaken the group's solidarity, so they use poetry to express strong feelings: sadness, love, and anger. Among other Arabs, poetry is also a way to convey feelings that might be difficult to express in everyday life.

An important feature of the Arab world, a shared honor code dictates certain behaviors, especially to preserve a family's reputation and help its members in times of need. For men, honorable behavior means supporting their families economically and defending the reputations of relatives, particularly women. For women, honorable behavior translates into being loving mothers and wives, running efficient and generous households, and acting in modest and respectable ways.

Arab families are very close, and home life is seen as the reward for hard work. While the life of a family remains quite private, Arab hospitality is famous: The people are generally quick to invite strangers to join them for a meal. Their hospitality and generosity are sincerely offered; as a result, friendships are intense, with sincerity and kindness expected in return.

BERBERS Berber hospitality in action: A woman prepares refreshing mint tea to slake many thirsts in Morocco's Ourika Valley.

Berbers

Population: 23 million
Location: Morocco, Algeria, Libya, Mali, Niger
Language family: Berber

A diverse group, Berbers comprise the indigenous peoples of northwest Africa. They include the Irifiyen, Imazighen, and Shleuh of Morocco; the Kabyles of Algeria; and the Tuareg of Algeria, Libya, Mali, and Niger. All speak varieties of Berber, a language of the Afro-Asiatic macrofamily.

Though scholars are uncertain about the Berbers' exact origins, they think members of this group probably came from the Near East and settled in northwest Africa around 3000 B.C. Over the millennia, Berber territories were invaded by Carthaginians, Romans, Vandals, Turks, Arabs, and Europeans, but the Berber culture remained distinct against all these foreign powers. The Arabs exerted the greatest influence on their culture, and in the early 8th century, Berbers were the main force behind the Arab conquest of the Iberian Peninsula. Later, in the 11th and 12th centuries, two great Berber empires arose and dominated much of Morocco, Algeria, Tunisia, and southern Spain. Berber traders once connected the Mediterranean markets to those of West Africa, and even today some Berber tribes still link trade between North and West Africa.

Most Berbers today are sedentary agriculturists who also tend small herds of goats and sheep. A few tribes, such as the Tuareg of Mali and the Imazighen of central Morocco, are seminomadic, herding sheep and using camels for mobility. The majority of Berbers trace their ancestry through their fathers, and men govern tribal affairs. There is a strong division of labor, with the men managing lands and herds and the women performing housework. Among the Tuareg, though, women inherit and own property, control the herds, cultivate land, and bring in the harvest. Tuareg men herd, tend the gardens, and conduct the direct activity of trade.

In urban areas, the distinction between Arabs and Berbers is diminishing, but Berber dialects, tribal customs, and traditional handicraft—handwoven carpets, silver jewelry, leather, and pottery, for example—continue to prevail in such remote areas as the mountains and valleys of Morocco's interior. A persistent custom is the local tribal council, which is composed of clan heads representing extended families.

Berber religious beliefs and practices are derived from Islam and are no different from the orthodox ones of other Sunni Muslims. There is an emphasis, however, on *baraka,* the belief in the special blessings from God, and *marabouts,* individuals who are considered holy. Even in death, marabouts are thought to act as spiritual intermediaries and to bestow God's blessings on the living. These individuals once played key political, religious, and economic roles, and they are still honored at numerous shrines throughout northwestern Africa.

Egyptians

Population: 88 million
Location: Egypt
Language family: Semitic

For thousands of years, Egyptians have owed their lives and livelihoods to the Nile River and its delta, part of the world's longest river system, which transforms a narrow band of the eastern Sahara into a fertile and viable nation, now home to more than 88 million people. In the past, the yearly flooding of the Nile, brought on by rains closer to the river's source, deposited the thick black silt that the *fellahin,* or peasant farmers, planted with wheat, rice, and cotton, all still valuable crops in Egypt.

The construction of the Aswan High Dam in Upper Egypt—southern Egypt, since the Nile flows, unusually, south to north—ended the annual floods, but irrigation systems maintain agricultural fields and even expand them by claiming more of the desert for cultivation. Still, less than 3 percent of Egypt's territory can support agriculture, yet it remains the largest arena of employment.

EGYPTIANS An Egyptian Christian lights a candle in the church that commemorates the spot where Jesus, Mary, and Joseph took refuge in Egypt.

Although they are proud of their pharaonic past, Egyptians identify with their Arab Muslim present; about 90 percent of the population embraces Islam, and nearly all Egyptians speak Arabic. Seventh-century Arab invasions brought about the conversion of Egypt's Coptic Christians, a major religion since Roman occupation in the first century. Egypt resonates with the everyday rhythms of Muslim life, including the calls to prayer that ring out over communities five times a day, and with the yearly cycle of fasting and feasting.

The ropes can be stretched far enough for forgiveness.

BEDOUIN PROVERB

Some five million Egyptians remain Coptic Christians, descendants of Egyptians who did not convert to Islam. Copts share many cultural customs and general appearance with Egyptian Muslims, although they may tattoo a small cross on the inside of the wrist. Most reside in Cairo and in several villages in southern Egypt.

Coptic Christianity has its own patriarch, who is chosen by the community and believed to be divinely ordained. The church's sacred text and liturgy traditionally use the Coptic language, which was based on ancient Egyptian, with a strong Greek influence. Coptic no longer is spoken, however; most Copts today speak Arabic. The Coptic community is wary of conflicts that can arise from religious differences, especially in light of clashes with ultraconservative Muslims, and in general they try to keep a low profile. Recently, the government showed support of Egypt's Christians by making Christmas, which falls on January 7 in the Coptic calendar, a national holiday.

Egypt's deserts are mostly unpopulated, except for the nomadic Bedouin who occasionally travel through them, though most Bedouin now live settled lives. They are found in Egypt's major cities, such as Cairo and Alexandria, but some prefer to live in oases, places in the desert where water trapped in rock layers rises to the surface and can support plant, animal, and human life on a year-round basis. Small groups of Bedouin are found in the Western and Eastern Deserts and on the Sinai Peninsula.

Despite programs to increase farm capacity, Egyptians continue to move to the cities. A population of some 16 million makes Cairo the largest city in Africa and the Middle East, and one of the world's most densely populated. Its population grows each year with people from the countryside who seek employment and the possibility of a better life. Reality tends to be harsh, however, and many end up living in makeshift settlements in Cairo cemeteries or in shacks constructed on the roofs of high-rise buildings.

Overcrowding is only one of Cairo's modern ills. Pollution fueled by several million automobiles as well as factories and metal smelters manufactures a yellowish fog that hangs over the city. Tourists, who bring in a significant portion of Egypt's revenue, tend to make short stays in Cairo—long enough to see the pyramids and Sphinx on the city's outskirts and to visit its outstanding archaeological museum—before heading off to the clearer air of Luxor and the pharaonic treasures of the Valley of the Kings about 400 miles south.

Egypt has been making a difficult transition in recent years from autocratic rule to greater openness, driven at times by massive public demonstrations, while balancing sometimes opposing forces of Islamic conservatism and Western-oriented values.

Nubians

Population: 605,000
Location: Sudan, Egypt
Language family: Eastern Sudanic

More than 50 years ago, the pent-up waters of the Nile began to pour over the houses, shrines, graves, and palm groves along the river's banks in southern Egypt and northern Sudan, near the site of the Aswan High Dam. The dam was designed to prevent the annual flooding of the Nile, to make its waters more manageable and available to agriculture on a controlled basis—an important development for Egypt, a nation wrested from the desert that encroaches on both sides of the river. In order for this happen, a homeland and a way of life had to be obliterated: the fate that befell the Nubians.

The Nubians are a non-Arab people who traditionally lived in Nubia, an unbounded area about 180 miles long that encompasses the Nile Valley on both sides of the border between Egypt and Sudan. Descended from peoples who produced a pharaonic dynasty in the eighth century B.C. and later established the Kingdom of Meroë in Sudan, the Nubians differed from Arab Egyptians and Sudanese in a number of ways, including language. Most Nubians spoke Nobiin, a tongue from the Eastern Sudanic language family, although one of the four main groups commonly spoke Arabic. Today, Nobiin is spoken mostly by older people, while the young tend to use Arabic.

The Nubians converted to Coptic Christianity after the sixth century. But unlike ethnic Egyptians, they did not accept Islam in the first wave of Arab invasions in the eighth century A.D. They chose to pay a poll tax rather than convert. The Nubians eventually did convert, but not until the 14th or 15th century. The vast majority of Nubians today, both Sudanese and Egyptian, are Muslim.

NUBIANS A Nubian girl at home in Sudan represents her people's long and distinguished heritage that dates back millennia in the Nile Valley.

Traditional Nubian villages hugged the Nile's banks in an area where the river formed cataracts, or steep rapids. The villages contained separate hamlets, usually based on familial descent, and were separated from each other by low, rocky hills. Nubian farmers had two cultivation seasons a year, harvesting grains such as wheat, barley, and millet, and vegetables. The date palm, however, was the Nubian signature crop. They carefully tended the planted palm shoots and nurtured the tree, encouraging the properties they valued in the fruits. The Nubians used every part of the date palm, including its wood and fibers.

A squeeze on agricultural land sent Nubian men to cities in search of work even before the Aswan Dam was completed. Egyptian Nubians migrated to places such as Alexandria, Cairo, and Suez, while the women of the villages carried on with the agricultural work.

As part of the Aswan Dam project, both Sudanese and Egyptian Nubians were offered relocation in villages away from the dam area. In Egypt, the new homes were only about 30 miles from the old settlements, but were 2 to 6 miles from the Nile; in Sudan, they were some 480 miles from the old settlements, along the banks of a narrower river, the Atbara. In the new settlements, houses were built in rows and were no longer clustered by kin group. The agricultural fields were larger and had to be managed in a different way, and there were no date palms. Although the new villages bore the same names as the old, they were nothing like the traditional Nubian homeland. They had more amenities but neither the organization nor the ethos of the communities they replaced.

The displaced Sudanese and Egyptian Nubians still mourn their fate. They and various advocacy groups point out that more care was spent moving the temples of Abu Simbel and other Nubian archaeological treasures to higher ground (or, in the case of the Temple of Dendur, to the Metropolitan Museum in New York City) than was taken with the lives of the displaced Nubians. Recently, land near Aswan has been opened to settlement by poor farmers, but the displaced Nubians receive no preference. It is land inferior to what they had, some claim, and it would never recompense the life they have lost.

Tuareg

Population: 1 million
Location: Niger, Mali
Language family: Berber

The Tuareg, their origins likely in Libya, remain one of the few Berber tribes to retain the nomadic way of life they have practiced for centuries. In the past, they led camel caravans across the Sahara to trade in gold, ivory, salt, and slaves they captured from sub-Saharan tribes. They also absorbed and assimilated a number of different peoples into their group, with the result that the Tuareg as a whole resemble both light-skinned North Africans and the darker peoples south of the Sahara.

Tuareg men are known for the long indigo veils they wind around their heads and across their faces, earning them the names of "The Blue Men" and "People of the Veil." The turbanlike cloths keep out sun and sand and make it possible for them to ride their camels or walk alongside them for more than 12 hours a day on weeks-long journeys across the desert.

The Tuareg number about one million people, dispersed in the countries of the Sahara and the Sahel, the semiarid regions to the desert's south. They are found mainly in Libya, Algeria, Niger, Nigeria, Mali, and Burkina Faso. They speak a number of dialects of Tamasheq, a Berber language, as well as some sub-Saharan languages such as Hausa and Songhay, along with French, the lingua franca of much of West Africa.

Historically, the Tuareg society divided itself into a hierarchy of groups related to men's descent and occupations. These included nobles, vassal-like offerers of tribute, smiths and artisans, and a serflike group.

Tuareg frequently practice polygamy. As Muslims, they are allowed four wives as long as they can provide for them and their children equally. Large families are common and highly desired. Children help with the many tasks of agriculture and other aspects of daily life.

Tuareg women enjoy more freedom than other Muslim women in Africa, and they do not necessarily veil their faces. Traditionally, descent and inheritance are reckoned in the female line, although Islamic practices have tended to undermine this.

Some Tuareg intensively cultivate the oases of the Sahara, planting crops such as maize, millet, carrots, tomatoes, and onions and tending orchards of oranges, lemons, pomegranates, and figs. They use their camels to pull up water from the aquifers that they then dump into channels to irrigate the fields. Some Tuareg also farm in the Sahel; some live as pastoralists, maintaining herds of sheep, goats, cattle, and camels and moving in search of water and pasturage.

TUAREG An Algerian Tuareg woman's headdress and veil pale in comparison to Tuareg men's traditional indigo wrappings.

Tuareg caravans still traverse the desert, cameleers singing to pass the time, creating images that seem no different from those of centuries ago. Caravans tend to be much smaller than in the past, when thousands of camels may have stretched in a line several miles long across the sands. Most caravan trade today deals in salt.

From time to time, during the more than four decades since the West African nations of Mali and Niger achieved independence, Tuareg dissatisfaction with government policies regarding assistance and access to land has fomented violent uprisings. In addition, the dramatic acceleration of desertification as a result of severe drought has forced many Tuareg to migrate temporarily to Algeria and Libya. Tuareg people also have often been at odds with non-Tuareg farmers over issues of access and use of land.

GENOGRAPHIC INSIGHT

SEEKING GREENER PASTURES

At the top of the continent of Africa lies a thousand-mile-wide sea of sand that has stood for millennia as one of the world's most inhospitable places. While the fertile tropics and savannas to its south nurtured the emergence of our species, North Africa's seemingly impermeable desert barrier kept most of Africa tucked safely away from the rest of the world through the early part of human evolution. The second largest desert on the planet after Antarctica and home to the world's hottest recorded temperature (136°F), the Sahara has remained off-limits to all but the most resilient migrants willing to brave its interior.

But the Sahara has not always been as it appears today. The Serengeti plains of today's Tanzania suggest a landscape not all that different from what the desert would have been like 7,000 years ago, verdant with tall grasses and acacia woodlands, a wide variety of grazing herbivores, and rain. Difficult to envision given the expanse and hostility of today's Sahara, it was a place suitable for human occupation. We know this history of the Sahara to be true because of what we have found by digging up large cores of the eastern Mediterranean seabed and looking for signs of life in those ancient layers.

Shifting Sahara

Much of Africa's climate is controlled by the intertropical convergence zone, a belt of low pressure created by the upwelling of hot, moist air above and below the Equator, which eventually cools, condenses, and returns to Earth where the intertropics converge. The system produces both equatorial rainfall and incredibly arid conditions to its north and south, resulting in today's Sahara and Kalahari deserts. But the process is thrown out of whack as the Earth precesses, or wobbles, on its axis, causing the convergence zone to migrate as much as 20 degrees in latitude northward and bringing tropical precipitation with it. Increased rainfall throughout the region floods into the Nile River, which dumps the water, rich with organic material, into the eastern Mediterranean, killing the sea's normal convection current in the process. With no current to circulate the water, organic material falls to the bottom and hardens into a new seabed layer. As Earth's orbit shifts back, the convection current starts over.

This cycle happens like clockwork around every 20,000 years, and carbon-rich layers known as sapropelles are deposited by Nile River runoff on the Mediterranean seabed in patterns that look like a sedimentary bar code. By digging up cores from the eastern Mediterranean and dating the carbon layers thus deposited, scientists can glean an accurate idea of when the Sahara might have been passable.

A nomad walks the ridge of a dune in the Sahara, which covers an expanse of North Africa nearly the size of the United States.

Desert Escape

The analysis of DNA from populations throughout North Africa and the Middle East confirms that when climate shifted, our ancestors headed for greener pastures. The first modern humans to migrate northward successfully along this fertile passage did so about 50,000 years ago, capitalizing on a short window of opportunity before desert conditions forced their descendants back out of the region as the Sahara expanded once again. Humans were subsequently locked out of North Africa until the next shift in Earth's axis. Then, peoples who had been kept apart for almost 20,000 years came together from both sub-Saharan Africa and the Middle East.

The genetic lineages of these Paleolithic hunter-gatherers survive today in some North African populations, evidence that humans have learned to adapt to harsh and variable conditions, in part by depending on an underwater system of aquifers filled during the wetter periods. More recently, successive civilizations have left their genetic stamp on the landscape of today's North Africans—descendants of the Egyptians, Phoenicians, Garamantes, and Arabs all survive—and the gene pool today reflects a genetic tapestry as complex as the environmental interactions that have shaped this timeless land.

The Middle East

A crossroads of geography, history, and culture, the Middle East is the nexus among Europe to the northwest, Asia to the northeast, and Africa to the southwest. It stretches from the Black Sea and the Caspian Sea of Central Asia to the Red Sea, the Persian Gulf, and the Gulf of Oman. The Middle East includes the entire Arabian Peninsula and Asia Minor, or Anatolia, the peninsula containing Turkey. Much of the Arabian Peninsula is dry desert or steppe, irrigated only by wadis, or seasonal streams, while the terrain just north of it, regions of Iraq and Iran known as the Fertile Crescent, stay fecund thanks to the Tigris and Euphrates Rivers.

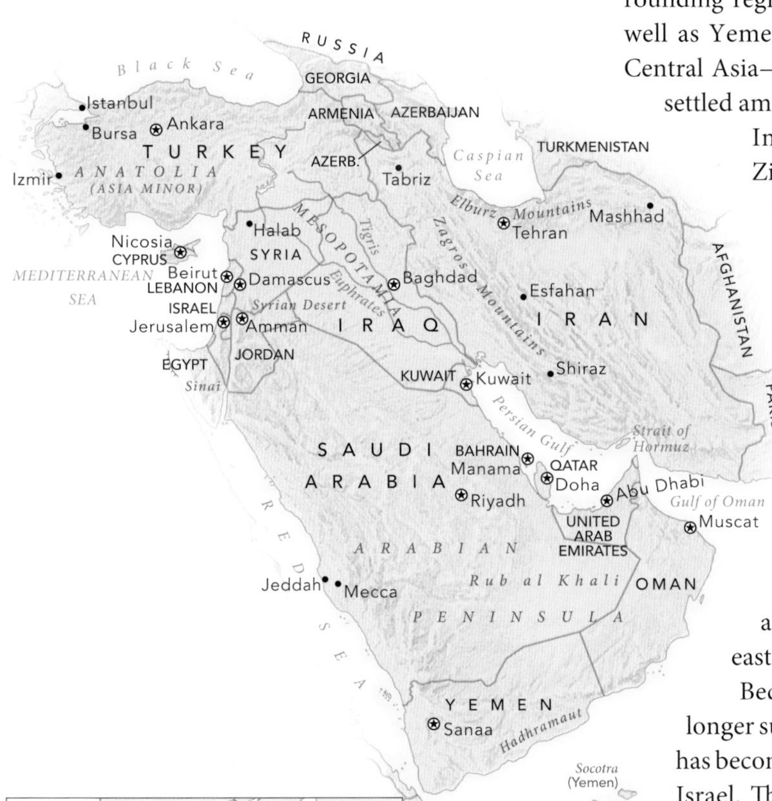

Jews

Population: 6 million (in Israel)
Location: Israel
Language family: Semitic

The more than six million Jews who live in the Middle East and North Africa share the same religion and have the sacred language of Hebrew in common; even so, members of this group are as diverse as the peoples they have dwelled among for more than two millennia. Until the state of Israel was formally established in 1948, almost all Jews also held in common the feeling of living in exile. Destruction of the First and Second Temples of Jerusalem—in the sixth century B.C. and first century A.D.—was followed by great dispersions of Jews from the city and the surrounding region. They went to what is now Iraq, as well as Yemen, Egypt, Ethiopia, Turkey, Iran, and Central Asia—and into Europe. Some diaspora Jews settled among the Berbers of northwestern Africa.

In the 20th century, tensions between Zionists and Arab nationalists, along with the realization of a homeland, led many Jews to move to Israel. As a result, the old Jewish communities of North Africa and the Middle East, once numbering in the thousands, were greatly reduced as Israel's cities and towns grew to include representatives from every Jewish group in the world. These groups are often placed in two categories: Sephardim—Jews from the Middle East, North Africa, and Western Europe—and Ashkenazim—Jews from central and eastern Europe.

Because they have a homeland, Jews no longer suffer as exiles, but their cultural diversity has become apparent during their resettlement in Israel. The country faces the enormous task of integrating the many languages, customs, occupational classes, and religious interpretations that have developed over the centuries. All Jews follow the same religious laws and offer up their prayers, but the words and styles may vary. The differences are especially obvious between Sephardic and Ashkenazic traditions.

JEWS Young Jewish army recruits take a break during a tour of Jerusalem's Old City.

Given the many Jews from the Arab world, Arabic is widely spoken, along with Hebrew, Persian, Turkish, Kurdish, Russian, and other Asian languages. Some African languages are also in the mix. Jews from Eastern Europe may speak Yiddish, a hybrid of Hebrew and Medieval German. People from the eastern Mediterranean may speak Ladino, a form of Spanish retained by 15th-century Jewish refugees from Spain.

Like Muslims and Christians, many Jews believe in pilgrimages and in mystics and holy people—saints—who can intercede in earthly affairs. And in several areas of the Middle East and North Africa, Jews and Muslims make pilgrimages to the same holy sites.

Some Jewish laws also have features in common with those of the surrounding Muslim majority in the Middle East and North Africa. Laws pertaining to inheritance, residence, and marriage, for example, have traditionally favored the father's side of the family, even though one major definition of Jewishness is birth by a Jewish mother. With immigration and urbanization, however, fewer Jews are following the practice of residing with the groom's family, and new families are setting up smaller, separate households.

Kurds

Population: 28 million
Location: Turkey, Iraq, Iran
Language family: Iranian

Residing for more than 2,000 years in contiguous mountain regions of Turkey, Syria, Iraq, Iran, and Armenia, the Kurds have remained reasonably autonomous, despite the fact that they have rarely enjoyed self-governance over the territory they call Kurdistan, "land of the Kurds." Some scholars believe these people are descended from Medes, an ancient Iranian group who settled in southwest Asia more than 2,700 years ago.

KURDS A woman in traditional Kurdish dress attends an International Women's Day gathering in Istanbul.

Today, Kurds live in Turkey, Iraq, Iran, Syria, Europe, and the United States. Because of their long, isolated existence in mountainous areas, they have been able to retain their own language—Kurdish, which is most closely related to Persian—as well as their own social organization and oral traditions. Almost all Kurds are Muslims, but a small minority known as the Yazidi practice a unique religion that blends Zoroastrianism, Manichaeanism, Nestorian Christianity, and Shiite Islam.

Though Kurdish society is patriarchal and only male siblings inherit property, Kurdish women nevertheless enjoy a more public life than do other rural women from surrounding Turkish, Arab, and Iranian societies. They have greater involvement in community decisions and more freedom of interaction with men. Women may also opt not to wear a veil, a choice that is much more difficult for women in some neighboring societies. A Kurdish woman traditionally wears a bright floral skirt, an embroidered jacket, and a turban or scarf on her head. A man's traditional attire consists of loose-fitting pants, a short-waisted jacket, a sash around the waist, and a turban similar in style to a woman's. Kurds share many culinary practices with their Turkish, Arab, and Iranian neighbors, and they are famous for their cheeses and yogurt.

Because of the Turkish, Iraqi, and Iranian policies of forced settlement, Kurds are now a settled agrarian people. Not long ago, more of them were nomadic and tribally organized, and some continue to organize themselves tribally, with loyalty building from family to lineage to clan to tribe.

Other Kurdish farmers have organized instead around village leaders, who usually own the land and lease it out to farmers in exchange for labor or a portion of the harvest. Unlike Arab nomads, whose leaders are selected, Kurdish nomadic leaders inherit their titles. Many aspects of the Kurds' social life are changing dramatically due to warfare, political instability, and sectarian tensions in their home areas. Some Kurds have migrated to cities and now work in urban trades.

Having been controlled by many outside forces and forced to witness the deaths of numerous members of their group, including children, Kurds have adopted a rather stoic attitude over the years. The Kurds suffered great chaos and upheaval during the Iraqi wars and saw the tantalizing prospect of autonomy. They remain part of the Iraqi polity, with greater self-governance, and in a world that is now more aware of their rich culture.

Lebanese

Population: 6 million
Location: Lebanon
Language family: Semitic

The ancient Phoenicians of Lebanon created a maritime culture that stood at the hub of the Middle Eastern interior and the Mediterranean cultures beyond. The modern Lebanese still play this role and also mediate the tension-filled distance between Syria and Israel, while trying to balance the interests of their own multicultural population.

Nearly all Lebanese are Arabs, except for a small Armenian minority, representing a number of different cultural traditions. An estimated 54 percent of Lebanon's six million people are Muslim, mainly Shiite and Sunni, but they also include the Druze, an Islamic offshoot that forms a closed community with esoteric

DEEP ANCESTRY

THE MOUNTAINEER

Fabrizio Scarselli heard about the Genographic Project through a friend who is a physiology professor. He had always been interested in genealogy and had even researched the University of Bologna archives to learn more about his family tree; he had found records that dated as far back as tenth-century Florence, Italy. Indeed, both sides of his family were Florentine.

DNA Results

Fabrizio's results showed a strong deep connection to Europe, with almost no mixture from other geographical components. Furthermore, his strongest component was Mediterranean, supporting his findings that his family had lived in southern Europe for millennia without much mixing. He did have a significant Southwest Asian genetic component (Middle Eastern), something typical of European populations and attributed to the Neolithic Revolution that began with the advent of agriculture in the Middle East and spread from there to southern Europe.

Fabrizio's Y chromosome lineage (R1b) is also quite typical of European males, and his branch was commonly found in southern Europe. His mitochondrial DNA lineage (K1a), however, was more atypical and is not associated with the agriculturalists who arrived in Europe from the Middle East in the last 8,000 years. K1a is closely associated with groups in central and eastern Europe and some of the earliest hunter-gatherer bands that settled the continent more than 30,000 years ago. Some of these early settlers may have interacted with the Neanderthals who were already there. Fabrizio's Genographic test showed that he carries 2.7 percent Neanderthal, a figure about average for Europeans based on calculations from nearly 200,000 Geno 2.0 participants.

Fabrizio Scarselli, mtDNA haplogroup K1a

Unlike his paternal lineage, which is found throughout Europe, K1a is not common in southern Europe; even in central Europe it occurs at frequencies of 10 percent or lower. A distinct branch of K1a is commonly found in Ashkenazi Jewish populations, and K1a is the lineage of the famous Oetzi the Iceman, a prehistoric man found preserved in the Alps in the border region between present-day Austria and Italy. However, Oetzi belonged to the ancestral type K1 and did not have any of the mutations that defines K1a or its sister branches K1b and K1c. This suggests that these old lineages have continued to evolve in the last 4,000 years.

In addition to having famous ancient relatives, Fabrizio shares his maternal haplogroup with comedian and *Late Show* host Stephen Colbert and news anchor Katie Couric. Colbert and Couric are among more than a dozen celebrities who have participated in the Genographic Project.

—Spencer Wells

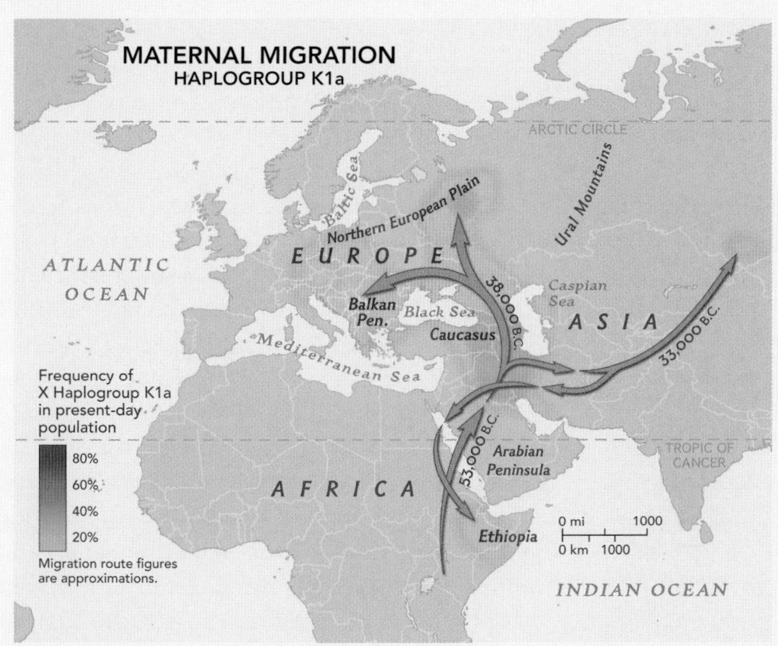

Maternal K1a: Early European migrants would have encountered glaciers and humanlike Neanderthals.

religious practices. Christians account for nearly all the rest, the largest group being the Maronites, the only Middle Eastern Christian group to wield considerable political power in the Arab world. Some Lebanese Christians currently reject the Arab label and refer to themselves as Phoenicians.

The smallest mainland country in the Middle East, Lebanon has accommodated its cultural diversity partly through geographical separation. In a space that averages only 35 miles from the Mediterranean to the eastern border, Lebanon spans a coastal plain, the Mount Lebanon range, the fertile plains of the Bekaa Valley, and another range, the Anti-Lebanon Mountains.

The Mount Lebanon range historically provided opportunities for cultural separation. The Maronites, who ultimately fall under the jurisdiction of the Roman Catholic Church while retaining their own rituals and local patriarch, occupy the northern end. Maronites have often allied themselves with the West, first with medieval crusaders and Rome and later with the French. After World War I, the French held a mandate over Lebanon and Syria. They gave considerable power to the Maronites, who continue to feel an affinity for French culture and often speak French in private life.

The central Mount Lebanon area is the traditional home of the Druze, divided religiously between the *ëuqqal,* who are initiated into the esoteric mystical teachings, and the *juhhal,* who live the Druze way but are kept ignorant of deeper knowledge. Generally known Druze beliefs encompass an omnipresent God, the return of a messiah, and reincarnation of souls into newborn Druze. Druze mountain communities usually were built toward the summits, near sacred shrines.

LEBANESE A group of Lebanese men gather to play board games in a Beirut park.

The southern end of the range is traditional Shiite homeland, characterized by a rural Arab lifestyle. Shiites tend to be more agricultural than Sunnis, who generally live in Lebanon's coastal cities and are involved in banking, trade, tourism, and other services that form the basis of Lebanon's economy and prosperity in times of political stability.

Play together as brothers but conduct business as strangers.

SYRIAN PROVERB

The Bekaa Valley, Lebanon's primary agricultural region, where crops of fruit, vegetables, and tobacco are grown, is also traditionally organized into sectarian enclaves, which include Shiites, Sunnis, Druze, and some of Lebanon's Christians, such as Greek Orthodox. Nevertheless, since the mid-1900s, there has been movement of all cultural groups to coastal cities, especially to Beirut.

The last official census taken in Lebanon dates back to 1932 due to the delicate nature of sectarian affiliation. Back then, Maronite Christians and Muslims were equally represented in the Lebanese population. When the 1943 National Pact was formulated, it allowed for a Maronite president, a Sunni prime minister, and a Shiite speaker of parliament.

But a higher rate of Christian emigration, along with a higher Muslim birthrate, shifted sectarian percentages. By the 1970s, with the population expanded by hundreds of thousands of Palestinian refugees—many of whom came following the 1967 Arab-Israeli conflict—the scene was set for the civil war that broke out in 1975. During the 16-year conflict, which also absorbed regional differences involving Israel, Syria, and the Palestinian Liberation Organization, some 100,000 people were killed and an estimated 250,000 left as refugees. Beirut and the whole Lebanese economy were in shambles. In July 2006, fighting erupted again between Israel and the Lebanon-based radical group Hezbollah, causing widespread devastation despite the conflict's short duration. Today, the impact of civil war in Syria has flooded Lebanon with streams of refugees, severely straining the country's resources.

Throughout the turmoil of the past 50 years, the Lebanese have retained customs that link them to other Arabs and have maintained values that make them some of the most skilled and educated people in the Middle East. Their complex cultural and political profile, however, continues to challenge notions of Lebanese unity.

Palestinians

Population: 2.7 million
Location: Palestinian Territory, Israel
Language family: Semitic

Perhaps the most highly disputed territory on Earth, Palestine exerts a sacred, symbolic, and very tangible hold on the members of the three great Abrahamic faiths. It is a name long used to describe the land between the Mediterranean Sea and the Jordan River, and it derives from the Philistines, an Aegean people who arrived there at about the same time as the Hebrews, 3,000 years ago.

After the Romans expelled the rebellious Jews in the years A.D. 70 and 135, Palestine eventually became a Christian land. The Arabs brought Islam in the seventh century, amassing converts and intermingling with the population. Later, four centuries of Turkish domination ended with World War I, replaced by a British mandate. Promises made to both Arabs and Jews gave both strong hopes of a permanent Palestinian homeland. Diaspora Jews began to return and form settlements.

In 1947, the United Nations voted to divide Palestine into an Arab and a Jewish state. Fighting began immediately and intensified first when the State of Israel was proclaimed in 1948 and then, with the exodus of more than 750,000 Palestinians from their homes, an event Palestinians call the Nakba, or Catastrophe. These events initiated the difficult situation that exists today—fueled by continual conflict and violence—with Palestinians now living in varied circumstances.

Some Arabs in the Jewish state were able to stay on and obtain Israeli citizenship. Others live in areas now partially overseen by the Palestinian National Authority—itself subject to internal contention—but mostly under Israeli control; these territories include the Gaza Strip, along the Mediterranean Sea, and the territory along the West Bank of the Jordan River.

PALESTINIANS At a workshop in the Gaza Strip, a Palestinian woman and her brother paint designs on ceramics.

Palestinians also inhabit Israeli-controlled East Jerusalem. Millions of Palestinians, however, are either refugees within the territories formerly occupied by Israel and in neighboring countries such as Syria, Jordan, Egypt, and the Gulf States or émigrés in other parts of the world. Recent conflict in Syria has made its Palestinians serial refugees.

Despite the turmoil of their lives and the fragmentation of their population, Palestinians continue to lead the lives that have defined them as a unique Arab people who are Sunni Muslim by a large majority but also Christian and Druze. Most speak Arabic; some speak Hebrew or English. They live amid economic circumstances ranging from abject poverty to amenity-filled comfort. Many Palestinians remain subsistence farmers, like their ancestors of long ago, and others thrive in business and trade. Refugees experience high rates of unemployment and underemployment. Although they have been torn from their land base, Palestinian families remain strong. Both Muslims and Christians continue to celebrate the holy days and festivals of their religious traditions.

For the Palestinian people, dress strongly conveys notions of ethnicity and identity. Men traditionally wear the black-and-white checked headdress known as the *kafiyyeh,* held in place by the *agal,* a black rope circlet. Often they don a long tunic and perhaps top it with a tailored suit jacket. Traditional dress for a Palestinian woman includes a *thobe,* a long, black, robelike dress with an embroidered bodice. The design and style of embroidery vary by region. In some areas, handstitching covers every inch of the bodice, whereas in others it forms decorative accents. While traditional dress often reinforces Palestinian identity, Western dress, including suits, slacks, and blue jeans, is very common, especially in towns and cities.

Palestinian foods have much in common with the rest of the Mediterranean world and feature wheat breads and cracked wheat, chickpeas, eggplant, rice, and stuffed squash and grape leaves, along with succulent dishes of lamb and beef. At feasts such as weddings, huge spreads of food are assembled for the guests.

Despite difficult circumstances, the Palestinian population is growing. Many refugee children receive a decent education through schools supervised by the United Nations. The handful of West Bank universities have large student populations who tend to be very politicized in their views, political strife and violence being all they have known since birth.

Persians

Population: 49 million
Location: Iran, Afghanistan
Language family: Iranian

The majority population in Iran, Persians are also well represented in Afghanistan, Tajikistan, the United Arab Emirates, and Bahrain. Since the Iranian revolution of 1979, many Persians have immigrated to western Europe and North America as well.

Persians are descended from waves of Eurasian pastoral nomads who arrived in southwestern Asia from 2000 to 1000 B.C. Ancient Persia was home to many different tribes, but as long ago as the sixth century B.C. these tribes were united under the Persian Empire and by the Zoroastrian religion. Only in the seventh century A.D. did invading Arabs bring Islam to the region and supplant the old religion.

PERSIANS Soccer fans react as they watch Iran play in a World Cup match at a coffeehouse in Tehran.

Despite many foreign invasions since then, Persians have continued to maintain their distinct culture, language, and group identity. Their uniqueness is in part reinforced by their religious beliefs: They represent the single largest group of Shiite Muslims in the world, in contrast to the Sunni Muslims who dominate in the rest of the Middle East. A minority of Persians still practice Zoroastrianism, and there are small communities of Jewish, Christian, and Baha'i Persians as well. Members of all faiths observe and celebrate the Persian New Year, Nowruz, an ancient Zoroastrian rite of spring and rebirth that begins with a thorough cleaning of the house and the setting of an altar rich with items that symbolize good luck and abundance. Beginning on the first day of spring and continuing for 13 days, the holiday is a time when families visit each other. On the 13th day, families go on picnics and spend time outdoors while ridding their homes of any bad luck from the past year.

A Persian family is dominated by the father, who tends to be formal and a disciplinarian, while the mother often plays the role of peacemaker and intermediary. The nuclear family is the most important domestic group, but extended families often choose to live near one another. Drawing artistic and poetic inspiration from their past, contemporary Persians have a cultural pride

GENOGRAPHIC INSIGHT

HARVESTS IN THE FERTILE CRESCENT

In 1971, Eldridge Niles and Stephen Jay Gould presented a new theory that changed the course of evolutionary thought overnight. Sifting through millions of years of fossilized arthropods, Niles and Gould found over and over again long periods of evolutionary stasis interspersed with short bursts of substantial morphological change. Known today as punctuated equilibrium, their theory explains the discovery that biological diversity tends to remain stable until something jolts the system, and new species emerge to replace the old ones in short, punctuated bursts of adaptation.

Sprouting Seeds

Niles and Gould's conclusions may hold true for recent human history as well. The Middle East has been an inhospitable habitation for a long time, and yet our ancestors have spent 50,000 years struggling to survive there. One reason for this was that it presented an inviting migration path, lying at the crossroads of Africa and Eurasia, the Mediterranean Sea and the Indian Ocean. Another reason is that their local climate was wreaking havoc.

Genetic evidence of migrations both into and out of the region of the Middle East suggests that early humans were often on the move. Then, around 10,000 years ago, humans changed tactics. Hunter-gatherers living in the so-called Fertile Crescent found that seeds they were collecting (and dropping on the way back home) began sprouting up close to their camps. They started planting systematically and, when conditions forced them to abandon a traditional hunter-gatherer lifestyle in exchange for a reliable food source, their nascent practice of sowing seeds and cultivating their products tied them to the land.

The Temple of Baalshamin, dating from the first century A.D., was part of the remains of the ancient Syrian city of Palmyra, until it was destroyed by ISIS in 2015 in the group's quest to wipe out the area's pre-Islamic heritage.

From Seed to City

This move to agriculture, often called the Neolithic Revolution, rapidly changed the course of human culture. It allowed higher population densities, stratified societal complexes, and warfare. Seemingly overnight, the age of civilizations was born. Though DNA from some populations indigenous to the region shows ancient ties to those first migrations, the predominant genetic patterns of today suggest that the early farmers of the Fertile Crescent moved south into both the Levant and the Arabian Peninsula, sticking to the lush Lebanese and Zagros Mountains until around 6,000 years ago. Then, when the region turned wetter, they began leaving the highlands to establish towns in lowland Mesopotamia. Urban societies quickly developed and, with the emergence of Sumerian, the earliest written language, the rise of the first great city-states was at hand.

that runs very deep: Mystical poetry, classical Persian music, calligraphy, and miniature painting continue to be practiced in their traditional forms, although the global pop culture also exerts an influence.

Perhaps the most dramatic differences among Persians in the Middle East are to be found between urban residents and those who reside in rural areas. Urban Persians live in a hierarchical society based on occupation and social class, while rural Persians occupy tightly knit villages and have a farming lifestyle. Whether urban or rural, Persians have formal social customs whose fluency is an important aspect of functioning correctly in their hierarchical society.

One of these customs is *taroof,* the social art of deference. Though a person may be offered much—in the form of concrete items such as gifts or abstract items such as social favors and friendship—by the subtle practice of taroof, the recipient knows how much to accept politely.

Qashqai

Population: 1.5 million
Location: Iran
Language family: Turkic

Largely nomadic or seminomadic, the Qashqai live in southwestern Iran, their traditional territory, and are famous for political prowess. They are among the few peoples in today's Middle East to organize themselves into a tribal confederacy with influence on national and local politics. Shiite Muslims, the Qashqai trace their origins to the 18th century, when several Iranian groups united under one tribal banner.

In the 1930s, the Iranian government forced the people to settle, and other periods of repression followed. But immediately after the founding of the Islamic Republic of Iran in 1979, many Qashqai made a dramatic return to their nomadic lifestyle, herding sheep and goats across the large territory between their summer and winter pastures in the Zagros Mountains. The city of Shiraz lies in the middle of this territory, and the Qashqai people have many ties to it.

Qashqai women enjoy social equality and hold roles complementary to those of the men in the tribe. Given the demands of nomadic life, every man, woman, and child participate in breaking camp, migrating, and setting up a new camp. Because of the policies of the Islamic Republic, Qashqai women now wear the veil more than in times past. Traditionally, however, women did not veil themselves unless they were going to non-Qashqai towns or into urban centers. Women's dress still retains its traditional style of bright colors, multilayered skirts, tunics, and ornate scarves. Men's attire is less colorful but consists of loose-fitting trousers, a tunic and vest, and the famous felt cap with its rounded top and side earflaps. This cap has become a symbol of Qashqai political unity. The Qashqai are also famous for woven carpets with designs depicting tribal life and lands in symbolic form.

QASHQAI Nomadic Qashqai of all ages participate in setting up and taking down these goat herders' encampments.

Turks

Population: 59 million
Location: Turkey
Language family: Turkic

From as far away as Mongolia and China, several migratory waves of different Turkic tribes came to southwestern Asia between the 11th and the 15th centuries. Each tribe created a distinct culture, identity, and way of life. The invaders who unified—first the Seljuk Turks in a large part of the Middle East, and later the Ottoman Turks in Anatolia (a portion of Turkey) and the Balkans—formed an expansive Turkish culture with a common language and historical experience. Islam was everyone's religion, and Turkish was the shared identity. Other Turkic peoples, speaking related Turkic languages, diverged and forged their own identities and cultures separate from the

Anatolian Turks. These groups live in Iran, Afghanistan, China, Azerbaijan, Kazakhstan, Russia, Uzbekistan, Kyrgyzstan, and Turkmenistan.

Nearly 60 million Anatolian Turks live in Turkey, with an additional 2.7 million in Europe, especially in Germany, Cyprus, and Bulgaria. Although most Turks speak Turkish, members of the large groups of emigrants now residing and working in Europe may also speak German or other European languages. A tongue unrelated to Arabic or Persian, Turkish is from the Turkic language family of Central Asia.

From the beginning of their settlement in Turkey, Turks rose in influence in the Middle East and North Africa. They ruled many Middle Eastern peoples during the time of the Ottoman Empire and integrated these groups into their culture. At their most powerful, around 1700, Ottoman Turks governed all the lands from the Persian Gulf, through the Balkans, and across North Africa to Algeria. The Ottoman Empire ended after World War I and was met by a rise in Turkish nationalism and pride under the leadership of Mustafa Kemal, or Atatürk, who advocated secular modernity.

A long time has passed since Turks in Turkey were nomadic or tribally organized, and today their society is a complex mix of opposites. People see their identity as both European and Middle Eastern, as secular while still devoted to Islam, and as Western while also honoring their Eastern traditions. Turks in western Turkey tend toward more cosmopolitan attitudes and more Western-influenced industries. Many Turks in the central and eastern regions of the country are closely tied to the land; they have maintained traditional customs, such as a division of labor where the women tend fields and weave rugs and the men govern public life. Turkish family relations, which are similar to those of other Middle Eastern peoples, follow marriage and inheritance practices that emphasize the father's lineage and are in accord with Islamic law.

A Turkish Islamic mystical order, the famed Mevlevi Sufis, are popularly known in the West as the whirling dervishes. While mystical Islam is expressed throughout the Middle East, in Turkey the Mevlevis uphold the philosophy that all humans can gain direct experience of the divine, and they combine this belief with the physical practice of whirling. Their aim is to achieve a trance state wherein their awareness is fully concentrated on God.

TURKS A Mevlevi Sufi, member of a mystical Islamic order, twirls in a dance that unites heaven and Earth.

ACROSS CULTURES

WHEN LANGUAGES DIE

The world is losing its languages at an alarming rate—and with them, we lose wisdom, diversity, culture, and knowledge.

Human languages are rapidly going extinct. In the year 2015, at least 7,120 distinct human languages were spoken worldwide. Linguists predict that by the end of our current 21st century—by the year 2101, that is—only half of the languages in current use may still be spoken. The voices of the last speakers of small languages are now fading away, never to be heard again.

There is another side to this picture. On a global scale, a few major tongues dominate. The biggest 83 languages, including Mandarin Chinese, Spanish, English, and Hindi, account for nearly 80 percent of the world's population. Most languages have far fewer speakers. More than half of the world's languages are spoken by less than one percent of the population. Languages such as Urarina (spoken by fewer than 3,000 people in the Amazon), Halkomelem (spoken by 200, in Canada), and Tofa (spoken by no more than 25 people, in Siberia) face a precarious and uncertain future.

Small languages are being abandoned by native speakers all over the world. Why is this happening? Native speakers stop using their original language for a variety of reasons. They may favor a different language because it is more dominant, more prestigious, or more widely known. They may be motivated by official state policies to suppress speech or by social pressure to speak differently. Many Canadian First Nations people and Native Americans who are now adults shared dismal experiences of childhoods spent at boarding schools where they were forbidden to speak their mother tongues. Children worldwide experience both subtle and overt pressures to switch to globally dominant languages.

The global biosphere faces grave threats as animal and plant species go extinct. But linguistic extinction is happening much faster as the human knowledge base erodes under the pressures of globalization. Scientists' best estimates show that since the year 1600, the planet has lost more than 1,200 plant and animal species. These numbers, while sobering, make up less than one percent of known species. Compared to this, the estimated 40 percent of languages now endangered is a staggering figure.

Our understanding is still quite imperfect as to how and why language death occurs. Children's decisions to learn a global language and abandon their ancestral tongues may ripple through societies to create a tidal wave of change.

We also lack a clear understanding of what exactly is being lost. Is it unique, irreplaceable knowledge, or merely commonsense knowledge easily translated? Could such knowledge ever be adequately captured in books and video recordings in the absence of any live speakers? Once it has vanished, can such knowledge reemerge spontaneously or be re-created, or is it forever extinguished?

The Multiplicity of Languages

How did we get so many languages in the first place? we might begin by asking. What is the global distribution of all those languages? The natural state of human beings, harking back to our hunter-gatherer past, was to live in small bands. This gives rise to language diversity, as speech habits are free to change rapidly within a group. If one group splits into two, the pace of language change is such that within the space of just eight or ten generations, descendants of the same ancestors may already begin to have difficulty communicating. Within two to three centuries, mutual comprehensibility can be lost: Where one language was, now there

Johnny Hill, Jr., of Arizona is one of the last remaining speakers of Chemehuevi, an endangered indigenous American language.

are two, and a small language family is formed. We can see similiarities that any two sister languages have inherited from their common ancestor; for example, the Irish *éan* and Manx *eean* both mean "bird."

We find the greatest linguistic diversity in places with small and sparsely distributed populations. Consider the 65 inhabited islands of Vanuatu, for example. Altogether, these islands equal Connecticut in size geographically, but their population is less than a tenth its size, around 250,000 people. Yet Vanuatu boasts 112 distinct tongues—one for every 2,200 people.

Linguists are working to get an accurate measure of global language diversity before it vanishes. The Living Tongues Institute for Endangered Languages, with support from the National Geographic Society, has developed a model called Language Hotspots. These zones, found around the globe in often remote and surprising locations, show the greatest concentrations of endangered languages. The Language Hotspots map helps to visualize this extinction crisis, focusing our attention and resources on the most critical areas worldwide, in order to better preserve languages.

When a Language Dies

What exactly is lost when languages vanish? Why should we care? One answer is that we lose unique human knowledge. When ideas go extinct, we all grow poorer.

Many medicinal plants are known only to local practitioners, their healing secrets encoded in unwritten languages that distill generations of experience.

Entire plant and animal species still awaiting scientific discovery by Western scientists are in fact already well known, described, and classified by the indigenous peoples who recognize, gather, and use those plants in the languages they speak. Losing those languages, we lose more than words; we also lose knowledge.

Societies that rely on nature for survival have developed technologies to cultivate plants, domesticate animals, and sustainably exploit local resources. These indigenous knowledge systems can be even more precise than our modern scientific ways of knowing. For example, the Musqueam people of British Columbia, fisherfolk who speak Halkomelem, have always used just one word to name the fish that we call salmon and trout. At first glance, it appears to be a simple mistake in classification. But in 2004, a genetic study confirmed the Musqueam classification to be more accurate genetically than our English one: Pacific trout and salmon belong to the same genus.

Much—if not most—of what humankind knows about the natural world lies completely outside of science textbooks, libraries, and databases, existing only in spoken languages. This unbroken oral tradition is always just one generation away from extinction, always in jeopardy of not being passed on. An immense knowledge base remains largely unexplored and uncataloged. The need for traditional knowledge becomes ever more acute as we strain our planet's carrying capacity. But we can hope to access this knowledge only if the people who possess and nurture it can be encouraged to continue speaking their languages and to pass them on to their children.

There is a second answer to the question of what is lost: our common cultural heritage. A wealth of traditional wisdom is found in oral history, poetry, epic tales, creation stories, riddles, and wise sayings. These products of human ingenuity and creativity may be found in all cultures. But the vast majority of human languages, never written down, exist only in memory and are especially vulnerable to forgetting as languages go extinct.

There is nothing so sacred in a culture that it cannot be forgotten. The Tofa people of Siberia are losing their language and no longer remember the following creation myth that they once believed:

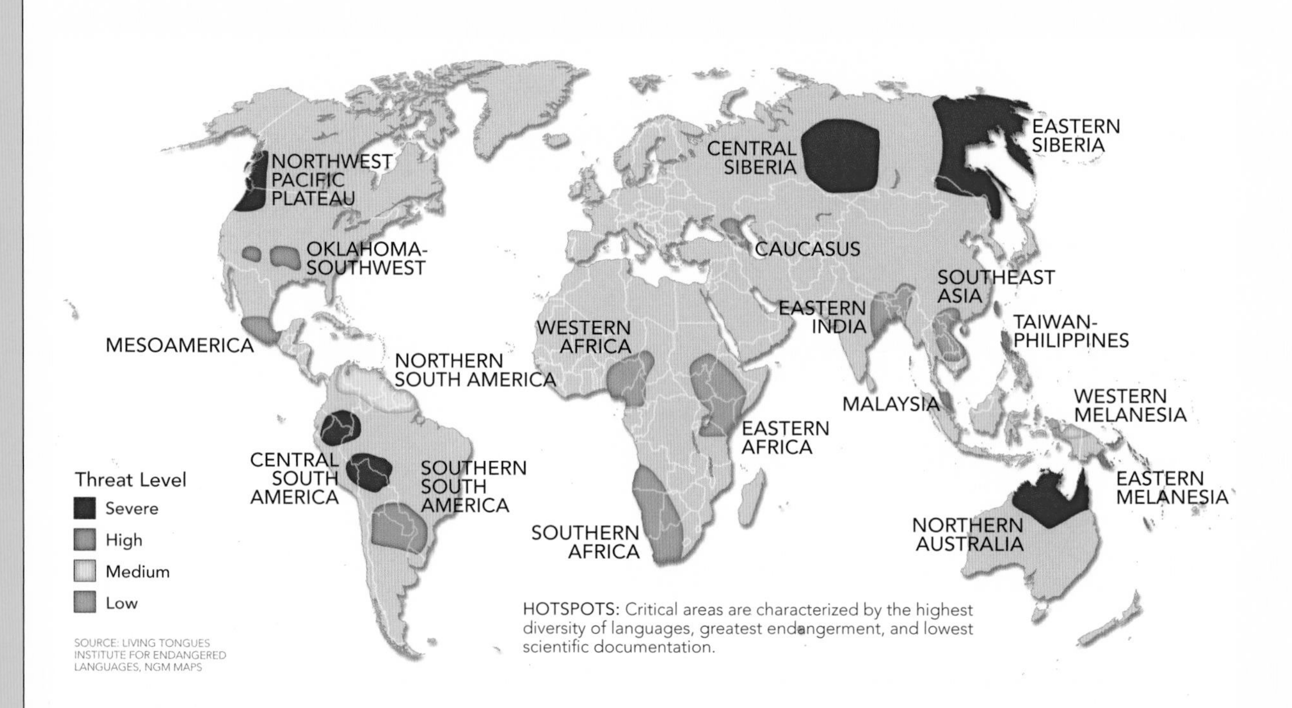

Linguists have identified global language hotspots where languages are in serious decline.

In the very beginning there were no people,
  there was nothing at all.
There was only the first duck,
  she was flying along.
Having settled down for the night,
  the duck laid an egg.
Then, her egg broke.
The liquid poured out and formed a lake,
  and the egg shell became earth.

We may be indifferent to the passing of this whimsical duck story, but all mythical traditions are attempts to make sense of the universe. Each one provides a piece of the puzzle of how humans understand life, creation, and the sacred. Without this Tofa creation myth, we are missing a small piece of that puzzle.

Young Monchaks, traditional nomads of western Mongolia, see more advantage to speaking Mongolian than Monchak.

A Conservation Challenge

A third answer to the question, "What is lost?" is our scientific understanding of the human mind. Languages reveal the limits and potentialities of human cognition—how the mind works. Underneath our global Babel lie deep similarities in the way our brains process speech and information. As a science, linguistics seeks the deep universal properties of all human languages.

But to advance their science, linguists need to examine the oddest, quirkiest, most unusual languages and words. If we had only major world languages to study, we would be severely limited in understanding human cognition. Linguists' theories have been challenged (or flatly contradicted) by the discovery of unique language structures not previously documented.

For example, a basic feature of a language is the standard order in which subject, verb, and object occur. In English, it's usually subject, verb, and then object ("He eats rice"). The Urarina tongue, spoken in Peru's Amazon jungle, places the object first, the verb second, and the subject last ("Rice eats he"). This pattern differs not only from English but also from almost every other known language in the world. Without Urarina and a few other Amazonian languages, scientists might not even suspect such a pattern to be possible. The loss of even one language may forever close the door to a full understanding of human cognitive capacity.

We must safeguard and document vanishing languages, first, because our human knowledge base is rapidly eroding. Most of what humans have learned over the millennia about how to thrive on this planet is encapsulated in threatened languages. If we allow them to fade, we may compromise our ability to survive as our human population strains Earth's ecosystems. Second, our rich patrimony of human cultural heritage, including myth, history, and poetry, is being forgotten. Allowing our own histories to be erased, we condemn ourselves to a cultural amnesia that may undermine our ability to live in peace with diverse peoples. Third, the great puzzle of human cognition remains to be solved. Much of the human mind is still a black box. We cannot discern its inner workings, and we often know them only by what comes out in the form of speech. Obscure languages hold keys that may unlock the mind.

For all these reasons, documenting endangered languages while they may still be heard, and saving or revitalizing tongues that still may be viable, must be viewed as the greatest cultural conservation challenge of our lifetime.

Adapted from the book *When Languages Die: The Extinction of the World's Languages and the Erosion of Human Knowledge* by K. David Harrison.

Chapter 3
ASIA

CAUCASUS
Armenians • Azeris • Chechens & Ingush

CENTRAL ASIA
Baluchi • Burusho • Kazakhs
Kyrgyz • Pashtun • Turkmen • Uzbeks

EAST ASIA
Ainu • Bai • Dong • Han • Japanese
Koreans • Mongols • Tibetans • Uygurs

SOUTHEAST ASIA
Balinese • Dayak • Hmong • Ifugao
Javanese • Khmer • Minangkabau
Thai • Vietnamese

SOUTH ASIA
Bengalis • Gujaratis • Nagas • Nepalis
Punjabis • Sora • Sinhalese • Tamils

THE ASIAN NORTH
Chukchi & Koryak • Evenk • Nenets
Sakha • Tuvans • Yukaghir

Asia's cultural makeup fully matches the geographic complexity and richness of the world's largest continent. From Japan to Georgia, from Mongolia to Malaysia, Asia is home to more ethnic groups than any other major region. These groups have evolved over a very long time. Some of them developed in isolation, while others have complex histories of interaction and sharing, either through peaceful migration or military conquest.

China, India, and Indonesia—the first, second, and fourth most populous nations of the world—are home to 40 percent of Earth's people, and in these three countries reside extraordinary numbers of ethnic groups. Language, a major measure of ethnicity, reveals in its diversity the range of a country's human wealth: Indonesia counts more than 700 languages, India more than 400, and China more than 200. In China, at least 90 percent of the people consider themselves to be Han. This term is commonly used interchangeably with Chinese and dates from the Han dynasty (206 B.C.–A.D. 220), China's first long-lasting period of unification. For more than 2,000 years, smaller ethnic groups have been driven into mountains and marginal areas by the Han Chinese, whose more than one billion people now constitute the world's largest ethnic group.

China's population has tens of millions of minority members constituting hundreds of ethnic groups, yet officially the Chinese government recognizes only 55 "national minorities." Some groups have been lost through a steady acculturation, which continues to threaten Tibetans, Mongolians, and others whose lands abut heavily populated Chinese regions. One group, the Manchu, ruled China as the last imperial dynasty (until 1911) but did so at the expense of ethnic identity: By assimilating Chinese ways, the Manchu lost their own language and culture. While millions survive, their language now has as few as several dozen speakers.

Recalling the era of Genghis Khan, a formation of Mongolian horsemen opens an annual festival in Ulaanbaatar.

RELIGION Throughout history, religion has given cohesion and meaning to ethnic groups, so much so that it often seems inseparable from an overall way of life. This is clearly true with Tibetan Buddhism, where belief and culture have evolved intertwined. Many parts of Asia have been touched by Islam, brought by Arabs from the seventh century onward. The religious, political, and social forces of this faith continue to dominate hundreds of millions of people every day. In Armenia, Christianity has been a foundation of society and a balm during the long and difficult history of its people. But religion can also inflame local hatreds. Among others, ethnic hostilities have erupted between Muslims and Hindus in India, Hindus and Buddhists in Sri Lanka, and Christians and Muslims in Indonesia.

Buddhism developed into a world religion from Indian origins 2,500 years ago. Eventually, it all but disappeared in its homeland, but by traversing the Himalaya and streaming along Silk Road trade routes, the religion reached China and from there spread to other parts of Asia. In Japan especially, Buddhism and contacts with China's mature civilization helped shape the character of the people.

While holding fast to their indigenous traditions, the Japanese borrowed and adapted foreign ways to great benefit. Landscape gardening, flower arranging, and the art of serving tea, for example, all began in China, with Japan bringing them to full bloom. The tea ceremony involves the graceful serving of tea with simple implements in impeccable surroundings, combining aesthetic and spiritual aspects to express the Japanese love of beauty and inner calm. Despite Japan's modern ways, this ritual survives and continues to grow in popularity.

INFLUENCES Through government and architecture, painting and sculpture, science and warfare, even the use of chopsticks, Chinese institutions also influenced the cultures of Korea, Vietnam, and other areas of East and Southeast Asia. Even so, local peoples created and maintained vital, thriving cultures of their own, while sharing a common written language and elements of Confucianism, a code of conduct as much as a religion that dates back to its founder, born in China in 551 B.C. In fact, the Confucian ethical system of social relations still underlies the modern state in China, Korea, Singapore, and elsewhere.

India has left its mark too, dramatically spreading its civilization and religions into South and Central Asia. Aspects of its culture reached into Sri Lanka, Pakistan, and Afghanistan, to name a few present-day countries, and traveled sea routes all the way to Bali, which remains Hindu to this day. Indian scholars and priests introduced many of Asia's peoples to the classical languages of Sanskrit and Pali, which were used in early writings and historical records.

A T-shirted young girl provides a sharp contrast to her traditional grandmother in Delhi, India.

RICE One of history's most important developments remains a shared legacy and a critical component of many Asian cultures: wet-rice cultivation. For millennia, Asians from Japan southward have shared the common inheritance of rice, a staple that is central to the survival of groups large and small. People eat it every day and rely on it for a hundred other uses. Versatile rice straw, for example, supplies material for thatch, basketry, floor mats, sandals, brooms, bedding, rough cloaks, and hats; it also serves as feed for animals.

People in all of Asia's rice-growing countries honor this grain through seasonal festivals and ceremonies. For them, rice has a sacred, even divine significance, so offerings of cooked and uncooked rice, rice cakes, rice stalks, and rice wine serve to ensure good harvests and to rejuvenate the vital forces of the Earth itself. Many Asian cultures esteem a god or goddess of rice. Among the Ifugao of the Philippines, a carved effigy of the *bulul,* or rice god, is placed in the rice granary to honor and protect the harvest.

Rice cultivation demands many hands and a spirit of cooperation to prepare fields, sow seeds, transplant shoots, maintain irrigation systems, and harvest, husk, and polish the grains that will be cooked and eaten. To grow this grain in submerged fields, farmers have reshaped entire landscapes with hand-built terraces, dikes, water channels, and carefully constructed beds stretching to the horizon. Stepped emerald fields, such as those of the Ifugao people in the mountains of central Luzon in the Philippines, stand as a lasting image of Asia.

Caucasus

The Caucasus region is divided into two major subregions: the North Caucasia, which is a part of Russia, and the Transcaucasia, which comprises Georgia, Armenia, and Azerbaijan. The topography here is defined by a 750-mile-long belt of mountainous land stretching from the mouth of the Kuban River, on the Black Sea, to the Absheron Peninsula, protruding into the Caspian Sea. For millennia, the Caucasus Mountains have been seen as the boundary line between Europe and Asia. It was on these heights that Zeus tied Prometheus, according to ancient Greek myth, to punish him for giving fire to humans. The region has served as the gateway linking those two continents with the Middle East. The political, cultural, and linguistic tableaus of these lands have shifted constantly through trade and travel, despite the daunting mountain range that slices though them.

Armenians

Population: 5.9 million
Location: Armenia
Language family: Indo-European

Landlocked Armenia, a mountainous country slightly smaller than Maryland, with many rivers and narrow, fertile valleys, counts some three million people within its borders. But this population represents about half of the world's Armenians. The rest—officially considered part of the nation—live in the former Soviet Union, India, the Middle East, Poland, Romania, Western Europe, and North America, where many have achieved success in business, crafts, and other occupations. Independent only since 1991, the Republic of Armenia faces major challenges: The economy is grim, many skilled workers are emigrating, and an ethnic war with Azerbaijan has killed tens of thousands.

There is no marksman who does not miss, there is no sage who does not err.

TURKIC PROVERB

Armenians remember a time during the Middle Ages when they were part of a major kingdom that also included what is now northeastern Turkey. Yet the greatest events in Armenia's collective memory are invasions, suppressions, and forced exiles over the past thousand years. The worst was the 1915 massacre, considered by many to be an act of genocide, when Turks drove Armenians from eastern Anatolia; more than 600,000 died.

Among diaspora Armenians, cosmopolitan attitudes coexist with a great love of home, family, and age-old traditions. The people strongly value their distinct language. Although it belongs to the large Indo-European family, it has no close linguistic relatives and is written in a unique 38-letter alphabet.

A distinctive brand of Christianity is the foundation that underlies the unity of all Armenians, who converted to the faith early and in the fourth century

ARMENIANS In Istanbul, Armenians gather to commemorate the more than 1.5 million Armenians slaughtered by Ottoman Turks in 1915.

declared that Christianity was their state religion. This tie has produced an invaluable corpus of religious literature as well as remarkable art and architecture. The people are renowned for their mosaics and tiles, and the circular dome of their churches is thought to have had a strong influence on Byzantine and Ottoman buildings. On theological grounds, primarily the interpretation of Christ's divinity, the Armenian Apostolic Church stands independently and proudly next to the Roman Catholic and Orthodox Churches. It has been a pillar of Christianity and a primary force in maintaining Armenian culture.

Azeris

Population: 8.8 million (in Azerbaijan)
Location: Azerbaijan, Iran
Language family: Turkic

The more than eight million Azeris of Azerbaijan form the largest group in the ethnic mosaic of the Caucasus. They also constitute an even larger, yet somewhat hidden, diaspora population of 15 to 18 million in neighboring Iran. A hundred years ago, their titular homeland was the world's top oil producer; today, it is again actively tapping the immense Caspian Sea reserves, with the oil industry advancing the trend toward urbanization. While most Azeris now live in cities, many are farmers and nearly all are Muslims, in contrast to neighboring Armenia and Georgia, which are Christian. According to local custom, Azeri Muslims may drink wine, and women are free to work and go unveiled.

Islam arrived with 7th-century Arab conquerors, the first of many waves that included Turks, Mongols, Persians, and Russians, who began dominating the area in the 19th century. A hybrid people, Azeris were ethnically and linguistically Turkic by the 11th century, yet their roots go deep into ancient Caucasia and Iran. To the south lies an Iranian region also named Azerbaijan, larger in both size and population but lacking a nation's cohesiveness.

Azeris also honor Zoroaster, who lived in the sixth century B.C. and founded Zoroastrianism, a religion with a central belief in the cosmic conflict between good and evil. Today, their Islam sets them apart. Of Islam's two main branches, the Shiites began to dominate after the departure of the Sunnis, who had allied themselves with Ottoman Turks in the 19th century. Later there was reconciliation as Azeri nationalism stressed a common heritage and downplayed sectarian differences. With improved relations came a cultural flowering from the mid-1800s to the early 1900s, bringing new libraries, hospitals, schools, art academies, and charities. A spirit of emancipation spread to Turkic peoples throughout the Russian Empire and elsewhere.

The Soviet era suppressed this awakening, closing most of Azerbaijan's 2,000 mosques, crushing intellectuals and artists, and forcing the people to abandon Arabic script and adopt a Cyrillic alphabet. Then, under glasnost and following independence in 1991, writers and historians returned to resurrect past glories. In recent years, Azeris have reopened or rebuilt their mosques, renewed ties to their Azeri cousins in Iran, and began a slow transition from Cyrillic to a Latin alphabet. After 70 years of Soviet rule, the Azeris are working to remake their polity into a modern, relatively secular Islamic state, with a putative democratic form of government, yet they are beset by corruption and limits on political freedoms.

AZERIS This Azeri man lived much of his life under Soviet rule, when Azeri culture was suppressed in favor of Russianization.

Chechens & Ingush

Population: 1.4 million (Chechens); 455,000 (Ingush)
Location: Russia
Language family: Nakh

The Chechens and the Ingush together share many aspects of culture and language, as well as a remote location in the northern Caucasus Mountains, on the southern fringes of Russia. They also unfortunately share centuries of history fraught with domination, violence, and repression, which have accelerated in the decades since World War II.

The names *Chechen* and *Ingush* are actually Russian designations; the two groups refer to themselves collectively as *vaj nakh*—our people. They live in adjoining territories that were formed by the Soviets into a single autonomous republic. Chechnya declared independence in 1991, and both it and Ingushetia are now republics within the framework of the Russian Federation.

Both groups are Sunni Muslims, having embraced Islam between the 17th and early 19th centuries. Chechen and Ingush represent two distinct but closely related languages of the Nakh branch of northeast Caucasian languages. Exposure to each other's language has allowed them to communicate easily; many Chechens and Ingush also speak Russian.

Their territory incorporates fertile, northern lowlands where farming is practiced, and high mountains and valleys where sheepherding forms the major occupation. Lowland villages tend to be larger than mountain ones, with compactly arranged one-story wood or brick houses, while mountain villages typically feature stone buildings, including tall watchtowers that also serve as homes. People from the highlands traded their livestock and eggs for grain from lowland farmers. In more recent times, both groups participated in oil production from the nearby Caspian Sea fields.

Traditionally, men and women led largely separate lives, with extreme deference shown to men by women: Women stood when a man entered the room, they kept eyes downcast when a man spoke to them, and they ate separately, in the kitchen, after the men had eaten.

CHECHENS Chechens cheer on their team at a soccer match in Grozny, the capital of Chechnya.

Women still dress modestly and cover their heads, usually with a wool or silk scarf. More traditional Chechen and Ingush men wear tall lambs' wool hats.

Like other people of the Caucasus, Chechens and Ingush place a high premium on hospitality. Visitors are entertained lavishly with the best the family has to offer in food and accommodation. Meals tend to include lamb or mutton; vegetables such as tomatoes, eggplants, and peppers; corn or wheat flatbreads; fruits such as apples, pears, and plums; and a large number of dairy products, including milk, butter, cheese, curds, yogurt, and sour cream.

Chechen and Ingush social organization is based on clans of paternally related lineages with no internal or external hierarchical relationships, other than the high esteem and respect accorded to elders and others who have earned merit. Marriages usually are contracted outside of one's clan. The egalitarian nature of Chechen and Ingush life has made it difficult for them to accept outside authority.

The Chechens, however, have demonstrated a more rebellious nature against Russia's many attempts to control the land and its access to main routes through the mountains, eliminate the people directly or eradicate their culture by forced Russianization, and, more recently, acquire more direct access to Caspian Sea oil. Chechens were fierce fighters during the nearly 50 years of the Caucasian Wars, which ended in defeat in 1864, with many northern Caucasians killed, repressed, or forced to emigrate.

The relative reluctance of Ingush to resist has not prevented them from feeling the effects of a kind of guilt by association. Along with the Chechens, the Ingush were accused by Stalin of Nazi collaboration during World War II and exiled for 14 years to Central Asia and Siberia. The journey and the harsh conditions of exile took their toll on both populations. During the exile, Chechen and Ingush lands were resettled by Russians and Russian-speaking Cossacks.

The Chechens' declaration of independence in 1991 caused the Ingush to withdraw from their union to avoid the hostilities between the Chechens and Moscow. A brutal war of succession with Russia followed from 1994 to 1996, followed by a long-term occupation by Russian troops. After most Russian troops withdrew in 2009, reconstruction began in earnest, yet progress is slow, and sporadic insurgency and discontent with Russian rule percolate.

GENOGRAPHIC INSIGHT

THE CENTRAL ASIAN PASSAGEWAY

Only a few thousand years after the second wave of humans migrating out of Africa had settled into life in the Middle East, climate shifted once again, drying out the vast plains of the Tigris and Euphrates River basins and forcing these early hunter-gatherers to pick up camp and search for a new homeland. The luckiest of these migrations headed northeast, crossing the Zagros and Taurus Mountains before moving directly into Central Asia, stopping finally to settle amid the giant granite upwelling known as the Pamir Mountains.

Genetic information from today's Pamiri populations shows that their DNA has been there continuously for the past 40,000 years, almost as long as the first lineages found in the Middle East. The finding is now well accepted by population geneticists, but it initially drew some skepticism. After all, it is difficult to imagine why humans, largely accustomed to life on the lowland African savannas, would make a move to settle in one of the world's more rugged landscapes.

There is an explanation.

On to Greener Lands

Around 45,000 years ago, the world was entering a dry spell, and the lush corridor that had made North Africa and the Middle East hospitable for life and fruitful for cultivation was becoming an inhospitable desert, poor in the food and resources necessary for survival.

Dependent on game and water, the human populations that had settled in the Middle East shortly after migrating out of Africa were now being evicted by climate change, forced to look elsewhere for green pastures. Fortunately for them, the mountains to their east provided a much needed refuge. In drier times, mountains act as atmospheric sponges, soaking up what little moisture is available in the air around them and providing a haven for wildlife. If the Garden of Eden had mountains, we can be fairly certain that they looked something like what central Asia did 35,000 years ago.

Staging Ground

Once in Central Asia, humans thrived, quickly expanding across the steppes, inhabiting the river valleys of the Pamir Mountains, and setting the stage for migrations to come.

As the climate turned again, groups began spreading out from the region, generations of rugged mountain living enabling them to inhabit all corners of the world successfully.

Europe, East Asia, India, and the Americas were all populated by independent migrations originating from the Eurasian heartland over a period lasting 20,000 years.

A Tehran woman makes a fashion statement with a fluorescent head scarf and matching sneakers.

Central Asia

Politically, Central Asia often refers to the cluster of former Soviet republics—Kazakhstan, Kyrgyzstan, Tajikistan, Turkmenistan, and Uzbekistan—encircled by the nation giants of India, China, and Russia. Geographically, the region combines bleak flatlands, including the Russian steppes, and nearly impenetrable mountain ranges: the Zagros of Iran, the Altay of Russia and Mongolia, and the Karakoram, in territory under historical dispute between China and Pakistan, which contains K2, the world's second highest mountain. The Silk Road snaked through these reaches, connecting Asia, Europe, and Africa through centuries of trade and cultural exchange.

Baluchi

Population: 7.4 million
Location: Pakistan, Iran, Afghanistan
Language family: Indo-Aryan

Fiercely independent seminomads whose origins may go back to Middle Eastern and Caspian Sea nomadic peoples, the Baluchi have for centuries integrated marginal groups from many cultures under their tribal name. Today, this broad ethnic group comprises dozens of tribes and occupies a vast area of Pakistan, southeastern Iran, and southwestern Afghanistan. Although arid and largely inhospitable, with dramatic landscapes and few roads, the region also encompasses green valleys, juniper forests, and lofty mountains with wildflowers, as well as irrigated gardens and orchards producing dates, apples, cherries, almonds, and cumin. Where the land meets the Arabian Sea, the 500-mile-long coastline has fine, sandy beaches.

About 7 percent of the Baluchi call Baluchistan—nearly half of Pakistan's land surface—their home. Worldwide, the group numbers some ten million individuals, including people who speak Baluchi, an Indo-Aryan language similar to Kurdish and Pashtu; people who consider themselves to be Baluchi but speak a different language, such as Urdu or Farsi; and even people whose disparate ancestors historically found solidarity among the Baluchi. Each tribal group follows a head chief, or *pir*, a Sufi mystic or Muslim holy man.

Nearly 80 percent of all Baluchi raise sheep and other livestock and rely on small-scale farming activities for part of their livelihood. True nomadism has almost disappeared, and the pastoral life has given way to cultural assimilation within larger nations. The Baluchi are increasingly entering the mainstream, frequently as migrant laborers. Even so, they retain a vibrant storytelling tradition that includes age-old songs with themes emphasizing courage, heroism, protection of the weak, and veneration of ancestors.

In Afghanistan and eastern Iran, the tribal name is still applied to intricate rugs with rich, dark reds and blues, highlighted sparingly with white. The colors originate from vegetable and fruit dyes used long ago but in the past century have been replicated synthetically.

BALUCHI A Baluchi woman wears her culture's distinctive slitted red mask to complete her veiling.

Burusho

Population: 87,000
Location: Pakistan
Language family: Language isolate

The Hunza region, within northern Pakistan's Karakoram Range, is home to the Burusho, a people who claim descent from soldiers in the fourth-century B.C. army of Alexander the Great. As fanciful as the story may be, many Burusho do appear European; some people say they look specifically Celtic. The Burushaski language is unrelated to any other world tongue.

Over the years, misconceptions about the Burusho diet and a romantic ideal held by gullible Westerners gave rise to stories about exceptional life spans in the Hunza region. While it is true the people have traditionally relied on apricots during the long winter, the stories connecting their diet to longevity are myths.

The Burusho once lived in a much larger area, but they were eventually driven into the mountains, where they established their virtually impregnable homeland. Hunza was ideally situated, allowing the Burusho to tax—or raid—caravans traveling between India and China.

Governed by the same family of *mirs* (rulers) since the 11th century, the remote outpost converted late to Islam; in 1904, it pledged loyalty to the Aga Khan and his Ismaili sect, a branch of Islam with no mosques and no imams. This sect defers, as do the Baluchi, to a *pir,* a religious leader who exhibits sanctity or special charismatic powers.

In 1974, Hunza became fully integrated with Pakistan, ending the long rule of the mirs, and four years later the Karakoram Highway was completed to relieve its severe isolation. Most important, the Aga Khan Foundation's programs for health, education, agriculture, transport, and small businesses have given

BURUSHO Burusho villagers in northern Pakistan sort sun-dried apricots, long a staple of the Hunza region.

new hope to the people. Money has slowly replaced barter, and schooling, especially for girls, is seen as the necessary way forward for Burusho society.

Kazakhs

Population: 10 million
Location: Kazakhstan
Language family: Turkic

The motif of Kazakh history has been the struggle and interplay between Mongol and Turkic tribes across the vast stage of Central Asia, from the Gobi Desert to the Caspian Sea. Kazakhs speak a Turkic language, but some scholars consider them to be Turko-Mongols; in appearance they seem more like Mongols than their Turkic brethren.

United under the Mongols in Uzbek lands, Kazakh clan leaders in the 1460s broke away to return to coveted pastures in the east, where they prospered as nomads. At that time they became known as Kazakhs, which means "secessionists" or "independents," and their ethnic identity firmly took shape.

The arrival of Russian soldiers and traders in the 1600s spelled a slow decline for indigenous culture. The distance from Moscow to Kazakhstan is a relatively short one, so Russian penetration came steadily. In the modern era, Kazakhstan played a main role in Soviet industrial, space, and military development, with its northeastern region subjected to more nuclear tests than any other place on the planet.

Today, Kazakhstan's ethnic Kazakhs do not form a clear majority in their own land; nearly as many Russians live there. Two and a half million more Kazakhs reside in China, Uzbekistan, Mongolia, and Russia. The largest country in Central Asia and the ninth largest in the world, Kazakhstan is sparsely populated and three-fourths desert or semidesert. Deep in the ground, though, lie rich reserves of oil, gas, and coal.

Kazakhs in all regions can communicate fairly easily, but their language is still developing a modern technical vocabulary and thus lives in the shadow of Russian, the language of education, business, and technology. The issue of language and ethnicity has come to the fore as the two official languages, Kazakh and Russian, vie for position and symbolize the deep rifts between two peoples living in entirely different spheres. Russians are skilled, technically capable, and international; Kazakhs seem unsure of the way forward as they struggle to balance the needs of a young nation, Islam, and modern mass culture.

Even in a modern state, the people find security by relying on traditional ties that go back 500 years. Each Kazakh holds allegiance to one of three ancient groups, or *zhuz* (hordes): the Lesser Horde in the west, the Middle Horde in the north and east, and the Great Horde in the south. Because clan and national politics, as well as economic decisions, depend on such affiliations, the ancient hordes remain important in the makeup and functioning of Kazakh society.

KAZAKHS A 13-year-old Kazakh boy brings his superbly trained hunting eagle to a yearly competition in western Mongolia.

Kyrgyz

Population: 3.8 million
Location: Kyrgyzstan
Language family: Turkic

The Kyrgyz represent an amalgam of nomadic peoples that reflects their movement through Central Asia for about 2,000 years and includes both Mongols and Turkic tribes. The earliest Kyrgyz state was a khanate that extended from south-central Siberia to the eastern part of present-day Kyrgyzstan, now Central Asia's second smallest nation.

Mountains dominate 90 percent of Kyrgyzstan's landscape. The Kyrgyz use the high pastures to raise livestock, primarily sheep, cattle, horses, yaks, and goats. They grow crops of wheat, sugar beets, potatoes, and fruit in the fertile Fergana Valley and elsewhere in the limited lowlands.

It is better to be a slave in your homeland than to be a ruler in a foreign country.

KAZAKH PROVERB

Ethnic Kyrgyz today form about 70 percent of Kyrgyzstan's total population. They speak Kyrgyz, which belongs to the Turkic family of languages and did not appear in written form until the 1920s. The Kyrgyz converted to Islam mainly between the 8th and 12th centuries. About 80 percent of Kyrgyz are Sunni Muslims, although the day-to-day expression of their faith is somewhat relaxed, and some Kyrgyz still observe practices from their culture's animist past. Relationships between men and women in Kyrgyz society also tend to be less rigid than among some neighboring Central Asian groups.

Under Soviet control throughout much of the 20th century, the Kyrgyz were forced to abandon their nomadic lifestyle and adapt to collectivized farming and mining. Some Kyrgyz still live mainly in yurts, circular portable homes of felt.

Yurt interiors traditionally are divided into a men's side on the left, which includes accoutrements for

KYRGYZ The family yak seems to comment on her vigorous milking by a tiny Kyrgyz girl.

hunting, fishing, and livestock, and a women's side on the right, where cooking utensils and sewing equipment are stored. A small stove used for cooking and heating sits in the center, and the bedding used at night is hung on a back wall protected by a decorative cover, forming a backdrop for the seating of an honored guest who might be served the delicacy of sliced sheep's liver and the equally prized fat of its tail. The traditional Kyrgyz diet features dairy products, including *kumys*—fermented mare's milk—along with mutton and vegetables.

Although yurts are no longer used as much for residences, even urban Kyrgyz may erect one on an important occasion, such as the birth of a child. So central is the structure to Kyrgyz identity that the national flag features a component of the yurt's roof at its center.

You will know the value of parents when you have children yourself.

KYRGYZ PROVERB

As Kyrgyzstan's mainly urban Russian population, concentrated in the capital city of Bishkek, has decreased, the population as a whole has become more rural. Now some two-thirds of Kyrgyz are involved in agriculture, including animal husbandry. Gold mining has become another important economic activity.

Proud stewards of a rich oral tradition, the Kyrgyz have maintained for a millennium an oral epic poem, *Manas,* that is twice as long as *The Iliad* and *Odyssey* combined—and it still has not been written down in its entirety. It relates the tenth-century saga of the folk hero Manas, considered the father of the Kyrgyz people. Many Kyrgyz know portions of the saga of Manas, his son, and grandson, which also is performed by a special class of storytellers called *manaschy.*

Like many other Central Asian peoples, the Kyrgyz are skilled and enthusiastic equestrians. Children learn to ride almost at the same time as they learn to walk. The Kyrgyz play many sports and games on horseback and show off their skills at shows and festivals.

An independent Kyrgyzstan strives to retain and rebuild Kyrgyz culture and language without making the country's non-Kyrgyz residents uncomfortable. After establishing Kyrgyz as the national language in 1993, the Kyrgyz added back Russian as an official language in 2001, partly to reassure Russian speakers. Despite the predominance of Islam, the Kyrgyz also have not attempted to establish a state religion.

Pashtun

Population: 38 million
Location: Pakistan, Afghanistan
Language family: Iranian

In southeastern Afghanistan and adjacent regions of Pakistan live the Pathan or, as they call themselves, the Pashtun, also spelled Pakhtun. The Persian name *Afghan* referred specifically to this group long before the word became a geographic designation. In multiethnic Afghanistan, where more than 45 languages are spoken, these people form the largest ethnic group, and Pakistan is home to even more Pashtun people.

PASHTUN A solemn Pashtun father poses with his equally solemn son in Helmand province, Afghanistan.

Pashtu, an eastern Iranian tongue with numerous loanwords from Persian and Arabic, forms one of Afghanistan's two national languages. Written in a type of Arabic alphabet since the early 1500s, this language has inspired many poets and writers, including Ahmad Shah Durrani, 18th-century founder of the Afghan nation.

Pashtun lands have witnessed tides of invasion by Persians, Alexander the Great, Turks, Moghuls, British, Russians (who invaded Afghanistan in 1979), and a continuing U.S. military presence since 2001. Undaunted and self-assured, the people have always resisted external control and have gained fame as fierce fighters.

The Pashtun make up some 60 tribes, all with separate territories, and together they form the largest tribal society in the world. Central to their lives as farmers, herders, traders, and warriors is the *Pashtunwali,* the Pashtun code of conduct embodying core values of pride, honor, vengeance, forgiveness, autonomy, and hospitality. The code's strict rules concern property, succession, marriage, the right to speak in gatherings, ethnic purity, and even slights.

Most Pashtun men carry guns, and blood feuds can erupt over land, water, injury, and women. A severe beating or death could be the price a stranger pays for talking to a Pashtun woman. While this behavior seems severe to outsiders, such strict adherence to rules has kept Pashtun society independent and intact.

Tribal ways do erode, though, as electricity, roads, and schools expand and government officials steadily encroach. Attempts by foreign powers and local governments to quash the age-old occupation of opium production have largely failed, and it continues to serve as a primary economic resource, even as the Pashtun connect to the larger world.

Turkmen

Population: 7.5 million
Location: Turkmenistan, Iran
Language family: Turkic

To be Turkmen is to claim lineage through a male Turkmen, be a Sunni Muslim, and speak the Turkmen language, related to modern Turkish and Azeri. In the past, Turkmen were famous as horsemen and warriors from the steppes of Central Asia. As nomads, they herded on grasslands that supported little besides ruminant animals, and they occasionally raided settled areas. Their transient lifestyle enabled them to avoid taxes and control by larger governing powers. Today, more than four million Turkmen live in northeastern Iran, Turkmenistan, Afghanistan, Turkey, Syria, and Iraq.

Though tribal ties are very strong, fewer Turkmen are nomadic now; most are either sedentary farmers who grow wheat, barley, and cotton or seminomads who divide their economic labors between raising crops and herding sheep, cows, goats, horses, and camels. The importance of nomadism and animal husbandry is seen in the higher status held by a *chorva* (herder) compared with the lower status of a *chomur* (farmer). It is widely held among Turkmen that a farmer is someone who does not have the wealth to remain a herder and is therefore no longer independent of larger political powers.

Using wool gathered from their own sheep—and occasionally adding silk to the weave—Turkmen

TURKMEN At an exhibition in Kabul, Afghanistan, a Turkmen woman shows off her weaving skills.

women create carpets that are internationally prized for their quality, design, and colors. Their geometric patterns make repeated use of polygonal shapes called *guls*. Each tribe has a distinctive gul that communicates the tribal affiliation of the weaver, and this design is passed from one generation to the next within the tribe.

The post-Soviet years saw Turkmenistan living under the personality cult and brutal dictatorship of Saparmurat Niyazov, dubbed Turkmen Bashi, "head of the Turkmen." Niyazov's death in 2006 began a gradual process for the Turkmen of establishing a viable political and social order, while tasting new freedoms.

Uzbeks

Population: 23 million
Location: Uzbekistan
Language family: Turkic

The Uzbeks occupy a Central Asian land famous through the centuries for its wealth, learning centers, and culture. Two thousand years ago, the Uzbeks' successful merchant ancestors traded along the Silk Road with Russia, China, India, and Afghanistan, creating a brilliant commercial hub whose riches inevitably attracted invaders.

Arab conquerors in the seventh century brought Islam and helped usher in a golden age, with Bukhara becoming one of the world's greatest cities. From the ninth century A.D. onward, Turks came from the northern grasslands.

Mongols arrived in the 13th century, but Turkic language and culture predominated because Genghis Khan's armies were in fact mostly Turkic soldiers, many of whom settled here and married local women. One Mongol chieftain, Öz Beg, left the legacy of his name, which survives as "Uzbek."

In the 1380s, the Turkic conqueror Timur (sometimes anglicized as Tamerlane) captured much of Asia and brought more scientists, artists, and architects to the region. But after 1501, as Tamerlane's empire was crumbling, a tribe from north of the Aral Sea invaded and became the core of people we now call Uzbeks.

In the 1900s, artificial ethnic and regional boundaries imposed by the Soviets dramatically altered Uzbek life. Until the Soviet Republic of Uzbekistan was set up in the 1920s, the people had seen themselves as members of tribes and clans, not as part of a nation. The Uzbek language, too, has only a short history as a distinct tongue. Before the 1920s, it was not considered a separate state language; instead, it was one of many Turkic dialects. All of the trappings of nationhood were achieved in 1991, when the Soviet empire ended and Uzbekistan became an independent country.

UZBEKS An elderly Uzbek man holds a baby identified as a Muslim by his crocheted taqiyah, or prayer cap.

Uzbekistan's population of 29 million, the largest in Central Asia, is more than 80 percent Uzbek. Sizable Uzbek populations also reside in Afghanistan, China, and Kazakhstan.

While its geographic position and ethnic clout lend Uzbekistan some claim to being Central Asia's leader, the country is beset with economic and environmental ills. Reckless expansion of cotton-growing areas and overuse of water have led to the vanishing of the Aral Sea, which now has just one-tenth its original volume. Fierce storms throw tons of salt and sand from the empty basin into the air, causing severe health problems. Many people worry about the future of the Uzbeks, who once lived so gloriously on the land.

East Asia

China, a country of more than 3.5 million square miles and the most populous nation of the world, forms the heartland of East Asia. It stretches from the 12,000-foot-high Tibetan plateau through highlands and plains of the east and northeast, to the 4,000-mile coastline along the Yellow, East China, and South China Seas. Within China's borders can be found the Tibetan plateau, 12,000 feet above sea level; the Gobi desert, 500,000 square miles; and two great rivers, the Yangtze, flowing into the East China Sea, and the Yellow, flowing into the Yellow Sea. East Asia also contains the mighty Himalayan region, including Mount Everest, at 29,035 feet above sea level the world's highest mountain; the rich and fertile Korean peninsula; and the densely populated, economically significant islands of Taiwan and Japan.

The People's Republic of China claims Taiwan as its 23rd province.

Ainu

Population: 15,000
Location: Japan
Language family: Language isolate

Among Asia's most enigmatic ethnic groups, the Ainu live on Japan's Hokkaido Island and on Russia's Sakhalin and Kuril Islands. For countless generations, these people led outwardly simple lives that relied on fishing, hunting, and primitive agriculture. They were once known as the "hairy Ainu," because men grew large beards and women adorned themselves with mustachelike tattoos around their mouths. Traditionally, they dressed in animal skins and beautiful robes of elm bark and feathers. Where the people came from remains a mystery.

Of the Ainu who survive today, few are considered true, or pure, Ainu. Into the 18th century, most lived on Honshu, Japan's main island, but they were so different from the rest of the population that the Japanese majority marked them for destruction. Official policy and population pressures eventually drove the Ainu people northward. Now, however, the Japanese hold their culture to be worth protecting, and the Ainu themselves have launched efforts to revitalize and reclaim their language, culture, and connections to the land.

For the Ainu, the focus of spiritual life has long been the bear-spirit ceremony, or *iyomante,* a ritual involving a sacred space and objects such as prayer sticks, carved staves, and elaborate shaved-wood animals. During the ceremony, a specially raised bear is sacrificed, to allow its soul to return to the spirit world. This offering affirms Ainu devotion to, and respect for, nature's hidden forces. The skin and meat of the bear are treated as gifts from the gods.

Traditionally, daily life centered on the village and its immediate territory, or iwor, which determined the limits of the Ainu world; within it, the people were largely self-sufficient, relying on salmon, deer, and the cultivation of millet. They excelled in

AINU Traditional dress and facial hair mark the distinctive look of Haruzo Urakawa, an Ainu elder originally from Hokkaido Island.

the construction of bows, arrows, spears, traps, and harpoons, and they retained a deep knowledge of the forests and waters of their homeland. Now many Ainu work in the Japanese construction industry, live in urban areas, and struggle to continue their tradition as a minority ethnic group within Japan.

Bai

Population: 1.2 million
Location: China
Language family: Tibeto-Burman

The 1.2 million Bai of Yunnan Province are among China's most assimilated, prosperous, and successful minorities. For centuries they have maintained close ties with the majority Han Chinese, and because they lacked a written language of their own, they have always relied on Chinese characters. While their name means "white" and they call themselves "Speakers of the White Tongue," the actual origin of the Bai and their language affiliation are not known for certain.

Dali, the Bai capital, once stood at the center of a kingdom that endured for nearly 500 years, separate from dynastic China until the mid-13th century. This well-documented independent history is the reason for the people's status as a recognized ethnic group.

The geography of their homeland has greatly affected the Bai, who dwell along the shores of sparkling Er Hai, beneath a range of snowcapped granite peaks. A 35-mile-long lakeside plain seems specially designed for a protected and bountiful life. Rich, productive rice fields and dazzling yellow rapeseed flowers contrast with the blue lake and red hills of the far shore, planted with peach and pear orchards. Towns and villages of weathered gray stone look across the entire plain, once covered with hundreds of pagodas standing as beacons to the Buddhist faithful. A trio of thousand-year-old pagodas still stands today as the emblem of the Bai.

In the past, horses were very important in this remote, mountainous region, and the Bai became known for the strong, small ponies that they supplied to members of various dynasties. Only with the coming of the Burma Road in the 1930s did the style of Bai life and transport truly change, and today's airports, highways, and trains have brought the Bai people into the mainstream of modern Chinese life.

Still, the old love of horses lives on in an annual fair hosted by the Bai. The event has grown enormously, with mountain dwellers and minorities from many parts of China coming not only to trade animals and traditional medicines but also to race horses, eat and drink, and enjoy the mountain scenery.

BAI Five exquisitely dressed Bai women stand at the edge of a pool near a gate to the Old City in Dali, Yunan.

Dong

Population: 1.4 million
Location: People's Republic of China
Language family: Kam-Sui

Dotting the border regions of South China's Guangxi, Hunan, and Guizhou Provinces are the fascinating villages of the Dong, an official minority of the People's Republic of China. Considered the north-easternmost extension of the greater Thai peoples, the Dong speak a language that, requiring more than 15 distinct tones, is one of the world's most difficult to learn.

In this culture, new parents commonly plant a grove of pines called "18-year trees," which will later supply their son or daughter with timber for a home. This concern for building needs extends beyond individual houses—generally two stories tall, with firewood and farm animals downstairs, belongings and sleeping quarters upstairs—to communal projects and public architecture. All major settlements proudly raise one or more drum towers, elaborate constructions up to 13 stories high, with tight, overlapping pagodalike roofs. People paint the eaves with scenes from myths, legends, and daily life: cooking, farming, weaving, hunting, and the playing of musical instruments. Statues of beasts and birds decorate the roofs. From afar these towers resemble stylized fir trees, sacred to the Dong.

In ancient times, a huge drum suspended inside each tower helped warn of danger; now, its beating calls people together for meetings, celebrations, and festivals, during which songs of praise may be sung for the towers. Dong architecture also includes "wind and rain bridges" of elegant proportions, each bridge having up to five separate pagoda structures. Originally built for shelter, these covered bridges and their special roofing now add beautiful touches to the landscape. Local residents have discovered yet another use for them: Leafy vegetables are hung from the railings to dry.

DONG A village street serves as a kitchen for these Dong soupmakers in Guizhou, China.

This attention to creative design can also be seen in Dong embroidery. On hot days, women may be found sitting in the shade of wind-and-rain bridges, dabbling their feet in cool streams while painstakingly adding bright stitches to bags, collars, aprons, or baby carriers.

Han

Population: 1.2 billion
Location: People's Republic of China, Taiwan
Language family: Sinitic

One in every six individuals on this planet is Han, a member of China's largest ethnic group. Today, about 1.2 billion Han share a country about the size of the United States with more than 100 million people representing 55 other "official" ethnic groups, most of whom live at China's fringes. Han also form the majority of the population of Taiwan.

The Han descend from the people who were settled in the middle and lower basin of the Yellow River 4,000 years ago. The name *Han* came into being as a dynasty in 206 B.C., as those people took an already somewhat unified northern China and solidified it, ruling what they called the Middle Kingdom. Then the Han people moved south, spurred by warfare and famine, spreading into the area below the Yangtze River in at least three great migrations and many smaller ones.

*Sugar may be sweet,
but barley meal is more filling.*

TIBETAN PROVERB

Most Han speak a northern dialect of Chinese. Commonly known as Mandarin, it has become the *Putonghua,* or standard language of China.

Chinese writing is based on a system of some 60,000 characters known as pictographs and ideographs. Unlike hieroglyphics, Chinese characters no longer present any visual clue to their meaning. The mastery of written Chinese requires a degree of education that has not been accessible to everyone. In the 20th century, attempts, not always successful, were made to simplify a script, applicable to all dialects, to make it more widely available.

HAN Bowls with reddish eggs symbolize good fortune for a new start in life for this Han bride and groom.

Since mountains, high plateaus, and deserts make up much of China's west, about 95 percent of the population crowds into the eastern third of the country. Millions live in the cities there, including more than 21 million in Beijing, the capital, but most Han still live in the countryside. Farmers in the west and north grow wheat, while rice is the staple grain in most of the country. A surplus of several hundred million agricultural workers is one of the causes of *liudong renkou,* or floating populations, that by law are not allowed to settle permanently in other locations. Some 24 percent of the Chinese work in industry; low wages, an industrious workforce, and the possibility of the cheapest unit cost of manufactured goods contribute to China's industrial explosion.

Despite China's official communist atheism, many Han still observe traditional religious traditions. Each of the three main traditions—Taoism, Confucianism, and Buddhism—offers a separate approach. Taoism, founded by Lao-tzu in the sixth century B.C., looks to balance and order in nature. The teachings of Confucius of the same period extol proper workings of society through hierarchical relationships.

Buddhism, a religious import, aims for enlightenment and the dissolution of earthly bonds. Many Han observe elements of all three, as well as honoring various gods and goddesses and taking measures to increase good luck.

The most important Han holiday is the Lunar New Year, or Spring Holiday, which usually occurs in January or February. People travel to be with family at this time, eat festive meals, and watch fireworks and parades featuring dragon- and lion-costumed dancers.

Rural Han families traditionally consist of several generations living under one roof. In the patrilineal and patriarchal Han society, women move in at marriage with their husband's family. Urban families often are nuclear and tend to enjoy a higher income and more amenities and luxuries than their rural counterparts. A growing, mostly urban, middle class widens the gap between rich and poor.

Worried about China's burgeoning population, the government enacted a one-child policy several decades ago that mostly affects Han Chinese; China's minorities are exempt from this rule. The prevalence of only children among the Han has contributed to a very child-centric culture and—as boys are favored because they carry on the family line—to discrimination against girls. Recently this policy has been somewhat relaxed due to an aging population and shrinking labor force, allowing a second child if either parent is an only child or in a rural family the first child is a girl.

Japanese

Population: 127 million
Location: Japan
Language family: Language isolate

Japan occupies four major and thousands of smaller islands spread out in a 2,000-mile-long arc in the western Pacific. More than 125 million Japanese crowd a country the size of California. Steep mountains run through the center of the islands, leaving flat terrain only along the coasts and river valleys. More than 50 percent of the Japanese live along one 265-mile stretch from Kobe to Tokyo on southern Honshu.

Japan is one of the world's most ethnically homogeneous nations: Some 99 percent of the population is Japanese, a group with origins on the Asian mainland. The remaining minority population includes people of Korean, Chinese, and Okinawan ancestry, as well as the Ainu, an indigenous people of northern Hokkaido. All speak Japanese, a language with many regional dialects and a written form derived partly from Chinese characters.

One uses a torch to look at others' feet.

VIETNAMESE PROVERB

About four-fifths of all Japanese live in large conurbations, packed into tiny apartments, although many have more rural roots to which they return for family celebrations during the year. With only 15 percent of the terrain suitable for agriculture, Japanese farmers have painstakingly constructed terraced fields on sides of mountains where they grow rice, the Japanese staple, along with vegetables and fruit. Japan's large fishing fleet ranges far and wide to provide the major protein component of the Japanese diet.

Most Japanese today follow the religious traditions of Shinto, Japan's ancient animist religion, which emphasizes connections with nature, and Buddhism, imported from China. Many families blend observances: They may marry at a Shinto shrine but hold funerals at a Buddhist temple. Christianity has taken a small but significant hold in Japan, and while celebrating Christmas has gained popularity, it is almost entirely a secular observance.

For several centuries, until the end of the shogunate in the mid-19th century, Japan was isolated from the rest of the world by choice. Since then, and especially since World War II, Japan has embraced Western ideas and technology and built the world's third largest economy, dominating the international market in cars and consumer electronics.

From obtaining a high-paying job to marrying well, success and status in Japan revolve around the credentials a person acquires, especially those achieved through education. Japanese education is a high-pressure experience from the earliest years on. Mothers

JAPANESE A busy Tokyo street captures the crowded feeling of Japan's densely populated capital.

closely oversee their children's academic efforts—the most hands-on are known as "education mamas"—including long hours of homework, as well as additional time in *juku,* or cram schools, to improve their chances of doing well on the very competitive entrance exams to upper-level secondary schools and universities. Many firms recruit only from the most highly regarded institutions, such as Tokyo University.

On landing that first job, the Japanese worker is folded into the company ethos in a strong, familial relationship that goes way beyond the nine-to-five realm. Male employees known as "salary men" often are expected to socialize for hours over drinks after work before returning home. As a result, Japanese fathers have little time to share with their children. Women increasingly work outside the home, but they hold fewer responsible positions and earn significantly less than men, and many quit their jobs to raise children. Japanese men tend to spend their entire working lives with one company. The Japanese business values of loyalty, conformity, and the collective good over individual expression permeate the culture.

Where there is no fire,
there is no smoke.

KOREAN PROVERB

The Japanese live in a state of perpetual preparedness for earthquakes. Some 20 percent of the world's earthquakes, with up to 10,000 tremors a year, occur in Japan, the result of riding the boundaries of at least three tectonic plates. Elaborate public safety plans have been devised to minimize effects of future earthquakes.

While heavily influenced by Western trends, especially those from the United States, the Japanese have made a number of contributions to worldwide popular culture, including manga (comic books or graphic novels) and animé (graphic animation). With a rapidly aging population and a falling birthrate, Japan may have to look outside for workers to fulfill its employment needs and hence accept the reality of more ethnic diversity in its future.

Koreans

Population: 74 million
Location: South Korea, North Korea
Language family: Language isolate

The robust Koreans occupy East Asia's strategic Korean Peninsula. Mountainous North Korea has more than half the land of the peninsula and yet less than half the population of neighboring South Korea. The 1945 division of the peninsula along a 150-mile border led eventually to the Korean War, which raged from 1950 to 1953. Tensions still remain high.

Despite their separation, Koreans historically have had a feeling of shared identity, reinforced by geography: water bordering them on three sides and mountains to the north. Their location between China, in the north and west, and Japan, in the east, has played a major role in shaping their national character. Early colonization by China encouraged Korea to form institutions based on Chinese models of administration, writing, and the arts, yet the Koreans never relinquished their own identity. They excelled in arts, crafts, and construction, and they have contributed significantly to the aesthetics of East Asian painting, calligraphy, sculpture, and pottery, with an artistic spirit that is natural and spontaneous. Many collectors prize the pale green porcelains known as celadon, created during the 10th- to 14th-century Koryo dynasty.

KOREANS Korean commuters pack a subway car at rush hour in Seoul's Gangnam Station.

DEEP ANCESTRY

THE STEPPE DWELLER

Aleena Torres heard about Genographic from a class project she was involved in at Cornell University. She knew her ancestry was complex, so she was curious about what the test would reveal. Her mother is half Japanese and half Finnish, and her grandfather is Finnish. "However, he has a darker complexion and dark brown hair, as opposed to being blond with blue eyes," she noted, hinting at possible Asian genetic connections. Aleena's father is from the Philippines, and like many other Filipinos, he has a Spanish surname. With such far-flung ancestry stretching to opposite corners of the Eurasian continent, she wasn't quite sure what to expect from her Genographic results.

DNA Results

Her results were consistent with her genealogical mix. She is 45 percent Northeast Asian and 24 percent Southeast Asian. These figures add up to more than the combined eastern Asian components that would be expected if these genetic markers were coming simply from her Japanese grandmother and Filipino father. But the answer to her high Asian affinity is likely to be partially influenced by her Finnish ancestry—Finns can have as much as 10 percent Northeast Asian because of their connection to the Saami reindeer herders, steppe- and tundra-dwelling peoples who migrated to northern Scandinavia from Central Asia's Altai Mountains in the past several thousand years.

Consistent with her maternal Japanese ancestry, Aleena belonged to haplogroup D4p. This lineage is a rare branch of the large D4 haplogroup that originated in Central Asia some 45,000 years ago. From the steppes of Central Asia, some of its branches dispersed throughout the Asian continents, and a few branches (D4h3a and D4b) made it to the Americas, migrating across the Bering Land Bridge that connected eastern Siberia to Alaska for several thousand years.

Aleena Torres, mtDNA haplogroup D4p

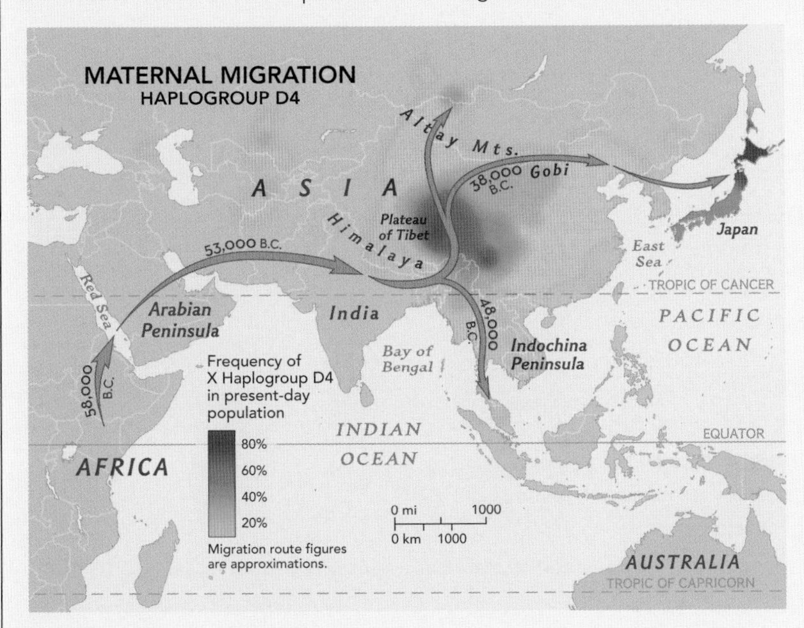

Maternal D4: This group was one of the most successful to spread throughout Central and Eastern Asia.

Ancient Asian Branch

Aleena's branch, D4p, is found at low frequency in populations from both Central and East Asia, specifically those from the Altai region in southern Siberia and Mongolia, as well as on the island nation of Japan. Researchers have long believed that the Altai region of Central Asia was a birthplace of many ancient Asian branches of the human family tree, including the widely distributed D4. Furthermore, ancient cultures from the region on northeastern Altai, which includes the shores of Lake Baikal of southern Russia, have shown archaeological similarities to some of the oldest cultures on the island of Hokkaido in northern Japan. Aleena's haplogroup D4p may be a remnant of this genetic link between these two distant Asian regions.

—*Spencer Wells*

The language used today evolved from the time of Korean unification in the seventh century and, like Japanese, has hierarchies of use and a sensitivity to social position. Koreans used Chinese characters for writing into the 15th century, when they invented a script called *hangul;* its 40 symbols are extremely accurate in representing spoken sounds. This break with Chinese tradition allowed Koreans to develop their own popular literary forms to deal with life, love, humor, and even social criticism; at this time, the people began to record colorful folktales possessing funny, ribald qualities. Today, Koreans enjoy one of the highest literacy rates in Asia, and South Koreans have one of the world's highest rates of university attendance.

Korean religion, diverse and complex, has roots in the old traditions of Confucianism and Buddhism, later overlaid with Christianity. The indigenous Ch'ondogyo faith, with elements of Confucianism, Buddhism, Taoism, shamanism, and Catholicism, has been joined in recent times by hundreds of new charismatic sects, such as the Unification Church of Sun Myung Moon and his millions of followers.

Diet is a point of pride for Koreans, who enjoy grilled meats and vegetable dishes such as kimchi, a fiery pickled cabbage that accompanies every meal.

Mongols

Population: 8.8 million
Location: China, Mongolia
Language family: Mongolic

The Mongols burst onto the world scene in the 13th and 14th centuries and left a short but audacious legacy that is remembered to this day. After Genghis Khan unified diverse clans in the heart of Asia's steppe, swift mounted warriors known as the Mongols went on to conquer much of Eurasia. Their empire—from the Pacific to the borders of Egypt and western Europe—lasted only 175 years. The stereotype of cruel hordes has survived much longer, however, even though peaceful herding has been the Mongol norm since that time.

MONGOLS A young Mongol horseman leads his mount up a grassy slope. Mongols traditionally excel at riding.

On the vast grasslands where they lived, Mongols found success as nomadic pastoralists, horsemen, and herders. They made their homes in circular tents (yurts, or *gers* in Mongolian) of felt and canvas, supported by wooden frames. Women and children took an active part in the pastoral life, caring for animals and collecting dung to burn for fuel.

A toughness born from their exposed, windswept life prepared the Mongols well for survival, which required the skillful handling of small horses while tending sheep, goats, cattle, camels, and yaks. Today, such animals continue to support a diet made up of dairy products and meat; milk, cheese, and yogurt appear in many different forms, and there is even a distilled "milk vodka" for festivities. All parts of the animals are used, with the sheep-tail fat considered a delicacy offered to esteemed guests.

Mongol culture has been strongly influenced over the years by Buddhism, introduced from Tibet, combined with indigenous animism. The nomads learned to use portable altars and devotional scroll paintings that could be rolled up and easily transported. In the 16th century, the bond between Tibet and Mongolia became so strong that a Mongol khan bestowed the title of Dalai Lama, metaphorically meaning "ocean of wisdom," on the Tibetan leader.

Today, Mongols have their own country in north-central Asia. Known as the Republic of Mongolia, it encompasses a high, rolling plateau nearly three times the size of France and embraces a population of 2.6 million. Even more Mongols live across the southern border, in China's Inner Mongolia and other Chinese provinces; additional clusters thrive in Kazakhstan, Siberia, and the Kalmyk region of Russia's lower Volga River.

Mongolia achieved its status as a republic and as the world's second communist country, after Russia, in 1924. Antireligious revolutionaries murdered many thousands of monks and forced changes in lifestyle, limiting nomadism and setting up collectives. Most Mongols now reside in towns and settlements, though many prefer to live in gers, even in the suburbs of Ulaanbaatar, the capital.

The popular revival of wrestling, archery, throat singing, and horse racing reveals the Mongols' deeply rooted desire to maintain their age-old traditions.

Tibetans

Population: 4.3 million
Location: China
Language family: Tibeto-Burman

With their numbers estimated at between four and five million, the Tibetan people inhabit an immense plateau, a high-elevation region of grasslands, alpine meadows, wetlands, and permafrost, bordered by the world's highest mountains, the Himalaya.

Cultural Tibet is far larger than the territory within China called the Tibet Autonomous Region. This "Greater Tibet" includes Bhutan, Sikkim, northern Nepal, and sections of Kashmir in northern India. Significant parts of China's Gansu, Qinghai, Sichuan, and Yunnan Provinces are ethnically Tibetan as well.

Groups indelibly influenced by Tibetan culture are the Sherpas of Nepal, the Drukpas of Bhutan, and the Ladakhis of northern India; dozens of smaller groups speaking diverse dialects contribute to the complex Tibetan world. Within China, Mandarin Chinese has become a lingua franca for many Tibetans who otherwise could not converse with one another.

Harsh physical barriers, along with official isolation, allowed Tibetans to maintain an ancient way of life into the mid-20th century. Their culture developed in the relatively well-watered southern part of the plateau, where raising barley proved possible, while in the great north, nomads living in black wool tents tended livestock, primarily the ubiquitous yak.

TIBETANS A backpacked baby responds to his caregiver's familiar voice in Lhasa, Tibet.

In the seventh century, the Buddhist religion was introduced into Tibet from China. Early word of the religion was soon followed by the arrival of Buddhist missionaries from India. Spiritual and secular institutions blended together, touching all aspects of life. The land eventually supported 6,000 monasteries, which required most families to have at least one son as a monk. Although Tibetan Buddhism has several branches, the Gelukpa sect has been dominant since the 17th century, when the fifth Dalai Lama reached the apex of Tibet's theocracy.

Buddhism fostered a sophisticated civilization of advanced learning and philosophy that survives to this day, despite political repression and exile. Tibet's de facto independence from 1912 to 1950 ended when China occupied the country and, in 1959, forced the exile of the Dalai Lama and 80,000 of his followers to India. Suppression of religion and destruction of monasteries followed, and a steady influx of Chinese immigrants has diluted the indigenous population. Tibetan culture enjoys worldwide fascination and respect like few others, yet is under threat in the Tibetan homeland due to political, economic, and demographic shifts.

UYGURS Uygurs ride a donkey cart along a poplar-flanked road in Xinjiang, China.

Uygurs

Population: 10.1 million
Location: China
Language family: Turkic

The Uygur people of western China's Xinjiang region inhabit desert and mountain expanses that were once known as eastern Turkestan. More Uygurs live in Pakistan, Afghanistan, Saudi Arabia, Turkey, Europe, and North America.

The Uygurs' origins date from the third and fourth centuries, when Turkic tribes roamed as nomads in search of pastures and oases. In time, the people created empires extending east to the Yellow River, north to Siberia, and south to Pakistan. Positioned at the Silk Road's midpoint, Uygurs slowly abandoned their nomadic lifestyle, took up farming and trade, and enhanced their culture with religious and intellectual exchange. They became scholars, administrators, inventors of a versatile script, and creators of magnificent cave art. Their contributions to architecture, music, printing, and medicine have enriched the whole region.

Uygur life, long tied to oases, was centered in two cities: Buddhist Turfan in the northeast and Muslim Kashgar in the southwest. Both lie in today's Xinjiang region of China. The Muslims conquered their brethren to the northeast at the end of the 14th century and were able to remain independent from China until 1759, when the Qing dynasty took over. The Chinese government has dictated the larger fate of the Uygurs ever since.

Although Uygurs have putative autonomy in their homeland, ethnic unrest is prevalent. For decades, large numbers of Han Chinese have been migrating to Xinjiang to relieve population pressures and develop the region. For the Uygurs, this has meant curtailment of their religion and dilution of their communities.

Today, the markets of Xinjiang are a mix of Arab souk, Chinese bazaar, and Central Asian trading station, filled with animals, tack shops, pungent spices, furs, carpets, brilliantly hued silk dresses, and vendors selling kebabs and huge flatbreads. Beyond the markets, farmers still irrigate their fields with an ingenious system of underground channels that bring water from the distant Tian Shan mountains.

Southeast Asia

Southeast Asia encompasses the mainland nations of Myanmar, Thailand, Cambodia, Laos, and Vietnam, as well as the island nations of the Philippines, Malaysia, and Indonesia. East Timor, once part of Indonesia, gained independence in 2002. Other islands and cities maintain distinct cultural and political identities, such as Bali, an Indonesian island with a defining Hindu presence; Hong Kong, now part of China but still steeped in its tradition as a British-owned port of trade; and Singapore, an independent city-state and a world-class center of trade and industry.

Balinese

Population: 4.3 million
Location: Indonesia
Language family: Malayo-Polynesian

The tiny, volcanic island of Bali, east of Java, has long enthralled visitors and residents alike. For the three million Balinese, this tropical island is an entire world, with physical and spiritual attributes important for existence. The Balinese believe that because their cosmos is so rich, the psychic, unseen forces constantly spill over into the mundane. Daily life becomes a vibrant expression of propitiation and praise for gods and nymphs, demons and witches. Hardly a day goes by without a procession or a temple festival. At night, villages come alive with the music of gamelan percussion orchestras.

In the otherwise Muslim expanse of Indonesia, Bali has remained a Hindu holdout. Hinduism arrived with Indian traders and found such a firm foothold in the seventh and eighth centuries that Bali was later able to resist all attempts at conversion, even after Islam dominated Java and other islands in the 16th century.

Beneath a holy volcano known as the Gunung Agung, "navel of the world," spread the rice fields and villages forming the heart of Balinese life. Here, sophisticated artists continually create, remake, and teach their traditions. Myths, rituals, painting, carving, acting, and dancing all seem to come together, where more than 5,000 dance troupes testify to the importance of creative and religious expression.

Bali's most powerful drama revolves around

BALINESE Wet rice cultivation is back-straining work for this Balinese farmer.

the clash between two mythical creatures, Barong and Rangda. The lion-beast Barong—representing sunshine, medicine, life, and light—provides the antidote for evil. He likes to dance for sheer joy, but his playfulness is interrupted by the appearance of Rangda, hideous witch and queen of death. In a climactic struggle, trance dancers rush to Barong's aid, only to have their own daggers turned against them by Rangda's evil magic. But Barong's power prevents the daggers from piercing the flesh. Slowly, the dancers are brought out of their trances by Barong's beard, his power center. Neither creature wins this cosmic struggle; goodness gains only a temporary victory.

Since the 1930s, Bali has seen growing numbers of tourists, and now more than a million each year swarm through temples and villages. The threat of cultural breakdown and commercialization of the entire island is always present, yet Bali's dynamic, adaptable people keep a firm view of themselves and their unique culture.

Dayak

Population: 3.5 million
Location: Indonesia, Malaysia
Language family: Malayo-Polynesian

Living along large, meandering rivers flowing through the forests of Borneo, the Dayak are a people whose name covers as many as 50 non-Malay ethnic groups, speaking a variety of languages. Most of the Dayak people live in southern and western Borneo, in both East Malaysia and Indonesian Kalimantan.

Traditionally, Dayak communities were made up of several hundred people who maintained few ties with other groups. Family clusters lived in longhouses and led a largely self-sufficient way of life that took raw materials and animals from the forest and fish from the rivers. Slightly back from the riverbanks, they practiced simple agriculture, growing dry rice for a few seasons in one field, then abandoning the site for another. Ritual warfare, including head-hunting, was also part of their lives.

In recent decades, the Dayak have been confronted with major changes. Beginning in the 1960s, the Indonesian government poured resources and people into Kalimantan in an effort to exploit the forests and relieve overcrowding elsewhere. A program called *trans-migrasi* relocated ethnic groups from other parts of the densely populated nation, especially the Madurese from Madura and East Java. The harvest of hardwood timber from Dayak regions has paralleled the growth of farms, plantations, and towns. Placed near native Dayak animists, Muslim immigrants caused economic disparity and distrust, fueling tensions. The Dayak feared newcomers would take tribal lands and jobs.

Fighting broke out periodically and became fierce in 2001. Hundreds were killed, and, during the frenzied bloodletting, head-hunting reappeared. Many Madurese left Borneo or fled to refugee camps; counterattacks in towns caused more deaths. Historic neglect and insensitivity to Dayak needs had set the stage for the misunderstandings and conflicts that exploded into full-scale ethnic trauma.

Hmong

Population: 7.7 million
Location: China, Vietnam, Laos
Language family: Hmong-Mien

The mountains of southern China and northern Laos and Vietnam are home to the Hmong, a tribal people known for independence and self-sufficiency. In China, the Hmong are called Miao or Meo—names that mean "barbarian" and that they consider pejorative.

The Hmong language, both tonal and monosyllabic, had no written version until Christian missionaries devised a Roman-based alphabet for it. Consequently, the Hmong have a rich oral tradition that includes sung poetry, proverbs, and aphorisms and combines embroidered figural cloths with storytelling.

The Hmong practice swidden agriculture, a technique that creates temporary growing plots. After clear-cutting an area, they burn off the rest of the vegetation before planting crops. The fields remain fertile for a few years; when they are spent, the Hmong

DAYAK Nonhuman skulls form part of the Dayak garb worn yearly at a festival on Borneo that promotes agricultural bounty.

HMONG A brightly embroidered dress and tasseled headdress make up this Hmong girl's traditional outfit.

move on. Hmong grow dry rice, corn, and vegetables. They eat their grains steamed or boiled and heavily seasoned with hot peppers. They also cultivate opium poppies for their own use and as a cash crop. They were urged to grow the poppies during the French colonial period, but current national governments discourage the practice.

Organized around patrilineal clans, the Hmong have no formal organization above that level. Their mountaintop villages consist of small, simple houses that contain few furnishings. Extended patrilineal families tend to live together, with sons bringing in their wives whom they have married outside their clan. Traditionally, Hmong marry young, in their teens. Couples may court during singing events. They form groups and sing to each other, sometimes passing a ball back and forth, hoping to kindle romantic interest.

The Hmong believe that spirits inhabit objects in nature and in the material world. Humans have multiple souls, and some Hmong rituals are attempts to call spirits back into the body. When illness occurs, a person known as a Txiv Neeb—"master of spirits"—may be called in to discern the cause from the vantage point of a trance. The spirits of ancestors also must be appeased to ensure the well-being of the living.

Beginning in 1975, many Hmong fled Laos due to the civil war that involved them as either members of the communist Pathet Lao party or as guerrilla operatives for the U.S. Central Intelligence Agency. Over the next decade or so, some 100,000 Hmong made their way as refugees to the United States. They live in communities, in Fresno and elsewhere in California, Minnesota, Wisconsin, and Rhode Island. Some work in farming. Quite a few of the unskilled older generation are unemployed, but the young are finding success through education. Many Hmong women make decorative squares of appliquéd embroidery called *paj ntaub,* which they sell at craft fairs.

Ifugao

Population: 168,000
Location: Philippines
Language family: Malayo-Polynesian

High in the mountains of Luzon, the largest of the Philippine islands, stands a monument to human ingenuity. Massive stone walls, hand-built by the Ifugao people, carve the steep mountainsides into thousands of terraced fields, while sophisticated hydrological systems supply water. The graceful stone terraces, designated a UNESCO World Heritage site, are served by a vast network of channels, pools, and aqueducts. Hardly a drop of rain falls that is not captured, controlled, and put to the service of growing rice. Boulders larger than a person and structures as small as a single leaf pinned down with twigs can serve to divert rainwater into an Ifugao pond field.

The worldview of the Ifugao has traditionally revolved around rice, a crop that they have domesticated intensely, producing dozens of distinct varieties. In a myth now mostly forgotten, ancient Ifugao were first given rice and instructions for planting by unearthly beings they call "skyworld people."

Many Ifugao villagers still reside in thatched houses supported on tall posts, with an open space underneath

IFUGAO An Ifugao man in full regalia represents his mountain community on the island of Luzon in the Philippines.

for cooking and relaxation. After a day's labor, villagers sit under their houses drinking fermented rice beer and gazing out over their fields.

In each hamlet, once a rice field is designated as ritually important, it must be harvested first by the joint efforts of the entire community. On harvest day, the family may invite a shaman to sacrifice a chicken, drain its blood, and read the entrails for signs of an auspicious year, season, and crop. The shaman sits for hours, chanting the family's genealogy to invoke the spirit world of deceased ancestors in support of a successful harvest. On festive occasions men may dress in traditional red loincloths, playing flutes and beating gongs to accompany dancers.

Down in the pond fields, women harvest rice by hand, using small metal blades. Out in front of the harvesters works a lone woman, the seed selector, an expert trained to identify and collect the plumpest and healthiest grains to use as seed for replanting in the coming season.

Occasionally women hum a song as they tread the rice paddies. In the past, epic chants called *hud-hud,* telling of hero-ancestors, were sung by women while harvesting rice. Hud-hud chants, designated a Masterpiece of Oral and Intangible Heritage of Humanity by UNESCO, have fallen into disuse and are now remembered only in fragments.

The Ifugao language is still spoken vigorously by people of all ages. It includes an intricate vocabulary for rice technology, with 27 different names for pottery vessels for storing rice wine, 30 names for types of woven baskets used to carry foods, and 130 phrases describing payments for the use of rice fields. Ifugao has uniquely expressive verbs like *tiw-tiw,* meaning "to frighten animals, birds, or chickens away from drying rice."

Ifugao rice knowledge—including genetics, hydrology, and the cultural belief system that fostered these technologies—is now eroding. As fertilization and genetically engineered rice varieties, promoted by international agrobusiness, gain popularity, Ifugao farmers are coming under increasing pressure to abandon traditional growing methods and strains of rice bred by their ancestors over millennia.

The Ifugao now face the challenge of nurturing their culture as they change and modernize. Many young Ifugao people have migrated to Philippine cities or other countries for work and education, and electricity and television have come to the villages. Few young people now aspire to grow rice, and the rich traditional knowledge of the Ifugao is being forgotten as terraces erode from disuse.

Javanese

Population: 102.6 million
Location: Indonesia
Language family: Malayo-Polynesian

An Indonesian island smaller than Florida, Java is home to more than 100 million people, the great majority of whom are ethnic Javanese living in the central and eastern parts of the island. Beneath Java's backbone of volcanoes and crater lakes, people have engaged in agriculture for 4,500 years. Here, rich soil and abundant water have allowed industrious villagers to turn their homeland into a wet-rice miracle, able to support the dense population.

Throughout their history, the Javanese have evolved an ability to adapt and transform outside forces into a profound inheritance. Layers of cultural history, vital and varied, have created a complex and sophisticated people. Invaders, immigrants, traders, and colonists from Polynesia, India, China, Arabia, Portugal, the Netherlands, and Great Britain left their imprint on the culture, bringing along many of the world's main religions: Hinduism, Buddhism, Islam, and Christianity.

Indian civilization on Java culminated between A.D. 750 and 900 in a great age of temple building, the legacy still visible in Hindu and Buddhist monuments sprinkled across the countryside. Borobudur, a massive carved stupa, is Java's most magnificent structure. The refinement of court life during this era also encouraged a language hierarchy that remains to this day. With separate vocabularies reflecting the social positions of its speakers, the Javanese language displays varying levels of speech: coarse, ordinary, modestly respectful, and fully respectful.

> *One parent can look after many children, but it takes many children to look after one parent.*
>
> THAI PROVERB

In the 15th and 16th centuries, maritime merchants introduced Islam, but the ascendancy of this faith was less a formal conversion than an adaptation of Islam to earlier Hindu-Buddhist institutions. The Javanese also practice animism and ancestor worship, acknowledging a world of spirits and ghosts, magic, and malevolence. Their religious epics, mytho-heroic stories, and comedies—presented as plays for *wayang kulit* (shadow puppets), *wayang golek* (wooden puppets), and *wayang wong* (human actors)—find enthusiastic audiences.

JAVANESE Two Javanese girls huddle under a rice basket during a chilly monsoon downpour.

In arts and crafts, the kris, a double-edged sword with a wavy blade, has an aura of holiness. Many believe krises can fly or that the original owner's soul lives in the blade. Javanese masters of batik, a wax-and-dye process on silk or cotton cloth, create exquisite fabrics based on themes that are secular and sacred, ancient and modern. Batik can be a fine art of subtle styles, colors, and symbols, or it may result in practical productions of shirts, sarongs, and bags.

Khmer

Population: 14.2 million
Location: Cambodia
Language family: Mon-Khmer

Cambodia's Khmer live in a fertile land watered by the Mekong River and Tonlé Sap ("Great Lake"), Southeast Asia's largest lake. This remarkable body of water shrinks in the dry season and then doubles in size in the rainy season, and is home to thousands of people who live in floating villages.

For more than a thousand years, the Khmer have been animated by Buddhism—influenced by earlier Hinduism—as both a personal faith and state religion. Under a line of Buddhist kings who combined just administrations with huge building projects, the people established a civilization of profound creativity and grand accomplishments.

The carved stone complex at Angkor, not merely a temple but a city, represents the flowering of the Khmer culture from the 9th to 13th centuries. This high point focused on a courtly life of ritual and classical dance, a slow, graceful art form that portrayed Indian epics, romances, and Buddhist themes. Temple dancers served as the earthly counterparts of celestial nymphs. Angkor's stone carvings record this tradition, which not only entertained the gods but also embodied Khmer ideals.

Incredibly, this land of beauty and refinement was transformed into a place of utter darkness between 1975 and 1979. The Khmer Rouge, a communist military and political contingent, embarked on a bloody revolution that killed more than 1.5 million people, brought the Khmer people close to cultural suicide, and almost destroyed all traces of the past. The revolution took the lives of artists, dancers, and musicians

KHMER Grilled, skewered meat is on the menu for this Khmer woman in Cambodia's Kampong Phluk floating village.

and threatened to erase the nearly thousand-year-old classical dance tradition. Every Khmer was changed by the mass death, forced labor, disease, and famine. Many fled, with nearly 150,000 finding refuge in the United States.

When the nightmare came to an end, a handful of survivors revived the classical dance and its slow, rhythmic music. Discipline and training started anew, with children as young as eight learning how to dress in opulent, glittering costumes and bring to life the ancient court dance. In their struggle to survive, the Khmer have preserved intact their nation's great arts, the living pieces of cultural memory that provide links to the past and hope for the future.

Minangkabau

Population: 6.9 million
Location: Indonesia
Language family: Malayo-Polynesian

Vigorous and successful, the Minangkabau people inhabit the western section of Sumatra, sixth largest island in the world. Their region of verdant plateaus, forests, gorges, and escarpments supports many small villages where people farm or engage in handicraft, creating articles of silver and gold filigree and fine embroidered shawls.

The Minangkabau form one of the world's largest matrilineal societies. This means that titles, property, family names, and inheritance go through the female line, even under Islam, which normally favors men. The arrangement provides a degree of gender equality uncommon in Asia.

Traditionally, many families live together in a *rumah gadang* (longhouse), and every person stays through life as a member of the mother's original house. When a couple wishes to marry, the prospective bride's family usually makes the proposal, and the new husband, not the wife, changes homes to enter the woman's clan house. In some rural places there is even a groom-price.

When a daughter takes a husband, an annex is added to the longhouse for the growing family. Such extensions, with swooping saddleback roofs, have hornlike adornments that rise toward the sky and symbolize the horns of a water buffalo, sacred animal of the Minangkabau. By counting longhouse extensions, visitors can guess the number of daughters with husbands and children.

Because of the strong family ties and an emphasis on education, the Minangkabau have one of the highest literacy rates in Indonesia. This group shines in the world of Indonesian letters, having produced many well-known writers over the past century.

Overpopulation and a decline in traditional ways have caused many Minangkabau men to emigrate to other parts of the island nation. Most have gone to the capital, Jakarta, and become successful shop owners, teachers, or cooks, who serve up the tasty, spicy dishes for which the Minangkabau people are famous.

MINANGKABAU Traditional architecture frames a Minangkabau bride. In this Sumatran matrilineal society, the bride's family often does the proposing.

GENOGRAPHIC INSIGHT

GRAINS OF RICE

As early as 50,000 years ago, the first humans arrived in uninhabited Southeast Asia. Glaciers trapped water in the polar ice caps, dropping global sea levels by as much as 300 feet and exposing an overland passageway known as Sundaland. Short migrations across stretches of open water brought people to both Australia and the highlands of Papua New Guinea, which were connected at the time, and left a corridor of early human settlements along the eastward route they had followed. Encampments presumably dotted the coastline from eastern Arabia across southern Asia, but they became submerged as the global climate warmed and as water, once locked in ice, filled the oceans. The absence of an archaeological footprint of this early migration has left many historians perplexed, but new technologies are finally allowing archaeologists to delve underwater in search of villages submerged thousands of years ago.

South Asian Arrival

Genetic research among populations along the South Asian coastline has uncovered a continuous thread linking Australians with Africans. The DNA carried by native tribes living throughout the region tells the story, and genetic work with populations from the Andaman Islands, Malaysia, and Papua New Guinea has revealed clues into these early movements. Recent studies of both their maternally inherited mitochondrial DNA and paternally inherited Y-chromosome gene pools indicate that these populations appear to have arrived by a route along the South Asia coastline within a few thousand years of leaving Africa.

The effects of chance—in genetics, known as genetic drift—meant that the patterns of genetic diversity among these people are quite different from among their African relatives. There appears in these groups to be a reduced Y-chromosome diversity, compared with their mitochondrial gene pools, a gender bias that makes sense when one considers the traditional marriage practices of these people. Both patrilocality—the tendency for a family to live in the male's community—and polygyny—the taking of multiple wives, which increases male reproductive success—are common practices in the region and help to explain this sex-biased genetic pattern.

A Khmer man dances in full peacock regalia. The peafowl dance is a folk form found in South Asia and China, as well as in Cambodia.

The genetic data suggest that these groups predominated in Southeast Asia until about 10,000 years ago, when the advent of rice farming in East Asia led to a large population expansion. Archaeological evidence for the spread of agriculture to Japan, Taiwan, and Southeast Asia parallels the genetic patterns of the people living there. Sites in northern China show evidence for the cultivation of millet and rice beginning around 7,000 years ago; by 4,000 years ago, rice farming had reached the Indonesian islands of Borneo and Sumatra. As these Asian agriculturalists moved south, they encountered populations that had been living in the area for tens of thousands of years.

The Role of Agriculture

There is conflicting evidence over the degree to which agriculturalists' DNA spread as their new way of life was adopted around the globe. The process by which genes are introduced into an existing population, known as introgression, is determined by factors such as population size, duration of contact, and the survival value of features offered by the new DNA. In Southeast Asia, the impact of agriculture appears to have been significant, for the vast majority of the genetic lineages found here today (between 50 and 90 percent, depending on the population) trace their origins back to the expansion of agriculture.

Though the first Southeast Asian populations lived in an area where large farming communities were popping up throughout the countryside, dominating the landscape with terraced rice farms and high population densities, their cultural and geographic isolation kept introgression at a minimum. Today, these ancient populations still hold on to the land and the genes of their earliest ancestors.

Thai

Population: 65.4 million
Location: Thailand
Language family: Taic

In the daily bustle and roar of Bangkok, the capital of Thailand and one of Asia's megacities, all the elements of Thai society have been thrown together. Here coexist the four core groups making up the greater ethnic "nation": Central Thai, Thai-Lao, Northern Thai, and Southern Thai. They all share the faith of Theravada Buddhism, dialects of a common language (Tai), and a similar ethnic origin—and they all face the same summer monsoon downpours and the long, dry periods from autumn to spring.

Today's Thai descend from ancestors who came from southwestern China's Yunnan Province. From the 9th to 13th centuries, a large, non-Chinese kingdom proved to be a buffer that allowed Thai to escape Chinese assimilation. Within this kingdom, Thai held high positions and could expand and migrate steadily into Southeast Asia, even into areas that extend far beyond the present-day borders of Thailand. As a result, most Laotians are Thai by descent, as are the Shan, who make up about 10 percent of the population of Myanmar (Burma). Other Tai-speaking people live in southern China and Vietnam.

Most of the early Thai settled in the fertile zone of the Chao Phraya River, which dominates Thailand's central valley. They officially emerged as a people in 1238, when diverse groups were unified under the Sukhothai kingdom and freed from the formerly dominant Khmer. Buddhism helped to bind them, and this in turn inspired a distinct art style reflecting prosperity and self-confidence, as still seen in carved temples and giant Buddha sculptures.

Today, one of the country's most popular activities is Thai boxing, in which punches and kicks are used; in fact, it is a national pastime. For all its outward appearance of violence, this sport is layered with refinements and cultural meanings, which aficionados have come to recognize and admire.

THAI The bond between mahout and elephant in Thailand is a strong one. Here, mahouts playfully bathe their charges.

Many Thai follow the lunar calendar and look forward to the predictable and joyous yearly festivals. Perhaps the most popular festival is Loy Krathong. Under a November full moon, the country's rivers, streams, lakes, and canals become sites for honoring Mae Khongkha, goddess of rivers and waterways. People of all ages prepare little boats in the shape of lotuses or miniature temples, which are then blessed and decorated with candles, incense, flowers, and coins before being set afloat. All through the night, millions of tiny lights drift and bob on the waters.

Vietnamese

Population: 80.8 million
Location: Vietnam
Language family: Mon-Khmer [Viet-Muong]

After four decades of intermittent warfare against varied foes—the Japanese, the French, the Americans, the Cambodians, and the Chinese—and following the successful unification of the country in 1975 after a long civil war, the roughly 80 million Vietnamese are rebuilding their ancient land.

The Vietnamese people originated in the Red River Delta region of the north, but their early roots are hazy. Sometime between 500 and 200 B.C., they emerged as an ethnic entity from a melting pot that contained seaborne travelers from Oceania and various migrant groups from Asia, with northern Mongol characteristics predominating.

Once established, the Vietnamese marched slowly and steadily southward, settling plains and deltas along the way and reaching the southern Mekong region by the early 1700s. Today, their thousand-mile-long country extends like a huge, stretched letter S, reaching all the way from the Chinese border to the Gulf of Thailand and taking in an area half the size of Texas.

The Vietnamese language was deeply influenced during a millennium of domination by China, when classical Chinese stood as the official discourse of mandarins—imperial officials representing the Chinese government—and scholars. Much later, in the 17th century, European missionaries created a new script representing the Vietnamese language using the Roman alphabet, which replaced Chinese to become the common form of writing Vietnamese.

VIETNAMESE Despite her modern dress, this young Vietnamese woman must forage for the firewood she hauls back to her home.

Confucianism from China held society together through interlocking relationships and duties, helping to establish four major social groups: scholar-officials, artisans, merchants, and farmers. Then as now, the peasantry cultivated rice and secured protein by fishing. Today, nearly 70 percent of Vietnamese continue to reside in the countryside.

Both Mahayana and Theravada Buddhism took hold in Vietnam, the northern branch from China, the southern from India. Buddhist institutions functioned freely until communism arrived in the 20th century, at which point all aspects of religious life were curtailed. Catholics, now three million strong, flourished for a century after French colonial dominance in the mid-1800s. In the early 20th century, Vietnam saw the establishment of Cao Dai—a mix of Taoism, Buddhism, Confucianism, and Christianity—in which utopian and cosmological ideals merged with an all-inclusive theology. Cao Dai saints include Buddha, Confucius, Jesus, Muhammad, Victor Hugo, Napoleon, Einstein, Churchill, and Sun Yat-sen.

South Asia

Rimmed by the great Himalaya to the north, South Asia stretches from Afghanistan, Pakistan, and the Indus River in the west to Bangladesh, beyond the Ganges River, to the east. The entire Indian subcontinent is contained within South Asia. Second in population only to China and significantly less than half its size, India's landmass varies from steep mountains in the north and arable river plains in the midsection to subtropical swamps and deltas, especially in the east. Rainfall depends on annual monsoon cycles for many parts of India and the rest of South Asia.

Bengalis

Population: 268.3 million
Location: Bangladesh, India
Language family: Indo-Aryan

The great Brahmaputra River meets the expansive Ganges River Delta in the Bengali heartland, and here the people divide themselves into two main regions: the nation of Bangladesh and the Indian state of West Bengal. Though short on many resources, both areas have water and people in excess. Bangladesh, with nearly 167 million people, is the world's most densely populated country, while West Bengal, with 101 million, is India's densest state. The Bengali (or Bangla) language ranks seventh in number of speakers worldwide.

In 1947, the partition of India split its Bengal Province into Indian and Pakistani sectors; eventually Bangladesh (formerly East Pakistan) declared independence; in 1971, it became a nation. The division followed religious, not ethnic, lines, for all of the people in the area speak the Bengali language and consider themselves Bengalis. India's West Bengal has always been overwhelmingly Hindu, but Bangladesh, with the world's largest concentration of Muslims, follows a form of Islam that contains pre-Islamic Hindu and Buddhist elements.

Both sectors share a common history and acknowledge common successes, such as the 19th-century revival of literature, music, and the arts. Their shared language, an Indo-European tongue, preserves a rich literary heritage. The works of Rabindranath Tagore, beloved poet and 1913 winner of the Nobel Prize for literature, are still read and memorized by Muslim and Hindu alike. Poet and playwright Kazi Nazrul Islam, called the voice of Bengali nationalism, grew to become a champion of poor farmers; his plays and tales on political and historical themes gave hope to long-subjugated Muslims.

BENGALIS Pots full of water, procured at the village pump, make a heavy load for this Bengali woman in Bangladesh.

While many Bengalis live in huge cities such as Kolkata and Dhaka, most of the people are villagers. To the north, they inhabit the hills, but elsewhere Bengalis dwell in a largely watery realm where fishers, traders, and farmers depend on shallow inlets, fragile islands, and shifting fingers of land. Although the annual summer flooding rejuvenates the countryside with silt, every few years great inundations kill thousands of people.

Gujaratis

Population: 56.3 million
Location: India
Language family: Indo-Aryan

The western Indian state of Gujarat sits at the junction of historical overland trading routes and the maritime outlets to the Indian Ocean. No wonder then that the Gujaratis have long been oriented toward trade. Their language, Gujarati, has ties to both Sanskrit, India's ancient language and the major vehicle of priestly ritual, and Prakrit, vernacular languages.

The vast majority of Gujaratis, some 90 percent, are Hindus, but there are also a number of Muslims, Jains, and Parsis. Banias figure prominently among the Gujarati Hindu castes. In caste hierarchy they are ranked third after Brahmins (priests) and Kshatriyas (warriors), but as traditional merchants and traders, they often experience greater economic success. In fact, much of the commerce in bustling Mumbai (formerly Bombay) in the neighboring state of Maharashtra is controlled by Gujarati Banias.

In Jainism, an offshoot of Hinduism, adherents go to great pains not to injure any living being; some even sweep the streets in front of them as they walk in order to avoid stepping on any insects. Jains, as well as the Parsis, Indian Zoroastrians who came to the subcontinent from Persia more than a thousand years ago, also participate in Gujarat's large trading community.

In addition to trade, Gujarat supports a large manufacturing base as one of India's most industrialized states, producing goods such as textiles, chemicals, and plastics. Gujarati farmers grow crops of millet, sorghum, wheat, and rice, many at the subsistence level, and also crops of cotton, groundnut, sugarcane, and tobacco for export. The cutting of precious stones, especially diamonds, is a traditional Gujarati specialty.

Gujarat was the birthplace of Mohandas K. Gandhi, best known as the Mahatma, or Great Soul. The slight, bespectacled, British-trained lawyer led his country on the arduous path to independence in 1947 through the power of his charisma and the principles of self-sufficiency and nonviolence, which have been adapted by leaders and causes the world over.

Many Gujaratis observe the same life-cycle rituals and celebrate most of the same festivals and holy days as they do elsewhere in Hindu India, but they emphasize some more than others. Diwali, the Festival of Lights, marks a fresh start for everyone: Merchants close out the year's books and begin new ones. Navratri honors the mother goddess with nine nights (*navratri*) of worship and celebration that include typical Gujarati dances such as the *garba,* a circular women's dance accompanied by singing and vigorous drumming.

On an international basis, Gujaratis outside India are perhaps even better known than those who reside there. Beginning in the 6th and 7th centuries, Gujarati traders traveled to Indonesia and East Asia; by the 13th century, they were in Africa. More Gujaratis came to Africa in the 19th century, when Indian labor was used to build the continent's railroads. A large number of Banias stayed on after others returned to India, and they are ubiquitous as shopkeepers and businesspeople, many with the common Gujarati surname of Patel, especially in East Africa. Gujaratis

GUJARATIS In Durban, South Africa, Gujarati women dance during Navratri, a nine-night celebration honoring the goddess Durga.

also formed the vanguard of immigration to the United Kingdom and North America in the 20th century. Many who came to the United States got their economic foothold in small mom-and-pop motels on well-traveled routes. Now Gujaratis own tens of billions of dollars worth of national chain hotel properties throughout the United States.

Nagas

Population: 2.4 million
Location: India
Language family: Tibeto-Burman

A strip of rugged territory in northeastern India between Myanmar (Burma) and the Brahmaputra River Valley is home to the numerous tribal peoples of Nagaland. Although the Nagas share some similarities with Assamese and Burmese ethnic groups, they take pride in a unique identity. The land's isolation and rough terrain has allowed development of more than 50 dialects, requiring different Naga peoples to use either a pidgin, English, or Assamese to communicate with one another.

Where there is rhythm and alliteration, there lies a proverb.

GUJARATI PROVERB

Historically, the Nagas obtained all they needed from the land, the forest, and minor trading. Farming and hunting, supplemented by fishing, supported them well inside their walled villages. Outside, beautiful gateways with bas-relief decorations greeted those approaching along the winding, sunken paths leading up to them.

A variety of local organizations distinguish the Nagas, from strong chiefs with near total power to democratic councils to gerontocracies. Throughout the Naga tribes, women have always enjoyed high status, with equality in work and at tribal meetings.

The British, by establishing control over the Naga Hill District in the 1870s, stopped ritual warfare and head-hunting, a practice that had grown out of the belief that the human head holds a vital force, or soul power. By taking the head of an enemy, a warrior could gain status, win a bride, and bring to his village an increase in positive energy. Such acts also inspired dances and various forms of art, all with skull motifs. The last case of Naga head-hunting occurred in 1958, several years after the British had left the area. In the late 19th century, American missionaries began converting Nagas, and two-thirds are now Christian. Their religion has reinforced their identity in a country where the majority of people are Hindu.

NAGAS A Naga tribesman from northeastern India participates in a yearly festival at the Naga Heritage Village in Kisama, Nagaland.

Nepalis

Population: 31.5 million
Location: Nepal
Language family: Indo-Aryan

In the 1760s, the ruler of a Himalayan principality formed a unified Nepal from a number of independent states with populations as diverse as the terrain they lived in. Today's Nepalis represent three ethnic stocks: the Indian peoples from the south, the Tibetans from the north, and indigenous peoples who predate the first two. Each source contributed differences in physical appearance, culture, and language. Nepalis speak dozens of languages and dialects, but many can converse in Nepali, an Indo-Aryan language closely related to Hindi, which some Nepalis also speak or understand. Highly educated Nepalis and those who

NEPALIS Finding enough wood for fuel takes up a lot of the day for these villagers in the Khumbu region of Nepal.

come into frequent contact with tourists from other countries also speak English.

About three-quarters of Nepalis engage in farming, mainly at the subsistence level. Nepal's terrain sorts into three distinct horizontal bands that rise to the Himalaya. The hot fertile plains of the southern Terai provide most of the country's agricultural land, where rice is grown in the monsoon season and wheat and other grains are planted after the rice. For many Nepalis, rice is a daily staple along with dal—a lentil dish—and vegetables.

Hills climb from the Terai into a zone of rivers and valleys and the large, ancient lakebed that forms the Kathmandu Valley, home to some 700,000 Nepalis. Here, crops grow at lower elevations and livestock grazes in pastures higher up. Above them rise more than 200 peaks over 18,000 feet high, including Sagarmatha, or Mount Everest, the world's highest at 29,035 feet. Nepal extends in places to the north of the highest ridge; the Tibetans who live on that side raise yaks and dzos, their yak-cattle hybrids.

The Indo-Aryans brought with them Hinduism, Nepal's dominant religion, embraced by four-fifths of the population. In 563 B.C., Gautama Siddhartha was born in the Nepali village of Lumbini in the Terai. He traveled to India and attained enlightenment and then, as the Buddha, imparted his insights to his followers, who filtered back to Nepal and now make up about 10 percent of its population. In general, Hindus and Buddhists celebrate each other's festivals and revere each other's shrines and sacred sites, of which there are more than 2,700 in the Kathmandu Valley. The Nepali population also includes small percentages of Muslims, Christians, and other believers.

The Hindu Indo-Aryans also brought with them notions of caste, which have been widely applied across Nepal's diverse population. Until recently, caste ranking correlated fairly closely with wealth, which was based on landownership. Now a wider range of people have opportunities in trade, tourism, and light industry. Nepal even boasts a small-scale film industry, known as Kollywood (*K* for Kathmandu).

Nepal's great scenic beauty draws many visitors as well as adventure seekers and serious climbers who attempt to ascend Everest. They require the knowledge and skills of the Sherpas, a group that traditionally has acted as guides and porters in mountain-climbing expeditions. The Gurkhas, another specialized group, have earned renown as soldiers and for nearly two centuries have supplied crack regiments to the British Army.

Still, Nepal is one of the world's poorest countries, with about a third of Nepalis living below the poverty level. Many of them belong to marginalized ethnic groups and have become highly politicized in a quest for greater rights and opportunities.

Punjabis

Population: 123.9 million
Location: Pakistan, India
Language family: Indo-Aryan

The Punjabis, a hardy people with strong ties to the land, originated in the northwestern corner of the Indian subcontinent. Their name is tied to their region,

DEEP ANCESTRY

THE TRADER

Junaid Syed, originally from Saudi Arabia and now living in Australia, decided to test himself through the Genographic Project after hearing about it in a live debate. He was intrigued about what it might be able to tell him about his Middle Eastern ancestry; his mother is of Turkish origin and his father is Arab.

DNA Results

His results revealed some intriguing patterns that were not typical of either his Turkish or Arab sides. In fact, they suggested an origin at least partially from farther east. While his maternal line was consistent with a likely origin in Turkey or the Caucasus, his paternal line—H1a1—is largely restricted to the Indian subcontinent. In fact, it seems to have been one of the founding lineages there, originating during the Paleolithic, perhaps 30,000 years ago. Today, it is largely limited to the Indian subcontinent, where it is carried by as many as half of the men, depending on the population and region. Junaid's closest matches in the Genographic database are all from southern India.

H1a1 is also found in places that traded with India—along the Persian Gulf coast of the Arabian Peninsula, for instance, and in southern Iran. Although H1a1 is not common in Saudi Arabia, it could have reached there through movement back west from South Asia. It is also found farther afield, in Cambodia and Indonesia. It is carried by 3.5 percent of the male population of Bali, a Hindu island in the Indonesian archipelago. The Balinese connection is interesting, as extensive Indian settlements have been found in the archaeological record, dating back more than 2,000 years ago, rich with Indian pottery, beads, and other artifacts. The current distribution of H1a1 is consistent with a spread via maritime trade networks—men who came from India for commerce but also brought their DNA, mixing with the local population.

Junaid Syed, Y-chromosome haplogroup H1a1

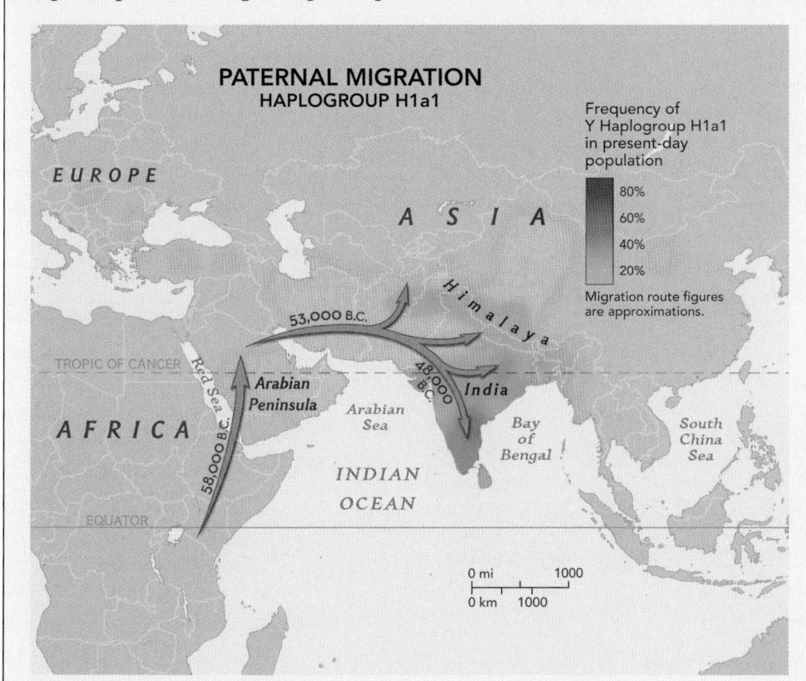

Paternal H1a1: An early migration from Africa took our ancestors via the southern route through India.

Surprise Connection

Junaid says that he "expected [his] maternal and paternal sides to be similar, and both be of Asian ancestry, so to see the South Asian connection was a surprise." His results confirm, though, that he has close ties to the Indian subcontinent. While he expected to learn more about his Arab ancestry, he never suspected that his DNA would reveal a story of ancient trade routes across the Indian Ocean, joining him to people living today in Mumbai, Chennai, and Indonesia.

—*Spencer Wells*

where five *(panj)* rivers *(ab),* tributaries of the Indus, drain fertile plains forming one of South Asia's most productive agricultural areas. The Punjab region was home to the flowering of the great Indus River civilization of the third millennium B.C., which included well-planned and -administered cities such as Harappa and suffered a lengthy procession of invaders—the Turks, Persians, Huns, Greeks, Afghans, and others. All left their marks on the present-day Punjabi people, who tend to be taller and more robust than other ethnic groups in the region.

Once a unified cultural and linguistic area accommodating members of the Hindu, Sikh, and Muslim faiths, the Punjab was divided during the partition that occurred at Indian independence in 1947, when the separate nation of Pakistan was created as a Muslim homeland. Hindus and Sikhs, who represent a distinctly Punjabi reformist movement within Hinduism with the city of Amritsar as its spiritual center, fled the western part of the region. Most Muslims went in the opposite direction. Violent reprisals took place on both sides of the border, leading eventually to almost a million deaths.

Today, the Punjabi population of Pakistan is about 97 percent Muslim, while in India, Punjabis are Sikh or Hindu, the ratio between them about three to two. Punjabi is spoken in both countries, although Pakistani Punjabis may also speak Urdu, Pakistan's official language, and those in India also often speak Hindi.

Punjabis in both Pakistan and India engage in agriculture as landowners or workers, although Pakistan's Punjabi population is much more urbanized, centered in towns and cities such as Lahore. Punjabi agriculture is carried out on a larger scale, with more mechanization and scientific advances, than elsewhere in the subcontinent, and even if a farmer provides his own subsistence, there often is a significant surplus to sell commercially. Wheat is the main food crop grown—together with lentils, vegetables, and maize in the summer—and cotton is the main cash crop.

PUNJABIS Brightly clad Sikh spectators watch mock battles at the spring festival of Hola Mohalla in Punjab, India.

Manufacturing, including food processing, other agriculture-related industries, and heavy machinery, also contributes to Punjabi prosperity.

The traditional Punjabi village consisted of a tight cluster of adjoining homes surrounded by a mud-brick enclosure fronted by a *darwaza,* or immense door; fields lay outside the village walls. Contemporary villages, especially in the Indian Punjab, often have brick houses with running water, electricity, and other amenities.

Rural Punjabi communities center on the intergenerational family: a man, his wife, his married sons, their dependents, and any unmarried children. Punjabis marry outside their clans and villages.

Muslim, Hindu, and Sikh Punjabis observe their own religious traditions, yet they also ecumenically celebrate festivals such as Vaisakhi, commemorating the Hindu harvest and also the founding of the Khalsa, the Sikh brotherhood. In Punjab, Vaisakhi features *bhangra* dancing, in which brightly costumed and elaborately turbaned men—in the traditional version—dance vigorously to a relentless drumbeat and vocal accompaniment, while women perform their own more subdued dances.

Bhangra has taken the international pop music world by storm, and this traditionally Punjabi music now appears singly and in fusion with hip-hop, disco, and reggae.

Sora

Population: 310,000
Location: India
Language family: Munda

The Sora are an indigenous people of India. Their ancestors settled in India long before the arrival of the Aryan and Dravidian populations that now dominate and outnumber them. They are ethnically and linguistically related to various Mon-Khmer peoples, including the Cambodians, who live a considerable distance to their east. To this day, the Sora, as well as other indigenous peoples of India, are referred to collectively as *adivasi,* "first peoples." Also popularly known as "tribals," they exist outside and beneath India's socially rigid caste system, lower even than the Untouchables.

SORA The Sora of eastern India, as this woman demonstrates, have transitioned from a hunter-gathering society to a more settled existence.

The Sora were formerly hunter-gatherers but are now rice cultivators, using yoked oxen and simple wooden plows to tend pond fields. Due to the alternating wet and dry seasons, they can grow only one or two crops of rice each year, and small family plots often do not yield enough food for a family to survive. Outside of the rice-growing season, the Sora work as day laborers and practice trades such as brickmaking, cashew processing, and tamarind growing.

In stark contrast to the Hinduism practiced by 81 percent of India's citizens, the Sora traditionally eschewed the worship of gods in temples, instead worshiping outdoors in nature, where some venerated stone megaliths that their ancestors built. Unlike Hindus, the Sora have no taboos against eating meat, and they used to sacrifice water buffalo ritually.

The Sora have traditionally valued communication with the spirits of recently deceased persons, for which they rely on mediums, who enter into trances and speak in the voices of the deceased. Sora will seek advice or protection from the dead and may also strive to console or appease them. This practice of dialoguing with the dead is now fading, however, under pressure from Christianity.

Sora women once tattooed their arms or faces and pierced and elongated their earlobes; Sora men once wore loincloths. These practices have now waned, and they may be seen only among the oldest members of rural communities.

In recent decades, many Sora have been converted to evangelical Christianity, which some now zealously practice, worshiping in elaborate churches that they have constructed. In the few small villages that have not yet been missionized, people retain more traditions. They brew palm wine, paint totemic designs on the interior walls of their mud-brick houses, and engage in ceremonial music and dancing. The vibrant sounds of musicians playing cymbals and horns can summon an entire village population out into the streets in minutes, congregating to dance frenziedly for hours at a wedding or important event.

The Sora language is one of the most important Munda languages. It has attracted the attention of linguists for its ability to build complex verbs. A Sora verb can incorporate the subject and direct object, effectively compacting an entire complex sentence into just a single word. The main concentration of Sora speakers is in the state of Odisha, India, but speakers are also found in adjacent Andhra Pradesh. Estimates of the total number of speakers vary between 150,000 and 300,000. In some areas with significant Sora populations, however, a language shift is occurring, and members of the younger generations no longer speak Sora. While not immediately endangered, Sora, like other similarly sized and marginalized minority languages of India, may be considered threatened. Sora remains a primarily oral language, not widely read or written even by speakers who may be fully fluent and literate in Oriya, a regionally dominant language, and in English. A rich story tradition helps impart values by cautioning against greed, family strife, and vanity. Though the Sora Bible may be found as a prized household possession in some Christianized villages, it is not typically read. In many areas, Sora remains a vital and thriving language; in other areas, it is being replaced by Telugu or Oriya.

Lost opportunities cannot be drawn back even by the might of elephants.

SINHALESE PROVERB

The Sora people are now eager to educate their young and join India's technological revolution. Though social mobility for tribals is quite limited and the value of their culture remains largely unacknowledged, schools for Sora and other tribal children are beginning to break down some social barriers and provide opportunities.

Sinhalese

Population: 16.5 million
Location: Sri Lanka
Language family: Indo-Aryan

In multiethnic Sri Lanka, a teardrop-shaped island off the southeastern tip of India, the majority Sinhalese number 16.5 million out of a population of 22 million. The Sinhalese have since earliest times coexisted with other ethnic groups on the island—first the Tamils, then Muslim traders, then the Portuguese, the Dutch, and the British. In recent times, though, contention with Tamils has grown into warfare because the Tamils, inhabitants of the northern and eastern reaches of the island, seek an independent homeland free of Sinhalese control. Points of division include religion, language, and ethnic identity.

Indo-Europeans (Aryans) from northern India moved to Sri Lanka about the fifth century B.C. and in time became the Sinhalese, with their own distinct language, Sinhala. As early as the third century B.C., they had fully converted to Buddhism, and their

SINHALESE A Sinhalese fisherman from southwestern Sri Lanka anchors himself on a stilt in the surf to cast his line.

mutual faith and shared language bound the people together. Sinhalese kings played important roles as patrons of Buddhism in the early centuries of the first millennium A.D. At a time when the Buddhist religion was declining and disappearing in India, the people of the island nation felt protective toward their faith, and the Sinhalese people stressed the need to preserve their special path to enlightenment.

The Sinhalese also strongly influenced the development of Buddhism in Southeast Asia through the missionary efforts of their teachers and artists, who traveled beyond their island home and used religion and a distinct writing system to direct the flow of education, literature, Buddhist chronicles, and common aspirations. Today, the village monastery or temple still stands as the local center for culture, and to become a monk is considered a noble goal.

Throughout history, successful kingship has been marked by massive irrigation works for the growing of rice and by impressive art and architecture: A period of exceptional flowering in the 12th century produced Polonnaruwa, a magnificent royal Buddhist city in north-central Sri Lanka that achieved greatness in art, sculpture, and architecture.

Tamils

Population: 76.2 million
Location: India, Sri Lanka
Language family: Dravidian

The Tamils largely inhabit Tamil Nadu, an Indian state created in 1956 to be a home for Tamil speakers, a group tending to be stockier and somewhat darker than northern Indians. Since before the arrival of the Aryans around 1500 B.C., these people have lived at the southeastern limit of the Indian subcontinent.

More than 75 million Tamils define themselves as Tamil-speaking Hindus who share in a great Dravidian culture. The term *Dravidian* describes an independent family of some two dozen languages and the special culture of the south of India. Except for classical Sanskrit, Tamil is the oldest written language in India. Going back nearly 2,000 years, its rich literary repository includes epics, religious and secular poetry, philosophy, and moral instructions.

TAMILS Tamil tea pickers work their way down the rows on a plantation in Sri Lanka.

Of particular interest are the verses and hymns of devotional literature dating from the sixth to ninth centuries A.D. Collectively, these works are considered to be among the great Tamil contributions to Indian civilization. Tamils believe that *bhakti,* meaning "devotion," elevates a devotee's personal relationship with a god to a level of ecstatic love and selfless praise. Above all, the path of bhakti leads to salvation.

Over the years, Tamils have vociferously resisted the encroachment of the Hindi language and northern cultural influences. They have done so even though they too are overwhelmingly Hindu, worshiping at more than 9,000 temples and celebrating a huge variety of riotous festivals.

Chennai (formerly Madras), the Tamil capital, was a weaving center when the British East India Company built a fort and trading station there in the mid-17th century. Later it became the administrative and trading capital of southern India. Because of British needs for labor elsewhere, emigration created Tamil communities as far away as Fiji, Malaysia, Singapore, South Africa, and Mauritius.

The Tamil people's presence in Sri Lanka consists of two groups: the Sri Lankan Tamils, who settled there more than 1,500 years ago, and the Indian Tamils, who came in the 19th and 20th centuries to work on tea and rubber plantations. Civil war in Sri Lanka has led Tamils to migrate to many other countries, but wherever they go, they build temples and develop strong local ties.

The Asian North

The area of the globe centering on the North Pole, the Arctic Circle is delineated by latitude 66° 30′ N. Among the Asian countries, it is Russia whose northernmost realms fall within the Arctic Circle. At its center is a large basin, the Arctic Ocean, ringed with a variety of landscapes. The Russian Arctic includes two mountain ranges, the Urals in the west and the Verkhoyansk in the east, and numerous islands in the Barents, the Kara, and the Laptev Seas. The northernmost realms of Scandinavia—Norway, Sweden, and Finland—belong to this region as well. Tundra vegetation characterizes the landscape year-round, with very cold, long winters and brief, mild summers.

Chukchi & Koryak

Population: 15,900 (Chukchi); 7,900 (Koryak)
Location: Russia
Language family: Chukotko-Kamchatkan

Across the Bering Strait from North America's Alaskan Eskimos live the Chukchi and Koryak, two peoples who have been in close contact over the years. The Chukchi inhabit the farthest reaches of northeastern Siberia, and the Koryak live on the Kamchatka Peninsula to the south. Both groups descend from Paleolithic hunters and fishers and are physiologically related to the Mongoloid northern Asians. They speak mutually intelligible languages.

Originally, the Chukchi and Koryak were split into coastal groups of maritime hunters and fishers and inland groups of reindeer herders. The coastal societies had more in common with each other and nearby Eskimo groups than they did with their inland relatives. They hunted on the ice in winter and from skin boats during the summer, relying on sea mammals such as whales, seals, and walrus for food. Since Koryak territory impinged on the taiga, or boreal forest, to the south and west, the Koryak also availed themselves of plant resources and freshwater fishing.

The inland Chukchi and Koryak maintained herds of reindeer for food, transport, clothing, and shelter.

KORYAK On Russia's Kamchatka Peninsula, a Koryak man stands in front of a traditional *yaranga,* a reindeer-hide portable shelter.

The semidomesticated deer had to be recaptured and retrained each season to pull the sleds. Watching over the animals were herdsmen who lived in family encampments of four or five large skin *yaranga,* or tents, made of reindeer hides. The families kept large herds, so concepts of wealth, private property, and social stratification became well developed among these people even before the first Russian contacts in the 1640s.

In both Chukchi and Koryak cultures, shamanism was highly developed, and their religious systems shared a number of beliefs and associated rituals, including the sacred family fire and the concept of Raven as a major cultural hero. Complex fertility ceremonies held at annual reindeer slaughters and sea mammal feasts were intended to ensure the continued availability of game.

The Chukchi now live mostly in villages established by the Soviets during the Siberian collectivization of the 1930s; only a few villages still keep reindeer herds. Among the difficulties faced by this nomadic culture as it adjusted to settled life were the pressures of industrialization and, later, nuclear testing, all of which—until recently—contributed to very high rates of infant mortality, alcoholism, and suicide.

Accidents will happen in the best-regulated families.

SAKHA PROVERB

These same general conditions affect the contemporary Koryak people, who were also formed into a national district by the First Congress of Kamchatka Soviets in 1930. The Koryak have suffered under the same deleterious effects of modern civilization as those that have affected their northern neighbors, and their average life span is now less than 50 years.

Evenk

Population: 30,500
Location: Russia, China, Mongolia
Language family: Tungusic

Among the most widespread of the Siberian Arctic cultures, the Evenk people are subdivided into two parts. The Evenki group inhabits the area extending eastward from the Yenisey River to the Pacific Ocean and northward from the Amur River to the Arctic Ocean; the Even, or Lamut, group lives along the coast of the Sea of Okhotsk.

The Evenk are reindeer pastoralists who keep large herds of completely domesticated deer, which are bigger and more uniform in color than the wild variety. For these people, reindeer domestication seems to be more closely modeled on herds kept by horsemen to the south than on dogs kept by the northeastern Siberians. Evenk reindeer are milked, ridden, and used as pack animals but, owing to their value, are not eaten except on ceremonial occasions or if starvation is imminent.

The Evenk derive the bulk of their food, clothing, and shelter from hunting and trapping, the main daily activities of the men. Like the Aleut, they have a great interest in animal anatomy, not hesitating to pursue and capture unfamiliar creatures to observe their behaviors and learn about their anatomies through dissection. Their knowledge of physiology and comparative anatomy astonished early Russian ethnographers, who also noted that they performed complex surgeries.

Shamanism was more highly developed by the Evenk than by any other Siberian group; in fact, the word *shaman* has an Evenki origin. Shamans could reputedly tell the future, predict the weather, cure or cause an illness, and ensure success in hunting or herding. An Evenk shaman frequently used the agaric mushroom to induce trance states. Bear veneration, another frequent circumpolar practice, was also common.

In the past, the Evenk lived near the taiga in winter and then moved their herds onto the tundra in summer. Because of climatic extremes, people were widely scattered in groups of only two or three families

EVENK An Evenk herder leads her domesticated reindeer to pasture in the early morning.

during the long Arctic winter, which lasts about three-quarters of the year; in summer, they formed larger camps.

The people continued keeping their herds and trading furs with the Chinese and later Russian merchants. Then, in the 1930s, during the Soviet period, they and other Siberian nomads were coerced into moving to permanent villages. Today, 30 to 50 percent of the population still herd and hunt for a living, while the rest of the people work in industry, education, administration, and health care or as unskilled labor.

The Evenk National Okrug (or District) was created in 1930, but it enjoys only limited autonomy within the Russian state. Nearly half the Evenk population lives in the People's Republic of China, where they are given the status of one of China's 56 officially recognized ethnic groups.

Nenets

Population: 44,600
Location: Russia
Language family: Samoyedic

Also known as the Nentsy, Samoyed, and Yurak, the Nenets inhabit northwestern Siberia, from the Kanin Peninsula on the White Sea to the Yenisey River Delta. This is a region of wet tundra underlain by permafrost, combined with tundra and taiga mosaic ecosystems to the south.

NENETS The world of these Nenets nomads revolves around reindeer herding.

Apparently the ancestors of modern Nenets moved into this area from the south during the early centuries of the first millennium A.D., replacing or assimilating the original hunting-and-gathering population with one that relied on reindeer herding. These newcomers practiced herding techniques that seem to have been adapted from ones used by horse and cattle breeders to the south.

Like other Siberian groups, the Nenets developed a migratory lifestyle fully adjusted to tundra existence. In autumn they moved with their herds from the coast to the taiga, and in spring they went back to the tundra. The Nenets became well known as breeders of reindeer large enough to be ridden like horses, and their animals were widely sought after by other groups. By comparison, the reindeer of the neighboring Evenk were so small that riders had to sit forward, above the shoulders.

Nenet contacts with the Russian Empire were hostile and fraught with uprisings, a situation that did not improve with the Sovietization of the 1930s; with a millennia-long history as successful nomadic herdsmen, they did not yield easily to collectivization. The area has since become the focus of the chemical and oil industries, and nuclear testing on Novaya Zemlya has posed grave dangers to the health of indigenous peoples. During the 2010 census, 44,600 Nenets were counted, about half of whom still spoke Samoyedic, a language related to Turkic. Their life expectancy is now reckoned at only 45 to 50 years.

Sakha

Population: 478,000
Location: Russia
Language family: Turkic

Although the early history of the Sakha (sometimes called Yakut) is obscure, the people emerged as a distinct ethnic group in the early 1300s, a fact that makes them the most recently arrived indigenous group in the Siberian Arctic. By the 17th century, they had peacefully assimilated with other northern peoples, particularly the Evenk, and had organized themselves into 80 independent tribes that were themselves subdivided into numerous clans ranging in size from a few to several hundred people.

SAKHA Vasily Atlasov interprets Sakha culture in an ice cave at his family's homestead in Siberia.

Like the Evenk, the Sakha based their subsistence on herding, but in this case, the herds were made up of horses and cattle. Because of the extreme cold, Sakha animals tend to be hardy but rather unproductive and have to be sheltered and fed for a large part of the year.

The far northern Sakha, known as Dolgan, adapted to reindeer herding and hunting learned from neighboring peoples, often hunting from the backs of oxen or horses. The people also hunted ermine, sables, otters, and ferrets, and their culture flourished as the Russian fur trade grew. The Sakha valued reindeer hides and furs as personal property, and families with hunters and large herds could accumulate great wealth. The society eventually became stratified, causing the old clan system to become nonfunctional and then fall apart. Shamans among the Sakha got paid for their services, and as shamanism developed a complex hierarchy of duties, the shamans too accumulated enough wealth and power to become threats to the social order.

From the 17th century on, the Sakha were best known as fur traders and blacksmiths; by the early 1800s, most had joined the Russian Orthodox Church. As in the rest of Siberia, Sakha pastoralism and agriculture were collectivized between 1930 and 1950, but in a particularly brutal manner: Thousands who resisted were arrested, killed, or forcibly deported. Sakha schools, publications, and organizations were officially banned by Joseph Stalin, who promoted the development of heavy industry in the region.

Sakha nationalism reemerged in the wake of Mikhail Gorbachev's reform policies of the 1980s, and in 1990-1991, the Yakut Autonomous Soviet Socialist Republic proclaimed its sovereignty as the Sakha Republic and established its own parliament. Despite political differences between Sakha and Russian leaders during the 1990s, President Boris Yeltsin in 1994 officially apologized to the people for their persecution by Stalin's regime and granted them greater control of their own resources, including diamond mines. In 1997, with the help of the Worldwide Fund for Nature, 270,000 square miles of Yakutia's Arctic tundra were set aside as an ecological preserve. In 2000, President Putin made the Sakha Republic part of the Far Eastern Federal District, a governing territory of more than 1 million square miles.

Tuvans

Population: 300,000
Location: Russia
Language family: Turkic

At the geographic center of Asia lies Tuva, a secluded and mountainous land in southern Siberia, adjacent to Mongolia. As a nation, Tuva enjoyed a few brief years of independence before being annexed by the Soviet Union in 1944. Since then, it has had limited autonomy within the Russian Federation.

The Tuvans are an Inner Asian Turkic people, related to the neighboring Altais and Kazakhs. They have cultivated a centuries-long, sometimes uneasy alliance with their Mongolian neighbors, sharing some physical and cultural traits though differing linguistically. Small Tuvan communities reside in the People's Republic of China and Mongolia, though they lack official recognition as ethnic groups in those countries.

Tuvans are known for their virtuoso throat-singing, also called overtone singing, in which performers produce multiple notes simultaneously, including high-pitched whistling and very low drones. An annual competition held in Kyzyl, the capital city, can launch the career of a young artist. Accompanying instruments may include a drum, bull's testicle rattle, bowed fiddle, goatskin banjo, and *khomus,* a metal mouth harp.

The Tuvan language boasts a rich oral tradition of songs, riddles, sayings, blessings, poems, and stories. Genres like the epic tale, once performed by itinerant storytellers over several evenings, are now seldom heard and in danger of being forgotten. Tuvan was first written in the 1930s, using a Latin alphabet, and later, under Russian influence, a Cyrillic script. Many books are now available, including elementary and high school curricula. Tuvan is spoken by people of all ages, and most educated Tuvans also speak Russian.

A minority of Tuvans practice nomadism and herd camels, yaks, sheep, goats, cows, and horses. In the far north of Tuva, a small population of reindeer herders thrives. Migrating seasonally to reach greener pastures, Tuva nomads live—as did their ancestors—in yurts, collapsible, round felt houses. Sheep entrails and tail fat provide a favored meal on special occasions, while salty tea with milk slakes their thirst daily.

TUVANS A Tuvan father and his strapping sons call a carpet-laden, round felt portable house their home.

Tuvans practice a form of Buddhism adopted from Mongolia and Tibet. Buddhist practice was suppressed in Tuva under Stalinism and communism, and many temples were destroyed. After the fall of communism, Tuvans rebuilt their temples with great enthusiasm, and they continue to look for guidance to their own Buddhist leaders, headed by their own Kamby Lama, as well as spiritual envoys sent by the Tibetan 14th Dalai Lama.

Most Tuvans also continue to practice animism, venerating Earth spirits and maintaining an elaborate cosmology of supernatural beings. Shamans are sometimes called on to heal the sick or mediate with the spirit world; they may drum, chant, and burn juniper incense to undergo a trance journey to another world. Tuvans revere the landscape they inhabit, making frequent offerings at sacred springs, rock cairns, and other sites believed to be inhabited by spirits. The Tuvan language has a rich vocabulary for naming topographic features and for imitating natural ambient sounds such as rivers, winds, echoes, birds, and game animals.

Tuvans enjoy some benefits of modern technology while also being shielded by geography. They are one of the few small ethnic groups of Russia that constitutes a majority in their home territory. As a result, they have managed to hang on to much of their language, lifeways, and beliefs. It is still possible in remote Tuva to meet elderly nomads who have never ventured more than one hundred miles from their birthplace, people who are still sustained by their connection to their land and animals. At the same time, many educated Tuvans have traveled abroad or served in the Russian military, and a few have risen to influential political positions.

Yukaghir

Population: 1,600
Location: Russia
Language family: Yukaghiric

Although the Yukaghir, or Odul as they call themselves, are among the smallest of the Eurasian Arctic populations, they are extremely important to the understanding of far northern cultures. These people probably represent most closely the ancient, original cultural adaptation to the coldest regions of Arctic Siberia. The area they occupy is generally north of the Verkhoyansk Range and is known for having the most severe winters in the world. Here, all of the water surfaces are frozen for seven or eight months of the year; the January temperature averages minus 70°F, as measured at the town of Verkhoyansk, and only 70 to 80 days each year are free of frost. During the short summer season, the land becomes a vast lake-studded marsh, because the meltwater from the considerable snowmass cannot seep down into the permafrost.

The most important subsistence animals in this region are reindeer. For generations, reindeer hunting furnished virtually all of the food, clothing, and shelter items needed by the Yukaghir, who were originally a nomadic people with an unusually large inventory of hunting methods. They used tame deer as decoys, crouching behind them to infiltrate a wild herd, and employed noose traps to catch the animals along their trails. From skin boats, they hunted deer at water crossings, using harpoons and compound bows. Elaborate, large-scale communal hunts called drives or pounds took place during calving season in the spring. At these times, men stampeded a herd into a keyhole funnel formed by waiting hunters, who then killed as many of the animals as possible.

The Yukaghir formerly lived in extended-family bands with fewer than a hundred people, occupying conical skin tents much smaller than the Chukchi's yaranga and somewhat resembling the tipi of the North American Plains Indian. Their shamans, like those of other Siberian societies, were powerful individuals, and elders and ancestors occupied places of particular prominence. At one time, the people used a pictographic writing system, the only remaining evidence of which is a love letter written on birch bark. Few today still speak their native tongue, a language related to Chukchi and Koryak.

Ethnologists knew the Yukaghir as a kind, mild-mannered, and honest people exploited and abused by Cossacks, Russian settlers, and local priests, leading to the rapid assimilation of their native culture. Russianization brought the Yukaghir to the brink of extinction. In recent years, the people have adapted to life in small permanent villages with rectangular wood houses and herds of no more than 30 to 40 domesticated reindeer.

YUKAGHIR Age and beauty in equal measure characterize this Yukaghir woman, a member of one of the smallest ethnic groups in the Siberian Arctic.

ACROSS CULTURES

A LIFE ON THE MOVE

Far from primitive, the pastoral nomads of today live according to sophisticated bodies of knowledge about their animals and their landscapes.

Nomads, contrary to popular belief, do not wander aimlessly, and they are never lost in their home territory. Their seasonal movements run like clockwork, responding to subtle cycles of vegetation, weather, and climate, as well as the need to fatten their animals and sustain themselves. Migration cycles may be long in both distance and time, sometimes cycling through a series of locations over a decade before returning to the same spot again. The outward simplicity of this way of life—owning only what can be easily carried on a few pack animals—belies an astonishing set of intellectual and scientific achievements made by nomadic cultures in order to survive.

Nomadism solved one of the great problems of human adaptation: How can people inhabit vast areas of land that are too poor to grow crops or forage for food? The answer lies in the many technologies of domesticating animals and migrating them across grazing lands. This advance allowed one of the greatest population expansions in human history, covering much of Central Asia and North Africa and persisting today. A significant human advancement, nomadism still has much to teach us. Though it is a way of life now under threat from overcrowding and globalization, nomadism is actively practiced in many parts of the world, from the high grassy plains of Mongolia and Tibet, to above the Arctic Circle in Finland and Russia, to the Andes of Bolivia and Peru.

Nomadic Technologies

Nomads have domesticated, or tamed, a wide range of animals around the globe: sheep, goats, reindeer, camels, horses, yaks, cows, and llamas. Living in intimate and constant contact with these beasts, they have developed expertise about the life cycles, habits, mental states, lactation, and useful by-products of the animals. Nomads often know more about their animals than do modern zoologists or veterinarians. Though their view is a more traditional one, we can learn much from their expertise.

For example, in Mongolia, herders of yaks and camels have developed a large repertoire of "domestication songs" and soothing sounds they produce while tending or milking the animals. They believe that these songs and sounds induce certain mental states in the animals that allow them to be better managed, more fertile, and more productive. In observing the calming effect of this singing on a mother yak or camel that is being trained to nurse its calf, it is impossible not to marvel at the effectiveness and genius of this sound technology.

Nomads are also pioneering geneticists, having made discoveries about animal breeding long before

Solitude is a way of life for pastoral nomads such as this shepherd in Qinghai Province, China.

A Wodaabe man of Niger gives his tall and streamlined mehari camel a rest. The breed is prized by nomads for its speed.

scientists knew about genes at all. They have learned to control breeding and reproduction carefully to maintain robust animals and in many cases to breed bigger and better animals. The Evenki people of eastern Siberia succeeded, through controlled breeding, in creating a bigger and hardier reindeer that could be saddled and ridden.

Early Genetic Engineering

The Bodi people of Somalia created an elaborate hierarchy of cow colors, markings, and patterns. By observing these outward morphological traits and using them to decide which animals to mate, they have effectively been successfully controlling genetic mutation centuries before scientists considered it.

Bodi herders give individual, descriptive names to most animals. The names may describe the coat color, pattern, horn shape, or other characteristic. And they can recite the ancestry of a particular cow, sometimes as far back as a dozen generations. The knowledge of folk genetics transmitted orally by the Bodi is astonishing: They give distinct names to more than 80 combinations of coat color and pattern. For example, a cow that is entirely black is called *koro.* If that black cow mates with a bull that has the pattern *ludi*—all white with a black head and rump—their calf will be a *kalmi,* all black with just with a narrow white stripe around its midsection.

By observing the cows across many centuries, the Bodi have mastered a highly complex system that names and keeps track of all possible colors and patterns for cows and tracks the patterns predictably produced by cross-matings. The Bodi people group certain colors and patterns together into cow clans that mirror the structure of their human clans.

Bodi involvement with cows goes far beyond sustenance of meat and milk. Though they are seminomadic—they also practice swidden farming (a technique that creates temporary growing plots)—their entire worldview centers on their symbiotic relationship with the cow, which represents fecundity, life, and survival. They sacrifice specifically patterned cows for marriage, for birth and death rituals, or when a human or cow falls ill, when they pray for rain, or when a lion is killed. Cows with certain patterns must never be sacrificed, lest plague or famine ensue.

Nomads invented veterinary medicine. They understand the health and illness of their animals. They also solve complex economic problems by using precise and time-tested formulas of mixed herding. For example, in the high Andes of Bolivia, herders combine llamas with sheep to make the most of the thin vegetation. Like the Bodi, the Aymara people of Bolivia use a complex classification system to name their alpacas and llamas according to sex, age, and fleece color or pattern. Though Andean pastoralists are now mostly sedentary, living in houses and keeping their animals in stone enclosures, the archaeological evidence suggests that nomadic pastoralists raised camelids for nearly 2,000 years in the Andes alongside sedentary peoples who grew potatoes and quinoa.

A nomadic Monchak boy of western Mongolia proudly displays medals he won in horse races.

An Azeri woman deftly prepares a large crepe-like dish in her open-air kitchen.

Naming of the Land

Nomads have an intricate knowledge of the landscape they inhabit. Typically they have names for every single topographic feature, at a far greater level of detail considering they may live in areas with no roads, towns, or addresses. The languages that nomads speak often encode features of the landscape into the verbs or other parts of their grammar. For example, in Tuvan, a language spoken by yak herders in southern Siberia, in order to simply say the word *go,* you must know the direction of water current in the nearest river, and you must choose one among three verbs that indicate "upstream," "downstream," or "cross-stream." In other words, the Tuvan nomads can orient themselves because their language requires them to pay attention at all times to the landscape.

The Work of Horses

Nomads' work is never done, and there are no vacations, because the animals never rest and must always be supplied with water and vegetation. In Mongolia, a typical horse-breeding nomad's day begins before daybreak. Horses in Mongolia are used for riding, carrying or pulling, milk, and sometimes edible meat. Horse racing is a national sport, practiced by both young boys who ride bareback and men who ride with saddles, competing for glory and medals. A common herd numbers up to 30 mares, 15 foals, 20 two- and three-year-old horses, 10 geldings (castrated male horses), and 1 breeding stallion.

Horses are a source of great pride, and their care and well-being is what the nomads—the entire extended family—devote their lives to. Early in the morning, the horses must be let out of the stockade to graze, and the foals must be let out of their separate pen. The foals will be allowed to nurse for just a few minutes, to get the mares' milk flowing. Then the foals are tied up and the mares must be milked, not once but every two hours during the daytime. The milk is processed to make a rich "white food" for young children. In three summer months, a mare can produce up to one and a half gallons of milk a day. Women and girls collect it by hand-milking.

Men work with colts to break them, train them to the bridle and saddle, and train them to carry a rider. Breeding must be carefully controlled to obtain a line of healthy offspring.

Leather must be worked by hand to make lassos, saddles, and bridles. Women collect horsehair from the tail and painstakingly twist it by hand into sturdy ropes. Hair may also be used to make stringed instruments, especially the *morin huur* (horsehead fiddle), decorated with a carved wooden horsehead and believed to embody the spirit of the horse.

Will nomadism survive our shrinking world? Even in the most remote areas of Central Asia or in the high Andes, nomads are feeling the pressures of globalization. Many lands now suffer from overgrazing, as well as periodic winter famines. As it becomes harder to fatten up the animals, young people may see herding as a dead end and prefer to migrate to the cities to work as day laborers instead.

Those who do stay in the countryside are acquiring technologies that change their way of life: Solar panels, televisions, radios, tractors, motorcycles, and cellular phones are all commonly seen. Despite these imports, many thousands of people in Mongolia, Siberia, Central Asia, North Africa, and the Andes still live as their ancestors did centuries ago, listening to the sounds of wind and rain, driving the animals on to greener pastures, and tending to the rhythms of their sheep, camels, llamas, yaks, and horses.

—K. David Harrison

On the move with their herd of cattle and flock of goats, Niger pastoralists transport their possessions on camels and donkeys.

Chapter 4

OCEANIA

AUSTRALIA

Arrernte • Australians • Cape York Peoples
Ngarrindjeri • Noongar • Pintubi • Tiwi
Torres Strait Islanders • Warlpiri • Wurundjeri • Yolngu

MELANESIA

Chimbu • Enga • Fijians • Middle Sepik
Motu • Tannese • Tolai • Trobrianders

POLYNESIA

Hawaiians • Maori • New Zealanders
Samoans • Tahitians • Tongans

MICRONESIA

Chamorros • Chuuk Islanders
Marshallese • Palauans • Yapese

In the 19th century, European scholars recognized three broad geographical, cultural, and linguistic zones across the vast Pacific Ocean: Melanesia, immediately to the north and east of Australia; Polynesia, in the central Pacific; and Micronesia, east of the Philippines in the North Pacific. This largely artificial division, which downplays important similarities and connections among the islands, is said to have originated with French explorer Jules-Sébastien-César Dumont d'Urville (1790–1842), who based it on the mistaken notion that three separate races of people dwelled on the widely dispersed islands.

Australia has always been viewed separately. Indeed, the Aboriginal languages and customs of this continent are not closely linked to those of other Pacific peoples, although archaeological evidence suggests that continuing migrations from Asia through Melanesia and onto the world's largest island began as early as 50,000 years ago.

RESISTANCE AND RECONCILIATION In the past few centuries, the indigenous cultures of Oceania—a collective name for the Pacific islands and Australia—have been greatly affected by colonialism; at various times, they were disrupted or even unified by their opposition to it. In New Zealand, for example, strong Maori resistance to settlement by Europeans resulted in an 1840 treaty establishing terms of coexistence.

Such a treaty was never contemplated with the Aboriginals of Australia. The British considered Australia *terra nullius*, a "land without people," despite the fact that at the time of James Cook's 1770 expedition, more than 500 tribes were living there. Because Aboriginals had no recognizable system of land tenure in the eyes of British law, the British believed there was no one with whom they could conduct negotiations for the transfer of land.

A diver explores a coral reef in the shallow, clear blue waters of American Samoa.

Following the tragic excesses of the chaotic frontier period, Australia's policy in relation to the Aboriginals ranged from protection to assimilation, integration, and finally self-determination. Not until 1992 was the terra nullius doctrine overturned by the Australian High Court. Now Aboriginal groups are reclaiming their homelands through federally enacted native title legislation, and Aboriginal reconciliation ranks high on the national agenda. Through it all, Aboriginals have held on to many of their traditions, continuing to follow rituals and practices associated with the Dreaming or Dreamtime, beliefs related to events at the "beginning of time."

Traditions are also important to Pacific islanders, providing links to the past and helping them maintain a sense of solidarity. Over countless generations and sometimes across great distances, islanders reached out and connected through trade and intermarriage to other Pacific peoples. Then, in the 16th century, European explorers sailed into their waters, and by the mid-1800s, most of their islands had been claimed by Germany, France, Spain, and England. The United States entered the colonial arena following the

Spanish-American War of 1898, with acquisitions in the Philippines, Guam, Hawaii, and Samoa. After World War I, Japan took control of many former German territories.

TOWARD INDEPENDENCE The invasiveness of colonial rule varied widely. In some cases, contact was infrequent and did not threaten the cultural integrity of local populations, while in others, the rights of self-determination were curtailed. Some groups felt compelled to develop strategies of avoidance or resistance that often grew into anticolonial and proto-nationalist movements. The most famous of these responses were the Melanesian cargo cults, which held that a supernatural "cargo" of deities and ancestors would arrive among supplies delivered to the islands, bringing about an age of plenty and a new, moral world order devoid of a ruling European presence.

Between 1962 and 1980, most colonial territories achieved independence. But the postcolonial period has been marked by deep division and civil unrest. In Polynesia, where chiefdoms are largely determined by genealogy and seniority, island chiefs have readily made the transition to political leadership; however, the justice of their continued rule is sometimes challenged. In much of Melanesia, political organization centers on "big men" who achieve leadership through oratory skills, gift giving, and marriage alliances. While Melanesians share a common religion (Christianity), they often resist forming alliances with strangers and former enemies.

Decorated to mimic birds of paradise, Huli tribesmen of Papua New Guinea await the start of a *sing-sing*, a gathering of tribes.

As indigenous groups achieve freedom from foreign rule, they sometimes find that the national unity in which they are newly participating threatens their traditional cultural distinctiveness. New states, often beset by local demands for autonomy, may actively resist the notion of land rights for cultural or ethnic groups within their boundaries. Governments may promote a sense of shared identity, particularly in regard to *kastom* (custom or tradition).

In Melanesia, people have a long history of making and remaking kastom, creating and re-creating songs, myths, dances, and other cultural expressions associated with constituent groups. Such expressions are sources of prestige for their creators, and new variations are traded and exchanged following traditional practices—as exemplified by the Trobriand Islands' kula ring. Members of this trading network travel in canoes around an island group off Papua New Guinea, and as they exchange arm rings, necklaces, and other items, they affirm ties between ring partners and their communities.

In some Oceanian states, population growth rates are among the world's highest; even so, survival of many groups is threatened by large-scale plantation agriculture, mining, migration and transmigration, and uncontrolled fishing and logging. Forced dislocation and illnesses stemming from atomic testing have been major problems in Micronesia, where many people are kept from their former homes by alarming levels of radioactivity. Traditional small villages with a strongly hierarchical social order remain, however, together with a deep sense of attachment to homes and ancient customs, a feeling shared by many Oceanians.

Australia

A continent and a country, Australia lies between the Indian and the Pacific Oceans, a landmass of nearly three million square miles that is by and large flat and dry. Its southwest corner collects moisture enough to be verdant and fertile, and its northeast corner, the Cape York Peninsula, supports a tropical rain forest. Much of the continent, however, is covered by desert. To the northeast, Australia's coastline is fringed by the Great Barrier Reef, the world's largest. Two major rivers, the Murray and the Darling, join and flow west into the Indian Ocean. The country is divided into five states—Queensland, New South Wales, Victoria, South Australia, and Western Australia—and includes several internal and external territories. The island state of Tasmania, to the southeast and separated from the mainland by the Bass Strait, is also part of Australia.

Arrernte

Population: 2,380
Location: Australia
Language family: Pama-Nyungan

Central Australia's Arrernte are perhaps the most famous Aboriginal Australians, thanks in part to the writings of such distinguished European social scientists as Sigmund Freud (1856–1939) and Émile Durkheim (1858–1917). Their works were classic ethnographic studies focusing on the concept and meaning of the Dreaming or Dreamtime—the traditional land-based Aboriginal belief system—and on the Arrernte kinship system, within which a man marries his mother's mother's brother's daughter's daughter (his second cousin).

In fact, the Arrernte are a collection of peoples from the well-watered MacDonnell Ranges region in the vicinity of Alice Springs. These peoples include the Alyawarre, Anmatyerre, and Western Arrernte. All suffered gravely after contact with Europeans, experiencing high mortality rates, physical and emotional dislocation, and ethnocide. Some groups became extinct.

In the 1860s, after 20,000 years of living in this region, the Arrernte first encountered non-Aboriginals. With colonization, their lands were converted into grazing lands, where they worked as ranch hands and domestic servants; in return, they received rations and pocket money. Since the 1970s, some of the lands have been returned to the traditional landowners through federal land rights laws and excisions on ranches.

The Dreaming is still relevant for the Arrernte, even though the beliefs and practices of Christianity are quite prominent in their lives. Large families are common among these former hunter-gatherers, who now reside in permanent settlements. Joblessness is a serious problem, and a large proportion of income comes from unemployment benefits and pensions.

Arrernte purchase most of their food from retail outlets, still supplementing with kangaroos, emus, bush

ARRERNTE An Arrernte woman reflects on the petroglyphs carved into stone at Corroboree Rock in Australia's eastern MacDonnell Ranges.

turkeys, goannas (a monitor lizard), rabbits, honey ants, grubs, and bush fruits that they hunt and gather.

Australians

Population: 22.5 million
Location: Australia
Language family: Germanic

Although Australia's indigenous peoples arrived on the island continent more than 40,000 years ago, the forerunners of its current non-native population date only to the end of the 18th century. In 1788, the first ship of British settlers, many of them convicts, landed at Sydney on the southeastern coast. Soon, free British immigrated voluntarily, and in the mid-19th century, the discovery of gold accelerated the process. Over time, people of many countries made the long journey, but Australia was to retain its British-based character, as seen in aspects of its government, laws, system of education, and many customs and traditions. Today, 22.7 million Australians, of whom only 2 percent are Aboriginals, live in an area the size of the contiguous United States, making Australia the least densely populated country in the world.

Alarmed by the rising Asian migration spurred by the gold rush, Australia passed an Immigration Restriction Act in 1901 promoting a "White Australia Policy" that was not totally dismantled until 1973. During this time British immigrants received top preference, followed by northern Europeans, and then southern Europeans; others were admitted as these sources dwindled. Australians now strive to be inclusive and supportive of the 200 different nationalities that call Australia home, but there has been conflict nevertheless. They also are attempting reconciliation with the Aboriginals, who for so long were marginalized and mistreated.

Encouragement of immigration in general did not work as hoped; newcomers still tended to settle in coastal cities, where some 90 percent of the population lives. Few have chosen the nearly empty interior, with its mainly arid conditions, as their destination.

Australia's cities tend to have compact centers fringed by far-reaching suburbs of single-family homes that contribute to one of the highest rates of home ownership in the world. The great majority of Australians—some 70 percent—now work in the service sector.

Australian English is based on British English, but with differences in pronunciation and a vocabulary that reflects Australia's unique contexts, such as the rugged bush life of the outback. As Australia's population of non-English origin has grown, so too have the number of languages heard routinely, especially in Melbourne, the country's most multicultural city, where Italian, Greek, Serbo-Croatian, Arabic, German, Vietnamese, and other languages are spoken.

There were many more men than women in the early days of Australian settlement, a situation that brought about a culture of "mateship." The hard life of remote sheep and cattle stations, coal and gold mines, and later the military fostered mutual reliance between and among men that has endured despite correction of the country's gender imbalance. Mateship also evolved with the Australian obsession with sports—as both participants and spectators. Girls and women also are referred to as mates now, but the concept generally describes institutionalized male camaraderie, with beer drinking often the common denominator.

AUSTRALIANS The work is hot and dusty for a sheep rancher at Anna Creek Station in South Australia.

Since Christmas falls in the Southern Hemisphere's summer, many Australians celebrate it as a beach day. At home, they may have a traditional Christmas meal—or a barbecue instead. Some counter the hot weather by observing "Christmas in July," feasting on turkey, ham, and Christmas pudding in more seasonal weather. Australians also observe the British custom of Boxing Day on December 26, taking advantage of more time for the beach.

Those who lose dreaming are lost.

ABORIGINAL AUSTRALIAN PROVERB

The national holiday, Australia Day, inspires parades, picnics, and fireworks. ANZAC (Australian and New Zealand Army Corps) Day on April 25 honors the contributions of all war veterans, especially those who fought at Gallipoli in Turkey in World War I, a military campaign that galvanized Australians and drew them closer to their New Zealand neighbors.

It appears that Australia's days as a member of the British Commonwealth may be numbered. In 1999, the population came close to a majority vote on a referendum, possibly paving the way for an Australian republic. In 1984, Australians changed their national anthem from "God Save the Queen" to "Advance Australia Fair," putting more distance between them and their colonial past.

Cape York Peoples

Population: 6,504
Location: Australia
Language family: Pama-Nyungan

Exhibiting considerable social, cultural, and linguistic diversity, as of 2001, the Aboriginals living in Cape York made up 35 percent of the cape's population and owned 20 percent of the land. Most of them dwell in Aboriginal-controlled townships and outstations. According to archaeological evidence, Cape York Aboriginals at one time tried farming activities, but the

CAPE YORK PEOPLES A Cape York boy participates in an Aboriginal dance festival held every two years in Laura, Queensland.

benefits of agriculture did not outweigh the costs in terms of labor. A migratory lifestyle provided much more time for leisure activities, rituals, and the forging and maintenance of clan alliances.

The Wik of the central west coast may be the best known Cape Aboriginals. Their participation in the momentous legal movement of the 1990s helped end the long fiction of *terra nullius,* "land without people," the policy that had facilitated theft of Aboriginal lands.

In the early 1990s, the Cape York Land Council was formed to pursue Aboriginal property rights on the cape. The council has embarked on a number of novel and groundbreaking regional comanagement schemes, connecting conservationists, pastoralists, and miners. Even so, Cape York Aboriginals continue to endure poor health standards and have a life expectancy that is among the lowest in the country.

Ngarrindjeri

Population: 160
Location: Australia
Language family: Pama-Nyungan

The Ngarrindjeri of South Australia's lower Murray River were once best known for the near equality of their men and women, something unique in Aboriginal Australia. Sacred rituals were the province of men, but Ngarrindjeri women also played important roles in the ceremonial life of the tribe.

Encompassing five different lowland zones, the tribe's rich environment was particularly attractive to colonists, and so the Ngarrindjeri became predominantly of mixed descent by 1900. Less than half a century later, few people retained knowledge of traditional ways and customs. This situation grew worse as

GENOGRAPHIC INSIGHT

AUSTRALIAN ARRIVAL

In 2005, a team of Genographic scientists undertook an ambitious expedition to Chad, in North Africa, in search of genetic evidence for the earliest Saharan migrants. It was the first foreign expedition into the remote Tibesti Mountains in more than 40 years, and the team relied on the expertise of local guides to shepherd them safely through scorching sands to the provincial capital of Faya-Largeau, an oasis town near the country's northern border with Libya.

On the third night, the team sought protection from summer winds at the base of a large dune and bedded down for a crystal-clear moonrise in the cold desert night. The following morning they awoke to find sleeping bags, tents, research equipment, and food covered in several inches of fine sand. Overnight, the sheltering dune itself had migrated as a result of the same winds they were trying to escape.

In the Sand

Shifting sands like those in the Sahara can both cover up and expose evidence of human occupation, and in 1974, such sands revealed an ancient fossil lying near Lake Mungo, one of several dry lakes in southern Australia's Willandra Lakes region. The modern human skeleton, named Lake Mungo 3, was found to be at least 40,000 years old. With evidence from other similarly dated sites in the region, it quickly became apparent that humans had been on the continent for quite some time.

These findings added fuel to a long-standing debate over human origins. Some anthropologists, citing morphological similarities between peoples today and extinct hominid species from similar geographic regions, argued a theory of multiregionalism, which proposed that humans descended from a number of ancestral lines of independent evolutionary origin. The implication of Lake Mungo 3, they argued, was that Australian aboriginals could be at least as old as African populations, if not older. Genetic research has since answered the question of origin resoundingly, showing that the DNA of all living humans directly descends from Africa, that Africa's indigenous populations are genetically older than others, and that they possess mutations ancestral to those found elsewhere.

As Australia redresses wrongs toward Aborigines, there is hope for younger generations.

But Lake Mungo 3's antiquity did present a difficult question: If they originated in Africa, how did modern humans make it to Australia, more than 10,000 miles away, so long ago? The first modern humans to leave Africa 60,000 years ago likely did so by exploiting a combination of terrestrial and marine resources, following the Nile River and Indian Ocean coastline northward until, unbeknown to them, they had left Africa and arrived on the doorstep to the rest of the world. This period marked a global dry spell, when much of the world's water was locked in the polar ice caps and continental Southeast Asia was connected to the nearby Indonesian and Malaysian islands as a single landmass called Sundaland, while Australia and New Guinea were similarly connected as a single landmass called Sahul.

The lower sea levels during this period meant that a continuous overland migration, with only a few minor open-water crossings, would have been possible from Africa to Australia. By traveling along the coasts of southern Asia at a rate of only two miles a year, people would have made it to within a boat ride of Australia in less than 5,000 years.

The Fires of Australia

Australia was then a lush landscape of lakes and rivers, full of large marsupials now extinct, like the buffalo-size *Zygomaturus* and a 450-pound short-faced kangaroo, *Procoptodon*. Evidence suggests that humans changed Australia's ecosystem by setting massive fires, and they transformed the region from forests and grassland to today's landscape of fire-adapted shrubs. The purpose of these fires remains a mystery. Nevertheless, roughly 60 species of the continent's large mammals went extinct soon after humans arrived, forcing them to maintain the mobile hunter-gatherer lifestyle that brought them to the continent.

Many of today's indigenous Australian people still practice the traditional way of life, and Lake Mungo 3 is a good example of this: The body was found sprinkled with red ochre in a burial ritual similar to that practiced by many of today's Aboriginals.

the government increased pressure on the people to assimilate into the dominant culture.

Currently, the Ngarrindjeri are perhaps more socio-economically integrated than other Aboriginal groups. Nevertheless, they embrace their Aboriginality and actively attempt to revive traditions and beliefs. They also hope to remedy problems associated with alcohol abuse and economic disadvantage.

One of their chief means of reinvigorating tradition is the clanwide, ongoing opposition to the destruction of sacred sites by developers, particularly in the vicinity of Hindmarsh Island. Led by Ngarrindjeri women, the protest has brought worldwide attention to the plight of this marginalized group and forced debate on a great national concern: Has the tide of history washed away not only knowledge of ancestral beliefs and practices but also traditional attachments and rights to the land?

Noongar

Population: 240
Location: Australia
Language family: Pama-Nyungan

For at least 40,000 years, the Noongar and their ancestors have occupied the Swan River area in the vicinity of Perth. The name of their language group, Nyungah, forms part of the collective label for a body of languages: the Pama-Nyungan language family, spoken by 80 percent of Aboriginals, from the tip of Cape York to the southwest corner of the continent.

NGARRINDJERI Meticulously painted and decked out, a Ngarrindjeri elder stands ready for the River Country Spirit Ceremony in southeastern Australia.

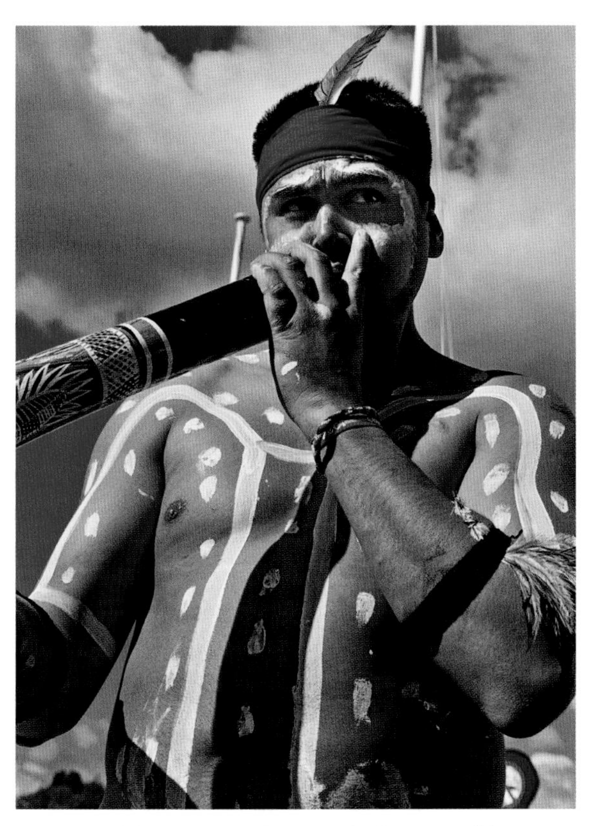

NOONGAR A Noongar man draws out the low, mournful notes on a didgeridoo, a traditional Aboriginal instrument.

Until 1829, when the British established a colony in the Noongar's Western Australia homelands, the people had led a seminomadic life; they moved freely from one area to another, depending on available water, fish, game, fruit, and root vegetables. But colonization struck the Aboriginal economy at its core. Depleted or destroyed were the Noongar's traditional food resources, and the people soon became dependent on introduced goods. The first few decades of the settlement at Swan River saw large numbers of Aboriginals dying after contracting measles, influenza, and other introduced illnesses.

Despite their violent 19th-century clashes with soldiers, convicts, and free settlers—and subsequent harassment by 20th-century bureaucrats—the Noongar have maintained a sense of themselves as a distinct cultural entity. As an urban Aboriginal collective, they face immense problems, not the least of which is how to reinvigorate a land-based culture when areas of greatest significance are covered by asphalt and concrete.

Aboriginality in Australia is a matter of self-ascription, but recognition from the state requires proof of descent and membership in a community of Aboriginals. In the case of the Noongar, various territorial and linguistic bands merged in the late 1800s, and new communities evolved that could claim descent from earlier groupings. Noongar associations now provide a collective voice for speaking out on topics of common concern.

Pintubi

Population: 1,600
Location: Australia
Language family: Pama-Nyungan

By 1965, the government of Australia's Northern Territory had removed the so-called lost tribe of Gibson Desert Pintubi Aboriginals from their homeland and settled them hundreds of miles to the east, in Papunya. About 150 miles west of Alice Springs, the settlement was one of the last Aboriginal communities to be established. It was intended as a place for "reeducating" people, to ensure their assimilation into mainstream society, but the Aborigines who were taken there managed to retain their traditional language and culture. The Papunya community, now one of the best-known autonomous Aboriginal settlements, instead became the focus of an extraordinary 20th-century success story. Ceremonial sand designs, as well as designs inscribed or painted on the body during sacred and secret Pintubi rituals, became the centerpiece of a local arts movement that eventually spread throughout Australia. One of the most successful businesses owned and directed by the Aboriginals is Papunya Tula Artists, a cooperative that began in 1971. Its art has done very well internationally, sometimes selling for record prices in New York and Paris.

Ferocious is the centipede,
but scared of the chicken.

SAMOAN PROVERB

PINTUBI The red rocks of Ngutjul, a site with sacred significance in northern Australia, dwarf a Pintubi boy.

The increasing prominence of Pintubi designs has had a dramatic impact on art and craft production throughout Australia's center. To meet tourists' growing demand for the art, other Aboriginals have been motivated to learn more about their Dreaming stories and designs and to seek out elders who could teach them, leading to a widespread cultural renaissance. Artists insist they are not selling their heritage; rather, they want the world to know about their culture. They keep the sacred knowledge for themselves, to be passed on to their children. Style may have changed, but the sacred message has not.

Since 1981, artists have established outstations at Warlungurra and Kiwirrkura, near the Northern Territory's border with Western Australia, and both are flourishing settlements. Kiwirrkura was the scene of a remarkable development in 1984, when first contact was made with perhaps the last of the traditional or "bush" Pintubi. Neither Aboriginals nor non-Aboriginals had believed that there were such desert dwellers who still followed a traditional lifestyle, and the reuniting of the nine bush Pintubi with family members, who had lost touch with them more than 25 years before, attracted international media attention.

Tiwi

Population: 1,700
Location: Australia
Language family: Tiwian

The Tiwi of northern Australia's Melville and Bathurst Islands experienced their first recorded exposure to Europeans in the 18th century; it was an encounter that would be followed by many more over the years. After a false start in the 1820s, European settlement of the Tiwi islands finally began in earnest in the early 20th century. Since that time, however, the Aboriginal Land Rights (Northern Territory) Act of 1976 has been passed, and the Tiwi have regained full control of their islands.

Today, most Tiwi live in three townships: Nguiu, the former Catholic mission; Pularumpi, which had been a school for part-Aboriginal children; and Milikapi, the former government settlement. Some of the people here go on weekend excursions, hunting kangaroos and turtles, but rather than paddling their traditional dugout canoes and carrying spears, they now steer outboard motorboats and wield shotguns.

The Tiwi concept of Dreaming distinguishes them from other Aboriginals. Divided into *murukupupuni,* or individual Tiwi "countries," their islands are "spirit-child" centers in which membership is determined by the burial place of one's father. Tiwi believe that when a new clan member is conceived, a spirit-child is transferred through a dream from a father to a mother of the same clan.

Celebrating the Tiwi culture, Nguiu's Tiwi Design Aboriginal Corporation is the base of operations for Tiwi Design, a collective of about a hundred artists involved in textile design, weaving, painting, printmaking, wood carving, and ceramics. Many consider it a good model for other Aboriginal art enterprises.

TIWI A pair of Tiwi women gather seafood along the shoreline in the Tiwi Islands of Australia's Northern Territory.

Torres Strait Islanders

Population: 52,600
Location: Australia
Language family: English-based Creole

Where once there was a land bridge, 90 miles of sea now separate Australia's Cape York Peninsula from Papua New Guinea. More than a hundred named islands and many thousands of islets, reefs, and sandbanks dot the Torres Strait, home to an indigenous population quite distinct from the original inhabitants of mainland Australia.

A good leader, like rain, stills the ocean.

PALAUAN PROVERB

Combining cultural and genetic influences from both Australia and Melanesia, the more than 52,000 Torres Strait Islanders trace their ancestry to 14 island groups in the South Pacific. Today, these people claim rights of ownership to the lands and waters where their ancestors have dwelled for more than 4,000 years.

Before colonization by Europeans, the Torres Strait's eastern and northern islands, such as Mer and Saibai, were home to permanent villages of bamboo-and-thatch huts. Elsewhere, settlements were transitory and populated according to the seasonal availability of resources.

In their large double-outrigger canoes, the Torres Strait Islanders engaged in extensive trading relationships not only with each other but also with Papuans to the north and Aboriginals to the south. They carried out pearl-shell harpoons, ornaments, and human heads, and in return they received sago, yams, coconuts, tobacco, bird of paradise feathers, and outrigger canoes. Subsistence came from both the sea and horticulture. Positioned on platforms constructed in shallow waters, islanders fished or hunted turtles and dugongs (a type of manatee); families also relied on defined garden plots, which they inherited from their ancestors along the male line. If they wished, the owners could transfer or lend their lands to others for gardening and foraging purposes.

Unlike inhabitants of the Australian mainland, Torres Strait Islanders have been able to maintain their traditional production, exchange, and subsistence practices. Today, the strait is largely an autonomous zone, a political configuration unique in indigenous Australia. Islanders manage their own affairs through the Torres Strait Regional Authority, the Island Coordinating Council, and the Torres Strait Council. In addition, each of the 17 most abundantly populated islands has a council that performs local government responsibilities.

Society is like a mat being woven.

TONGAN PROVERB

Major issues include a significant out-migration to the mainland, poor health, relatively low education levels, and overcrowding in the limited available housing. Ties to coastal Papua New Guinea communities, once strong, are now in decline or have been terminated. The international boundary limits communications, and there is growing disparity in wealth among the former trading partners.

Warlpiri

Population: 2,500
Location: Australia
Language family: Pama-Nyungan

Among the most populous Aboriginal groups in Australia's Northern Territory, the Warlpiri were traditionally hunter-gatherers living in small groups of up to 30 people. They led a hydrocentric existence in the Tanami Desert, their movements around water sources characterizing life in their vast, arid homeland.

Regular contact with Europeans began in 1911 with the establishment of the Tanami goldfields.

TORRES STRAIT ISLANDERS Children use a log as a tightrope to play in the water in Masig Island, part of the Torres Strait Islands.

Later, in the Coniston Massacre of 1928, as many as a hundred Warlpiri were killed by government forces; others fled northward from home territories to escape further depredations. Survivors were forced to settle in the 1940s, and today, they and their descendants inhabit the small communities of Yuendumu, Lajamanu, Warrabri, Willowra, and Nyirrpi, as well as outstations.

Following the 1976 enactment of federal land rights legislation, almost all of the country the Warlpiri owned prior to contact was returned to them. This has not restored the independence they once enjoyed, however. Most of the people are unemployed and getting by on various forms of government support, minimally supplemented by mining royalties and art sales.

The Warlpiri are known for their "dot paintings," which illustrate visions of their *Jukurrpa* (Dreaming) narratives. Artists re-create individual and collective Dreamings and associated places in the landscape, using customary U-shapes, concentric circles, and journey lines. With this art, the Warlpiri artists assert their rights and obligations to the land.

Wurundjeri

Population: 60
Location: Australia
Language family: Pama-Nyungan

Koorie is the popular name among Aboriginals for the indigenous people of southeastern Australia. The Wurundjeri, a Koorie group, live on the Yarra River plains, now encompassed by Melbourne, Australia's second largest city, after Sydney.

In the first decade of contact between English colonists and the Wurundjeri people, diseases decimated local populations. Survivors watched as their hunting grounds were cleared and drained and their sites of cultural significance were destroyed. Within ten years, only 300 to 400 Wurundjeri people remained, and they soon became totally dependent on the newcomers.

Very little is known about the Wurundjeri's traditional religious and spiritual life, except that clan members belonged to one of two intermarrying moieties, or halves, called Eagle Hawk and Crow. The Wurundjeri also were part of a confederacy of clans

WARLPIRI Renowned Warlpiri artist Michael Jagamara Nelson instructs his wife and assistant, Marjorie, about one of his paintings.

WURUNDJERI A Wurundjeri elder participates in Sorry Day, an annual event on May 26 that recognizes Australia's poor treatment of Aboriginal peoples.

known as Kulin; today, though, social life centers on the extended family. The Wurundjeri still maintain spiritual and cultural ties to the land, asserting their role as traditional caretakers of the territory created by the Dreaming entity known as Bunjil.

In response to the situation of urban Aboriginals, the federal government established the Indigenous Land Council, which has since helped the Wurundjeri purchase the old mission site of Coranderrk and the surrounding 200 acres, a small portion of their former tribal lands. The Wurundjeri have also established a heritage walking trail and a bush-food trail in Melbourne to raise awareness among the general population about the region's history and its significance to the Koorie. Downtown Melbourne was built atop Aboriginal burial sites, resource extraction areas, ceremonial and sacred areas, and former campsites. Markers placed along the trails describe the old Aboriginal way of life.

Yolngu

Population: 600
Location: Australia
Language family: Pama-Nyungan

The Aboriginal people of Arnhem Land, in Australia's Northern Territory, refer to themselves as Yolngu. They are well known for the richness of their cultural heritage, as expressed in traditional feathered ornaments and paintings on eucalyptus tree bark, and for their Christian beliefs.

The Yolngu community of Galiwin'ku, the setting for the landmark Adjustment Movement in Arnhem Land, is especially noteworthy for the innovative ways people have incorporated Christianity into their worldview without compromising the sanctity of the Dreaming. Affirming their belief in both traditional religion and Christianity, residents publicly state that neither would receive preference in their lives. Galiwin'ku is also the base of the touring Aboriginal Black Crusade, whose aim is to re-Christianize Australia's non-Aboriginals.

From the late 1600s to the early 1900s, the Yolngu traded with Indonesian fisherfolk who were seeking access to fisheries and trade beyond Europe's control. Together they gathered trepang, a sea cucumber prepared as food in China and Indonesia. Thus Australia's first international industry was established. These precolonial contacts probably enabled the Yolngu to better deal with the missionaries, miners, and bureaucrats who arrived later.

In the early 1960s, the Yolngu mounted a protest against bauxite mining by a multinational company. They presented a bark petition—a typed manuscript combined with a traditional bark painting. They did not win the case, but their efforts set into motion an inquiry into how land rights could be delivered to Aboriginal Australians and led to the Aboriginal Land Rights Act of 1976, which mandated that land in the Northern Territory be returned to Aboriginals. Today, the Yolgnu host Australia's largest indigenous culture gathering, the annual Garma Festival.

YOLNGU In East Arnhem Land, a Yolngu woman visits a memorial marking a suicide—an unfortunately common event there.

Melanesia

Melanesia is one of three divisions of the Pacific islands to the north and east of Australia. The name means "black island," although its origins are not clear. Perhaps the name applies to the rock of the islands, perhaps the dark skin of the people who live there. Melanesia is the westernmost of the three South Pacific island clusters, lying closest to Australia. It includes Papua New Guinea, at the east end of the Indonesian archipelago; the Bismarck Archipelago, a volcanic island group dominated by New Britain; the Solomon Islands, a chain that includes the historically significant Guadalcanal; Vanuatu, a double-chain of about 80 islands; and Fiji, a grouping of about 300 islands, 100 of which are inhabited. Most of the islands of Melanesia are ancient volcanoes, ringed with coral reefs and containing tropical forests on their windward slopes and mangrove swamps on their leeward stretches. Occasional hot springs give evidence of their volcanic past.

Chimbu

Population: 115,000
Location: Papua New Guinea
Language family: East New Guinea Highlands

Inhabiting the Wahgi and Chimbu River Valleys of the eastern highlands, the Chimbu are one of Papua New Guinea's largest populations. They were first contacted by Jim Taylor and Michael and Dan Leahy, Australian explorers and gold prospectors, on their patrol into the New Guinea highlands in 1933, when the region was controlled by Australia.

Soon after contact, the Chimbu were missionized by Lutherans and Catholics. They also experienced rapid cultural change in the post–World War II period, when Australia placed the Territory of Papua and the Trust Territory of New Guinea under a single administration. In 1975, Papua New Guinea became independent.

Originally part of the Eastern Highlands District (now a province), the Chimbu broke away and their lands later became the Simbu Province. Today, they and other eastern highlanders are successful coffee growers and a dynamic force in the economy and on the national political scene.

Along with western highlanders, the Chimbu were transformed by the agricultural revolution that

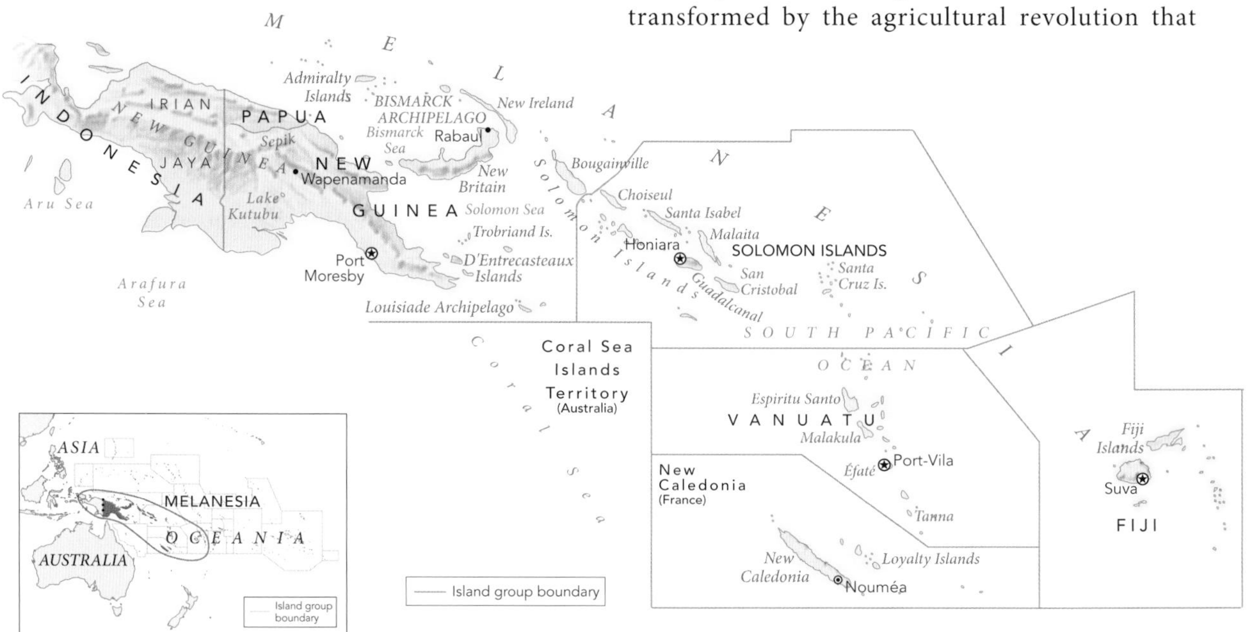

CHIMBU A Chimbu woman finishes up the makeup and other adornment of a young girl in Papua New Guinea.

introduced sweet potatoes into New Guinea almost 300 years ago. This crop, which allowed intensive cultivation, not only spurred growth of the human population but also supported a large domestic pig population. The Chimbu subsequently developed an elaborate ceremonial system based on the exchange of pigs, shell valuables, feathers, and other items.

Because of their expanding population, these people became chronically short of land. Early in the 1960s, population densities in some parts of Chimbu territory reached as high as 500 inhabitants per square mile. Endemic warfare ensued, and the Chimbu expanded into less densely populated, lower-elevation areas south of their original home.

Like the western highlanders, the Chimbu have no hereditary positions of authority. They are very egalitarian, with men exerting authority on the basis of personal characteristics such as strength, bravery, and generosity.

Enga

Population: 230,000
Location: Papua New Guinea
Language family: East New Guinea Highlands

The Enga are one of the most numerous peoples of the western highlands of Papua New Guinea. As is the case with the Chimbu of the eastern highlands, the Enga have their own province, Enga Province, with its capital at Wapenamanda. They survive on intensive sweet potato cultivation, which in turn supports vast pig herds. Other garden foods, including greens, sugarcane, bananas, and introduced vegetables, supplement the sweet potato.

The Enga do not include pork in their daily diet; instead, they use the pigs in a regional exchange cycle called the *tee,* which also involves trading ceremonial crescent pearl shells. The pigs are periodically slaughtered at highly ceremonialized events that are part of the tee.

ENGA Faces painted and ceremonial spears at the ready, Enga tribesmen get ready to perform a *sing-sing*, or intertribal festival.

Before Australian colonial pacification in the 1950s, warfare was endemic among Enga groups. Casualties claimed a quarter of Enga males between 1900 and 1955. The acquisition and competitive display of shell wealth and pigs during the tee is a peaceful substitute for warfare, and the ceremony has become enhanced and more important in local political life since the advent of the Pax Australiana.

The Enga are known for their strict separation of men and women. Like their Huli neighbors, Enga men feared contamination by women's menstrual blood and other sexual fluids. Cohabitation was not permitted, and women were not allowed to enter the men's space in the village during their menstrual periods. Feminine fertility, on the other hand, figured centrally in the Enga's precontact male cults, a general characteristic of groups in the western highlands.

Hereditary chiefs and other formal leadership positions are not part of Enga culture or, for that matter, part of the societies of most other large horticultural groups that make their homes in the New Guinea highlands. Male leadership is acquired through prowess in warfare, success in the various aspects of the tee exchange, skill in oratory and magic, and management of ceremonial wealth. Enga politics are thus mutable and dynamic, and in many ways they reflect an egalitarian society.

Fijians

Population: 516,500
Location: Fiji
Language family: Malayo-Polynesian

A cluster of more than 300 volcanic islands in the South Pacific, Fiji is home to the Fijians, a people of mixed Melanesian and Polynesian ancestry who have inhabited the islands for some 3,500 years. Since the late 19th century, Fiji has also absorbed an influx of East Indians, who came initially and involuntarily as indentured workers for British-run sugarcane plantations and later—voluntarily—as merchants and traders. Fiji's population is represented also by small numbers of Europeans, Australians, New Zealanders, Chinese, and other Pacific islanders, as well as people of mixed European and Fijian ancestry.

Native Fijians insist on exclusive entitlement to designation as Fijian, for which they now officially use the term *iTaukei;* currently, they make up some 60 percent of the total Fijian population of about 900,000. At the same time, the Indo-Fijian population has formed an Indian diaspora community that intersects with ethnic Fijians, mostly in commercial and political spheres.

Ethnic Fijians inhabit only about a hundred of the Fijian islands, which they call Viti, a word that means "east" or "sunrise" to the Fijians. Most live on the two largest islands, Viti Levu and Vanua Levu, which account for about 86 percent of Fiji's territory. In rural areas they form villages occupied by members of one or several patrilineal lineages or subclans headed by a hereditary chief.

FIJIANS A Fijian woman carries a harvest of kava, a plant used to make an intoxicating beverage.

When the British took control of the islands in the 19th century, they acknowledged the power and importance of the traditional chiefdoms and incorporated them into the administrative structure. They also allowed the Fijian system of communal land tenure to continue. Fiji has been independent since 1970 and a republic since 1987. The Council of Chiefs still meets at least yearly; it also appoints Fiji's president (a largely ceremonial position), vice president, and nearly half of the Fijian Senate. Fiji's government officially uses English, but Fijian and Fijian Hindi also are formally recognized.

Each bay, its wind.

FIJIAN PROVERB

Fijian villagers typically fish and cultivate taro, cassava, and other crops that meet their own food needs as well as providing some surplus to sell. Sugar production is still the largest export; the fields, worked mainly by Indo-Fijians, are on leased land.

Over the years Fijians have been moving from rural areas to the cities, so that 40 percent of them are now urban. Many ethnic Fijians work in the services tied to tourism, a major source of revenue. Every year, thousands of cruise ships stop in Fiji, disembarking tens of thousands of tourists, who also arrive by air, especially from Australia and New Zealand. Fijian women also work for low wages in the textile industry.

European missionaries, predominantly Methodists who created a writing system for Fijians in order to translate the Bible, successfully converted most Fijians to Christianity. Although churches have replaced native temples and Fijians embrace Christian practices, animistic beliefs live on, especially those concerning ancestral spirits

Some of the central practices of traditional Fijian life also persist and have been modified for presentation to tourists. One is the drinking of kava, also known as *yaqona,* an intoxicating beverage made from the root of a pepper plant that has been the focus of a Fijian ritual for centuries. Formal etiquette traditionally surrounds the preparation and consumption of yaqona, but it is also drunk socially—known as "having a grog"—and tends to loosen the tongue and encourage storytelling.

Middle Sepik

Population: 24,700
Location: Papua New Guinea
Language family: Several language families

Because of their highly developed art and architectural styles, the Middle Sepik River peoples are probably among the best-known groups in New Guinea. Their now famous *haus tambaran* (spirit house), a ceremonial longhouse with a high, overhanging roof, was the model for the Papua New Guinea High Commission building in Canberra, Australia.

The peoples of the Middle Sepik—Kwoma, Arapesh, Manambu, Iatmul, and others—share many cultural similarities, including a totemic clan system organized into larger groups and an elaborate system of initiation for boys, organized around sacred bamboo flutes. Secrecy is a major dimension of public life along the Sepik River, and as men grow older, they attain greater knowledge of mythological names and significances. Iatmul and Manambu men debate in public, challenging each other's knowledge of secret totemic names. Among the Iatmul, the journeys of ancestral creators are depicted by a knotted cord, each knot representing a place where an ancestor stopped; knowledge of these creators is a mark of ritual prominence.

MIDDLE SEPIK This Middle Sepik mother and child likely encounter visiting tourists who come to see the tribe's distinctive spirit houses and watch their ceremonies.

Among the Kwoma and other lowland Sepik River peoples, the staple yam is the focus of fertility ceremonies. In recent times, these ceremonies and the boys' initiations have taken on new forms—nonsacred versions performed for the many tourists who visit every year. Art, too, has become important for the tourist market and is an integral part of the local economy. Sepik carvers are renowned for their masks and depictions of local animals and birds.

East Sepik Province has long provided prominent leaders for the nation of Papua New Guinea, including Michael Somare, who in 1975 was inaugurated as the first prime minister. The province was also one of the first in the country to incorporate traditional landowning clans, allowing them to become modern business organizations suited to the demands of commercial development.

Motu

Population: 39,000
Location: Papua New Guinea
Language family: Malayo-Polynesian

Indigenous people of the coastal area around Port Moresby, in southeastern Papua New Guinea, the Motu inhabit a tropical savanna that changes dramatically between dry and wet seasons. Traditionally the people have been potters, traders, and fishers, with gardening and hunting as supplementary activities. They have lived closely with the Koita people, who depend on horticulture, and have intermarried extensively with them, exchanging pottery and fish for the Koitans' garden produce.

The Motu became most famous for *hiri,* their annual trading voyages along the Papuan Gulf coast in

MOTU In times past, Motu men undertook elaborate trading voyages in their outrigger canoes.

outrigger canoes. Under the leadership of a man who had claimed a vision in a dream told him to organize such a journey, as many as 600 men would sail forth on a voyage. They took pots made by Motu women to Elema and the home of the Namau people. There, they exchanged the pots for sago and for timber they would make into canoes.

When the traders got back home, the Motu held a *koriko* feast, during which the return gifts were ceremonially distributed. As a result of these voyages, Motuan became a trade language along the south coast and today serves as the other indigenous lingua franca of Papua New Guinea. Tok Pisin, a mixture of Tolai, German, and English, remains the country's dominant language.

One of the central ideas in the Motu religion is *irutahuna,* which can refer to the center of a house, an individual's heart or mind, the central part between two masts of an outrigger, or the base of a single-masted vessel. Clearly, the core or center of an important architectural or organic entity is a space of great sacredness for the Motu. This concept reflects a widespread preoccupation among Melanesians with the vital, germinal core of all things.

Tannese

Population: 21,000
Location: Vanuatu
Language family: Malayo-Polynesian [South Vanuatu]

The island of Tanna, in the southern part of the Vanuatu archipelago (formerly the New Hebrides), has a fascinating history as well as a rich ceremonial culture. As is the case with villages elsewhere in Vanuatu, a dancing ground called the *nakamal* is situated near each small group of huts. This site is where ceremonies, rituals, and social events occur, including gatherings of men who come to drink kava at the end of the day. The core of men who meet this way is called a "canoe," the local term for a small group related through the male line.

TANNESE A Tannese child walks along the black sands of Sulphur Bay on Tanna Island in Vanuatu.

DEEP ANCESTRY

THE SEAFARER

Dustin-Alan Kaiahua, born in Honolulu, was inspired to participate in the Genographic Project because he "wanted to know the migration path of the Hawaiian people." He knew his Hawaiian family tree back to the time of the Kamehameha dynasty. King Kamehameha I was born in the first half of the 18th century and was the first Hawaiian ruler to unite all of the islands under a single government. Prior to this, they had been separate, occasionally warring island-based chiefdoms. His descendants ruled Hawaii for nearly 80 years until it fell under the influence of the United States at the end of the 19th century. Clearly Dustin was expecting something to corroborate his Hawaiian heritage. What he found surprised him.

The Hawaiians are part of the Polynesian expansion, a mass migration across the Pacific Ocean that began more than 5,000 years ago in Asia. Small groups of ever more intrepid sailors known as Austronesians began to island-hop southward along the coast of Southeast Asia and the Philippines, eastward along the north shore of New Guinea, ultimately heading out into the open Pacific around 2,500 years ago. By 1,000 years ago, they had reached places as distant as New Zealand, Easter Island, and Hawaii. Along the way, the East Asian and near Oceanian cultures—and genes—blended, giving rise to the patterns we see in Polynesia today.

Dustin-Alan Kaiahua's seafaring ancestors traveled the Pacific waters in outrigger canoes.

DNA Results

Dustin was "expecting a mixture of Hawaiian and Japanese," which he knew about from his recent genealogy, but was "shocked . . . [to see connections to] Cambodia, Laos, and Vietnam." Upon looking deeper, however, the results started to make sense. Dustin's paternal ancestry was Hawaiian, and his Y-chromosome reflected this. He belongs to a lineage known as C-M208, where C is the main haplogroup and M208 is Dustin's terminal genetic marker, his own twig in the broader C lineage branch. It turns out that C-M208 is common in Polynesians, so the results weren't entirely unexpected. But his C lineage ultimately traces back to Southeast Asia, consistent with the migrations of the Austronesians through this region thousands of years ago. And before that, haplogroup C ultimately traces back to the first migration out of Africa around 60,000 years ago, along the south coast of Asia.

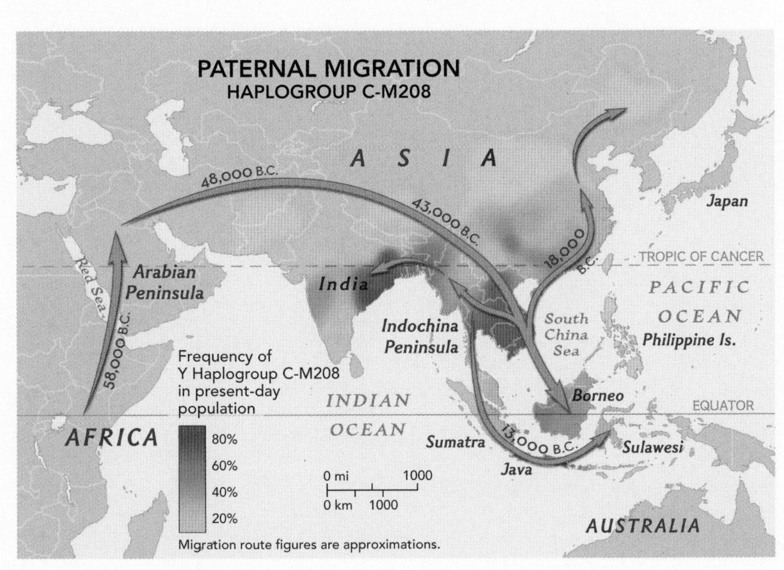

Paternal C-M208: The journey of early humans took them to Southeast Asia, Indonesia, and from there to the Pacific.

Mix of Ancestries

Dustin's results revealed a mix of ancestries, as he had predicted—both East Asian and Oceanian. While his terminal genetic marker identified him as a Polynesian, his ancestors had migrated to Hawaii from Southeast Asia over thousands of years. He had been expecting his results to confirm his Hawaiian-ness, and had certainly found that—but also his distant connection to people living thousands of miles away.

—*Spencer Wells*

Because the Tannese have a long history of contact with the European world, their culture has in many respects evolved as a reaction to, and an incorporation of, colonial encounters. Among the most important encounters were the notorious "black-birding" expeditions mounted in the late 1800s by seagoing Australian merchants. These expeditions transported thousands of Tannese—as well as indigenous Melanesian men from eastern New Guinea, the Solomon Islands, and other nearby areas—to Australia, where they were put to work in the cane fields of northern and central Queensland. During this period, more than 100,000 Melanesians were brought to Australia and to plantations and mines on Fiji, New Caledonia, and Samoa, leaving many of their home islands depopulated and their cultures weakened. The name *Tanna* survives as a common surname among descendants of the original laborers.

Tannese efforts to preserve their heritage and maintain distinctiveness in the face of dominant outside forces can be summed up in the term *kastom* (custom), from the Bislama creole of Vanuatu. Used throughout the archipelago, the word describes a fusion of old, rediscovered traditions with new or re-created traditions.

The island of Tanna was home to the famous cargo cult started by John Frum. In the 1930s and 1940s, Frum preached the return of traditional political power to the Tannese and prayed for the departure of the Europeans. The British acted decisively to crush the movement in 1941, but the legend of John Frum and the influence of his followers continue to this day.

Tolai

Population: 61,000
Location: Papua New Guinea
Language family: Malayo-Polynesian

The Tolai inhabit the Gazelle Peninsula at the eastern edge of New Britain Island—site of Rabaul, the most important town in Papua New Guinea and the location for a major Japanese naval base during World

TOLAI A trio of Tolai men prepare for a dance at a *sing-sing* in Papua New Guinea.

War II. This group experienced colonization relatively early in Melanesian terms. Before the Japanese, Germans came in 1884 and Australians arrived after World War I. As a result, the Tolai became one of the most literate, sophisticated, and vocal groups in all of Papua New Guinea, and they are a dominant force on the national scene today.

Tolai society is organized into groups called *vunatarai,* a term that can refer to either clans—descendants of a particular ancestor associated with a particular everyday leader—or groups into which the clans are placed. Members of vunatarai hold land, property, and ceremonial objects in common.

Even a small twig can bring the pot to boil.

PALAUAN PROVERB

Because of the early plantation activity on New Britain, cash cropping has long been a primary component of the village economy, with copra and cocoa the most important crops. The people are famous for the use of shell money known as *tambu,* which consists of nassa shells sewn into plaits about six feet long. A true local currency, tambu is used to purchase food and implements and is exchanged for ceremonial purposes such as bridewealth and mortuary payments.

Art created by the Tolai includes elaborate effigies and masks representing the female spirits of the *tubuan,* entities owned by the clans and considered the incarnation of ritual and sociopolitical power.

Trobrianders

Population: 20,000
Location: Papua New Guinea
Language family: Malayo-Polynesian

Inhabiting Kiriwina and a number of smaller islands off the east coast of Milne Bay Province, Trobriand Islanders are probably among Papua New Guinea's most famous groups—thanks in large part to anthropologist Bronislaw Malinowski, who did pioneering

TROBRIANDERS A Trobriand woman shows her delight before a dance at her island home off the east coast of Papua New Guinea.

fieldwork on them early in the 20th century. As he discovered and subsequently reported, Trobriand men and women engage in relatively uninhibited sexual activity from an early age. Sexuality is prized and admired as a desirable trait, with both men and women devoting time, attention, and resources to obtaining love and beauty magic.

The Trobrianders also are noted for their involvement in a regional interisland trading network called the *kula,* which encompasses all the coral and larger islands in the Massim group, off the east coast of Papua New Guinea. In the kula, valuable arm shells and shell necklaces are circulated among the islands by traders who travel in opposite directions around the island group in seagoing canoes. This trading arrangement, essentially a competition for prestige, is not limited to the valuable shell objects. Many islands also contribute specialty items such as clay pots to the trade.

Like most other Massim area people, the Trobrianders are organized into clans based on descent through their mothers. But fathers also bequeath important property rights to their children, and these are from their own matrilineal connections.

In recent years, Trobriander artists have become internationally known for their traditional carving of canoe prow boards, figures, and masks, an art form that has caught the attention of the tourist trade.

Polynesia

The boundaries of Polynesia, largest of the three divisions of Pacific islands, are roughly drawn as a triangle including Midway and the Hawaiian Islands at the north point, Pitcairn Island and French Polynesia at the east point, and New Zealand at the south point. Contained within that large boundary are numerous other islands and island groups, including Tonga, Tuvalu, Samoa and American Samoa, and Kiribati, a republic that includes the former Christmas Island. The well-known island of Tahiti is part of French Polynesia.

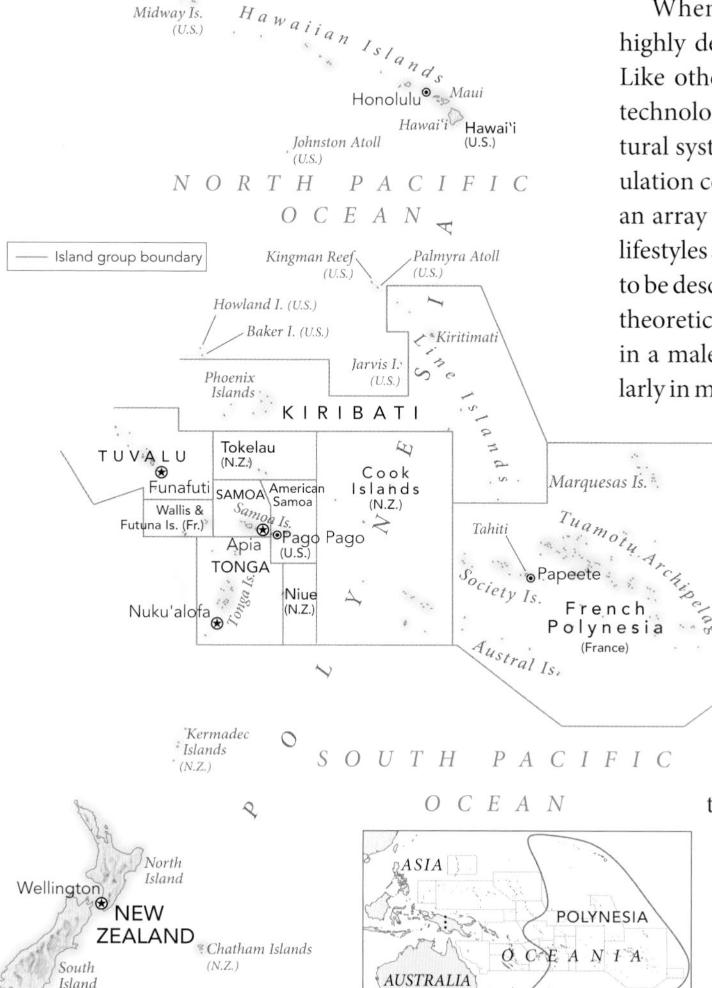

Hawaiians

Population: 141,000
Location: United States
Language family: Germanic (English), Malayo-Polynesian [East-Fijian Polynesian] (Hawaiian)

Native Hawaiians are the descendants of intrepid voyagers whose canoes reached one of the most isolated island groups in the world. Though scholars debate the details, the original Hawaiians probably arrived from the Marquesas Islands at least 1,700 years ago. There may have been a subsequent migration from Tahiti as well, long before Captain James Cook's 1778 "discovery" of the islands put them on maps made in Europe.

When Cook arrived, he encountered the most highly developed political system in all of Polynesia. Like other Pacific islanders, Hawaiians lacked metal technology, but even without metal tools, their agricultural system produced enough food to support a population conservatively estimated at 300,000. There was an array of chiefs, priests, and craft specialists, whose lifestyles set them apart from the rest of society. Thought to be descended from the gods, Hawaiian chiefs, or *ali'i*, theoretically inherited their positions as the firstborn in a male line. In practice, however, ability—particularly in making war—was a potent factor in establishing and expanding a chiefdom. Women too could enjoy ali'i status.

A Hawaiian chief controlled land, especially irrigated fields, together with the labor of commoners. The complex system of chiefdoms was supported by an ideology of supernatural power: Kamehameha I, for example, inherited a war god's title. He used his own skills and European weaponry to unite all the islands by 1810, an unprecedented feat in Hawaii. After the chief's death in 1819, his widow, Ka'ahumanu, became a powerful figure in her own right.

American Protestant missionaries arrived in the islands in 1820. Even before that time, native Hawaiians had begun to suffer severe population losses from measles, influenza, and

HAWAIIANS A drummer accompanies young dancers performing against the spectacular backdrop of Waipo'o Falls in Waimea Canyon.

other introduced diseases for which they had no immunity; by 1854, their numbers had been reduced by at least 75 percent. When foreigners began to establish sugar plantations, they found it necessary to import labor from Asia. These same commercial interests, as well as missionary descendants, began assuming power in the Hawaiian kingdom. In 1893, the last Hawaiian monarch, Queen Liliuokalani, was deposed by a group of mostly American businessmen, who established a republic in 1894. The United States annexed the islands as a territory in 1898, and in 1959 Hawaii became the 50th state.

Native Hawaiians, now mostly of mixed ancestry, began to increase in number in the 20th century; generally, however, they remained socially inferior not only to "white" Americans but also to the descendants of Asian laborers, who moved into higher economic and political positions after statehood. In the past half century, though, a new sense of Hawaiian identity has emerged. Hawaiians have self-consciously revived their language and customs such as hula chanting, dance, and traditional healing. In 1976, the sailing canoe *Hokule'a* was built according to traditional style. With the help of a navigator from the Caroline Islands, a crew sailed the vessel to Tahiti using only stars, wind, and waves to guide them. Subsequently, *Hokule'a* has voyaged to all the major islands in the Polynesian triangle, inspiring young Hawaiians as well as Tahitians and other Pacific islanders to assert their ethnic pride.

Maori

Population: 625,000
Location: New Zealand
Language family: Malayo-Polynesian

The Maori people were the last major group of Polynesians to settle their islands. Legend has it that Kupe, a fisherman and navigator, left his eastern Polynesia homeland of Hawaiki to pursue a huge octopus that had been stealing his bait. He caught up with the creature and killed it; then he began exploring what is now New Zealand. Kupe later sailed home to tell others of uninhabited land available for settlement. Regardless of this story's accuracy, modern scholars agree that Maori began settling in New Zealand about 1,200 years ago, after they had made the longest and last voyage of settlement in Polynesian history.

The new settlers, having come from a warmer climate, faced challenges in adapting to a different environment. In the early days, they depended in part on the moa, a large flightless bird that Maori hunters eventually hunted to extinction. The people subsisted by fishing, cultivating sweet potatoes, and gathering forest products. Between sweet potato harvests, foraged plant foods such as fern roots provided an important part of the diet.

The largest social unit in traditional Maori life was the *iwi,* often translated as "tribe." Each iwi occupied a distinct territory and was ideally based on descent through males or females from a founding ancestor. Within each iwi were several *hapu,* or sections, whose members also claimed common kinship. Because the Maori could claim descent through either sex, individuals potentially could belong to more than one hapu. Actual membership depended primarily on one's residence and cooperation in group activities.

Each hapu had its own chief, and the chief of the most senior hapu was the paramount chief of the entire iwi. Ideally, the tribal chief was succeeded by his eldest son, but in some groups, a senior daughter received special recognition. Conflict was frequent, even within the same iwi. When Europeans eventually reached New Zealand, the Maori quickly gained a reputation for ferocity. Maori tribes were known for enslaving their defeated enemies or perhaps even eating them as a final humiliation.

MAORI A young Maori woman prepares to dance on Waitango Day, commemorating New Zealand's founding treaty.

At planting time one labors alone; at harvest, friends are all around.

MAORI PROVERB

Like other Polynesians, Maori traditionally believed in an elaborate pantheon of supernatural beings, from which the highest-ranking individuals were thought to derive their power. Chiefs presided over major ceremonies, but a specially trained priesthood also existed. Most public rites were celebrated on special open-air platforms that were associated with a particular hapu or iwi.

Europeans were drawn to New Zealand at the end of the 1700s, when they decided to exploit the rich sealing and whaling grounds off the North Island. Protestant missionaries arrived in 1814, and soon other Europeans and Australians followed. In 1840, many Maori chiefs signed the Treaty of Waitangi, which gave sovereignty to England. But from 1860 to 1865—sometimes called the period of the Maori Wars—battles were fought over land rights and questions of sovereignty. These issues remain controversial; now, however, Maori struggles to obtain rights are usually played out in courtrooms.

People identifying themselves as Maori make up between 10 and 15 percent of New Zealand's population. Although they are a more cohesive ethnic group than their Hawaiian relatives, they have followed some of the same strategies for cultural survival. Exhibits of traditional arts have toured overseas, and younger artists are making their own statements about ethnic identity. Maori "language nests," learning communities that teach Maori language and culture, provide a language revitalization model for Hawaii and many other places.

New Zealanders

Population: 3.1 million
Location: New Zealand
Language family: Germanic

While the indigenous Maori have populated New Zealand, with its two main islands and some 140 smaller ones, for at least a millennium, settlers of European origin began to arrive in significant numbers only in the mid-1800s. They now represent about 70 percent of the nation's population.

New Zealand was first sighted by Dutch explorer Abel Tasman in 1642. Under attack from Maori, he caught only a quick glimpse before sailing off. Other Dutch explorers got a better look and named the islands Nieuw Zeeland, for the Zeeland province of the Netherlands. Within a hundred years, European and American seal hunters and whalers were stopping there, until both the seal and whale populations were depleted.

About the time of an 1840 treaty between the British and Maori, the first groups of British and Australian settlers came to the North Island, where 76 percent of New Zealand's population lives today. They began to farm and herd sheep, starting immense flocks that by the late 20th century had grown so large that the sheep in New Zealand outnumbered the humans, twelve to one. With refrigerated shipping, New Zealanders began exporting meat as well as wool, adding to their economic success.

Anglican missionaries settled the South Island and gave it the unabashedly British feel it still has today, with its fine homes, gardens, and the elegant neo-Gothic cathedral in the city of Christchurch. The South Island's population sorted itself into a mainly English northern end and a Scottish southern one, as clearly witnessed in the place names: Dunedin and Invercargill are the principal southern cities. Today, Anglicans represent almost half of the 40 percent of New Zealand's Protestants, including Presbyterians and Methodists. Some 13 percent of New Zealanders are Roman Catholic.

Other immigrants came to New Zealand. Polish, Lebanese, and Chinese flocked there during the gold rush of the 1860s. In recent decades, new immigrants have arrived mostly from South Asia as well as other South Pacific islands. All New Zealanders, including Maori, share the nickname "Kiwi," after the round, brown, flightless bird that has become the emblem of the nation. When the New Zealanders developed the Chinese gooseberry into a fuzzy-skinned fruit now popular worldwide, they naturally gave it their favorite name.

NEW ZEALANDERS Surf's up for champion New Zealand surfer Bobby Hansen. New Zealand is a prime surfing destination.

New Zealanders revel in their diverse landscape and take to its hills, glacier-carved mountains, rivers, lakes, volcanic thermal springs, and beaches as often as possible. Hiking—or, as Kiwis call it, tramping—is a favorite pastime, especially along the South Island's famous Milford and Routeburn Tracks. On weekends many people spend time in beach houses called "cribs" on the South Island and "bachs" on the North Island.

As the demand for wool has declined, so has the raising of sheep, and New Zealanders now focus more on lumbering and the export of dairy products. They also encourage adventure tourism, which brings in thousands of tourists each year, not only for the more staid possibilities, such as trekking, but for the bigger thrills such as bungee jumping off cliffs. New Zealand's spectacular scenery also draws filmmakers, providing the backdrop for recent Hollywood productions such as *King Kong, The Chronicles of Narnia, The Lord of the Rings,* and *The Hobbit,* as well as the country's own burgeoning film industry.

New Zealanders celebrate many holidays with the same traditions as the British, and their cuisine follows a number of British preferences, with the Sunday roast typically being lamb, not beef. ANZAC Day commemorates the sacrifices of Australian and New Zealand armed forces at the Battle of Gallipoli in Turkey in World War I, in which New Zealand's casualties were higher per capita than any other country's.

Samoans

Population: 231,400
Location: Samoa
Language family: Malayo-Polynesian

Since 1900, Samoans have been divided into two polities. Eastern Samoa is a territory of the United States, while the western islands—ruled first by Germany and then by New Zealand until 1962—are independent. Samoans settled here more than 3,000 years ago, and for centuries the people have regularly interacted with Tongans and Fijians. Today, their islands are home to the world's second largest concentration of full-blooded Polynesians: nearly 50,000 in American Samoa and some 180,000 in independent (formerly Western) Samoa.

Traditional Samoan life was based on horticulture, with taro and breadfruit considered the primary crops. Men carried out more strenuous tasks, including clearing and planting the land, fishing beyond the reefs, and even cooking food. Women collected wild plants and gathered small sea creatures in the lagoons and reefs. They also wove mats, and these continue to be important exchange items on ceremonial occasions such as weddings.

Samoans typically lived in self-contained communities made up of a number of kin groups related through descent from either male or female ancestors, or by adoption or marriage. Still basic to the Samoan social structure is the *matai* system, in which chiefly titles are assumed on the basis of birth, ability, or a combination of the two factors. Titles are of two kinds, *ali'i* (head chiefs) and *tulafale* (orators), and are ranked within the village and in wider political contexts. The degree of authority exercised by titleholders varies according to rank; for example, a council of matai exercises supreme authority over village affairs.

When Samoa became independent, only titleholders were allowed to vote and to be candidates in general elections. This changed in 1990, when a referendum created universal suffrage; only matai can be candidates for most of the parliamentary seats, however.

During aboriginal times, matai also served as religious practitioners, and they were responsible for the worship of family gods. But teachers from the London Missionary Society eventually began their work in Samoa, and today, about half the people of

SAMOANS A Samoan man performs a traditional dance often embellished with fire to entertain tourists.

independent Samoa are affiliated with the Congregational Church. About one-quarter are Catholics, and the rest are Methodists, Mormons, and members of other faiths.

Nearly as many Samoans live away from their home islands as on them. In New Zealand, there are more than 100,000 Samoans, and about the same number live in Hawaii and the 49 other states. Samoans have enjoyed disproportionate success on professional football teams and prize warrior-like athleticism. But no matter where Samoans live, they maintain ethnic identity through their own church congregations. They have also revived the practice of elaborate tattooing, which is done by using traditional instruments, such as toothed combs and mallets, to tap pigments into the skin.

Tahitians

Population: 128,500
Location: French Polynesia
Language family: Malayo-Polynesian

Many romantic images of the South Seas were created when French and English explorers first met Tahitians in the late 18th century. Before these early encounters, there may have been as many as 30,000 islanders; afterward, the population declined for a time. (In 1999, estimates exceeded 100,000, though, including a number of people with mixed ancestry.) Like other speakers of Polynesian languages, the early Tahitians settled their homeland after making long canoe voyages from Southeast Asia. These voyagers passed through Tonga and Samoa before arriving in what are now called the Society Islands.

Traditionally Tahitians divided their society into three classes: nobility *(ari'i),* commoners, and servants. Nobles inherited their rank from their fathers. Usually the firstborn male would become the heir, though women might also enjoy high status. Because people at the highest level of society claimed descent from the gods, the culture was a complex mixture of religion, status, and power. Chiefs received tribute from lower ranks, but competition among those who were claiming power in different parts of the island group meant that no ruler could rise above all the others. That situation changed when Europeans arrived and introduced new weapons and different political ideas to the islands.

Though Tahitians were among the first Polynesians to be converted to Christianity, they originally recognized a ranked series of gods. Some of these deities were worshiped by an entire region, while others received the devotions of families or individuals. Drawn from the ranks of the ari'i were the priests who officiated at major ceremonies involving whole districts; many of them were thought to have the power to practice sorcery.

People believed that divine anger could cause untimely death, but they also recognized death through aging as a normal occurrence. Funeral rituals depended on the deceased's social status. For example, the body of a high-ranking Tahitian was exposed on a covered platform, or ghost house, before burial, and lengthy and elaborate expressions of mourning were a high chief's prerogative.

Tahitians used to support themselves by gardening and fishing. The men fished, did construction work, and made tools, weapons, and canoes. Women made bark cloth, an important item for trade, formal gift giving, and clothing. They also wove mats, hats, and baskets from the leaves of the pandanus plant.

Both sexes shared the gardening chores, raising a variety of crops that included yams and sweet potatoes. The numerous coconut trees provided not only food but also craft materials. Today Tahitians participate in

TAHITIANS Canoe paddles play a part in this dance performed by Tahitian men and women.

modern economic activities including tourism and the cultivation of black pearls. Traditional dancing, featuring rapid hip movements accompanied by frenetic drumming, has become a staple of Pacific island tourist performances and a potent symbol of identity.

Tahiti has long been tied to France, which made it part of a protectorate in 1842 and today includes it within the Overseas Territory of French Polynesia. In recent decades, France was criticized for carrying out nuclear tests in other parts of the Society Islands. The tests ended in 1996, and now funds for compensation and other French government subsidies provide much of the income for Tahitians. For more than 70 years, the people of Tahiti have pressed France for greater freedom, and they have recently achieved some increase in local autonomy.

TONGANS Tongan women arrive at the residence of the new king with food for a banquet for the royal household.

Tongans

Population: 96,300
Location: Tonga
Language family: Malayo-Polynesian

In all of Polynesia, only Tongans can boast that their islands were never ruled by a foreign power. Settled some 3,000 years ago—like Samoa, whose chiefs Tongan nobility once married—the Tongan archipelago is the largest group at the western end of the Polynesian triangle of islands. Of the Tongans there today, almost 70 percent of them live on the main island of Tongatapu.

Tongan society seems always to have been extremely centralized and stratified. Traditionally, the supreme ruler held the title of Tu'i Tonga; he claimed descent from the gods and had power that was originally both sacred and secular. As the population grew, however, the Tu'i Tonga delegated more responsibility to his close relatives, who became chiefs in their own areas.

The political instability that ensued led to a separation of powers, with the Tu'i Tonga remaining sacred ruler and someone else becoming secular leader. Later, the secular leader delegated authority to his son, creating a new royal lineage and title. The Tu'i Tonga's sister, Tu'i Tonga Fefine, was also a powerful figure, but her oldest daughter, the Tamaha, was the most sacred being of all. Theoretically all Tongans, including the reigning monarch, trace their kinship affiliations and relative social rank from one of these four chiefly titleholders.

A significant feature of Tongan life was the superiority of women as sisters over men as brothers, meaning that brothers were expected to defer to their sisters and their sisters' children. The higher status had a material basis, in that women created valuable property such as mats, baskets, and bark cloth, to be exchanged at important events. They also prepared the pepper plant–based kava drink featured on social and political occasions.

Christian missionaries brought major changes to traditional Tongan life, including an increase in the political instability that was created by newly arriving European commercial interests. In 1826, a Wesleyan Methodist mission was established, and by 1845, a Tongan newly converted to Protestantism had brought political defeat to the Tu'i Tonga, who was supported by Catholic missionaries.

Proclaimed King George Tupou I and guided by Methodist missionaries, he modeled his rule after British constitutional monarchy. A constitution was created in 1875, giving the sovereign ownership of all the land while designating a class of hereditary nobles as trustees and guaranteeing all adult males a farm and village allotment.

Today, the independent Kingdom of Tonga remains centralized and traditional. The king, sixth in the Tupou dynasty, appoints a cabinet that is the highest governing body. Although a pro-democracy movement emerged in 1994, the monarchy and its associated nobility appear firmly in control. Tonga relies heavily on foreign aid and on income that is earned elsewhere.

GENOGRAPHIC INSIGHT

FAST TRAIN OR SLOW BOAT?

More than 3,000 years before Captain James Cook first encountered the Polynesian islands, a remarkable group of seafaring explorers landed on their uninhabited shores, completing the greatest migration in human history. These proto-Polynesians had learned to harness the Pacific winds and currents to power their exploration, and they did so without European navigation technologies like the compass, astrolabe, and sextant. But while these open-water voyages landed humans on the shores of paradise, the conditions forcing their migrations were anything but.

Calibrated radiocarbon dating from early Holocene human encampments in continental East Asia shows that by about 10,000 years ago, people had begun domesticating a few wild types of a tall grass found throughout the region. Over the next few millennia, the practice grew, and cultivated rice spread through South and East Asia.

Agricultural intensity led to a surplus of rice, and new domesticates led to denser populations. This was not much of a problem as long as these early farmers were able to expand through peninsular Southeast Asia and the nearby islands of Indonesia, the Philippines, and Taiwan. Eventually, however, expansion became a problem: Fertile land was at a premium. Humans had effectively invented the notion of real estate.

Space Race

The competition for space fueled intense rivalries. Rather than going to war, some turned their gaze east toward the Pacific, emboldened by a novel double-hulled outrigger canoe that had been perfected sailing through island Southeast Asia. Packing every known plant and animal domesticate into their newly invented vessels, these daring agriculturalists set off to colonize islands they didn't yet know existed. Along with breadfruit, taro, and banana, the tropical farmers brought with them breeding pairs of pigs, dogs, and rats, fast-reproducing food sources capable of thriving in unknown conditions.

Tahitians await the start of an outrigger canoe race in French Polynesia.

The exact route of those first migrations is still hotly debated, and historians tend to fall into one of two camps. The "express train" theory proposes that the Polynesian ancestors left from Taiwan speaking Austronesian languages and rapidly colonized the South Pacific in a straight migration eastward, depositing their people and culture along the way. The "slow boat" theory argues that the expansion originated in Asia and proceeded slowly through Melanesia before continuing to Polynesia.

Trade Networks Widen

It is fairly certain that around 3,500 years ago, the first settlers came into the region: a people known as the Lapita, distinguished by a pottery style first discovered in the 1950s. Using Fiji as a home base, the Lapita people systematically colonized the Tongan and Samoan islands, the Marshall Islands, and much of Micronesia, depositing their characteristic pottery style across western Oceania. The oral histories of people living today throughout the region largely corroborate this account and serve as a good instance of the genetic and archaeological data supporting ancestral storylines.

The Lapita seafarers established a trade network. Some of the best evidence for this comes from obsidian, a naturally occurring glass formed by the rapid cooling of volcanic lava, which can be pinpointed to its original source based on geochemical analysis. Obsidian, widely used in Oceania for tool construction, has been found on some coral atolls and has been determined to have been imported from islands hundreds of miles away. People certainly continued east, extending the Melanesian trade networks to distances greater than ever before. They colonized the remote Marquesas, Hawaiian, and Easter Islands, and they probably established contact with coastal South America, bringing its sweet potato back with them. The Lapita made it as far east as Tonga and Samoa—but suddenly, in what remains one of the great mysteries of the Polynesian expansion, they disappeared. None of their famous pottery can be found in the more remote archipelagos of Oceania.

Micronesia

Lying primarily north of the Equator, Micronesia spreads to the northwest among the three divisions of Pacific islands. It includes the Caroline Islands, the Marshall Islands, the Mariana Islands, the Gilbert Islands, Palau, and Nauru. Some of these islands—the Carolines, without Palau—have formed the Federated States of Micronesia, which are self-governing entities in free association with the United States, bringing U.S. investments. Kosrae, Pohnpei, Chuuk, and Yap are the four states within the federation. Deposits of bauxite, iron, and phosphate represent the mineral riches of these islands, especially the Carolines. Nauru, an island republic within Micronesia, also holds significant phosphate deposits. The Marshall Islands have less economic potential; they are coral islands and atolls from which the major exports are fish and coconut products.

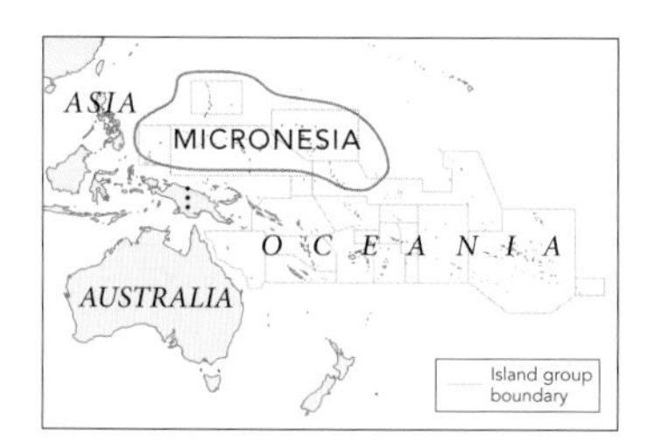

Chamorros

Population: 72,800
Location: Guam, Northern Mariana Islands
Language family: Malayo-Polynesian

The Pacific people having the longest history of contact with the Western world are the Chamorros of Guam and the Mariana Islands. In 1521, Ferdinand Magellan landed on their islands while in command of the first expedition to circumnavigate the world. Spain claimed the islands in 1565 but established settlements there only in 1668. Three decades of war and disease followed, and by 1710, fewer than 4,000 Chamorros remained. Today, an estimated 160,000 people live on Guam, while 52,000 reside in the Commonwealth of the Northern Marianas. Many are of mixed ancestry, with forebears who may have been Spanish, Filipino, American, Asian, or perhaps Micronesian (from the Caroline Islands). Regardless of their ancestry, though, Chamorros are linked by a language believed to be related to ones spoken in the Philippines.

The original Chamorros wore very little clothing, and early European observers were impressed with their physiques. They were fishers and gardeners, as well as the only Pacific islanders who, when first contacted, cultivated rice as a staple crop. Men did most of the gardening and some of the fishing, built houses and canoes, and created stone structures. Women

CHAMORROS Chamorro dance has undergone many transformations, often influenced by colonial powers that controlled Guam and the Mariana Islands.

made pottery, cooked, gathered food in the jungle, and fished with hand nets on the reefs. They were particularly skilled at working with pandanus fiber to create fine mats, sails, hats, baskets, and boxes, some of which were traded in places as far away as the Caroline Islands. The betel nut, chewed with pepper plant and lime, was the Chamorros' preferred stimulant.

Traditional Chamorro society was based on clans whose membership was inherited through females. These clans were ranked to form separate social classes, with women of the upper class holding high positions in society. Within the clans themselves, rank was determined by seniority, and the senior member of each group would use clan property for the benefit of all.

Today, Chamorros are overwhelmingly Christians; the majority are Catholics, reflecting 17th-century Spanish missionization. A form of ancestor worship seems to have prevailed originally, and at that time, some individuals were thought to have been professional sorcerers.

Chamorro history has taken many turns since Europeans first made contact. Under early Spanish rule, the Marianas population was moved to Guam and not returned until 1815, when other Micronesians were also settled on the islands. After the Spanish-American War, the United States took over Guam, and Germany purchased other Mariana Islands. After World War I, Japan ruled Germany's old possessions as a League of Nations mandate. World War II brought bloody fighting to all the islands. Today, Guam is an unincorporated territory of the United States, while the Northern Marianas have been an independent commonwealth in association with the United States since 1986. Hence, the Chamorros operate in a modern political and economic world, while their language links them to the lives they once led.

Chuuk Islanders

Population: 51,800
Location: Federated States of Micronesia
Language family: Malayo-Polynesian

Chuuk (formerly Truk) Islanders live on a group of islands surrounded by coral reefs enclosing a large lagoon. After sailing there in outrigger canoes, their ancestors settled these islands by the beginning of the Christian era. They eventually established and maintained trade relationships with nearby atolls, obtaining pandanus mats and sennit cord made from plant fiber and occasionally receiving canoes built to order. In return, they supplied the atolls with a plant dye used as a cosmetic.

Chuuk was divided into small districts, within which households spread out to form loose neighborhoods. These households were made up of extended families formed by a group of related women; husbands generally lived on their wives' land. Clans and smaller kin groups, such as lineages, were also based on descent through females. In each district, certain lineages claimed the right to particular lands, and for each district, this claim was the basis for chieftainship. Though lineages were based on descent through women, the senior male in the line of senior women usually served as chief. The oldest man in the lineage generally acted as a kind of symbolic chief, receiving gifts of food as his right. Conflict often occurred over just which lineage member had the right to succeed as chief and thereby control land and other resources.

While coconuts provided food, drink, and fiber for cord, breadfruit was the traditional staple food, preserved by fermentation in pits. Gardens also included taro, plants for cosmetic dye, and sugarcane. Men gardened, prepared food in bulk, fished in deep waters, engaged in war and political affairs, built canoes and houses, and worked with wood, shells, and stone. Women wove with hibiscus and banana fibers on looms, and they plaited mats and baskets. They also prepared routine meals, did inshore fishing, and took the main responsibility for child care.

Since the middle of the 20th century, the people have been either Protestant or Catholic Christians, but Chuuk Islanders traditionally believed in a wide variety of spirits, some of whom lived in the sky and were especially important. Chiefly clans derived part of their power from association with these spirits and the magic lore they controlled. There were spirits who controlled particular crafts and others who brought illness. Sickness was commonly believed to be caused by malevolent spirits, who might be invoked by a sorcerer to harm the infirm. Every person was thought to possess two souls—one good, the other bad—and at death, the good soul went to the sky world, while the bad one became a potentially dangerous ghost.

CHUUK ISLANDERS This young family of Chuuk Islanders probably lives more independently than recent generations, who resided in large matrilineal groups.

Like many other Micronesians, Chuuk Islanders became the subjects of successive colonial regimes. Between World Wars I and II, Japan administered Chuuk as a League of Nations mandate and developed the islands as a military base. U.S. troops mounted an invasion in World War II, bringing much suffering and loss of life. Chuuk later became part of the Trust Territory of the Pacific Islands, administered by the United States for the United Nations. After two decades of negotiation, in 1986, the Federated States of Micronesia became self-governing in free association with the United States. Chuuk is one of the four states making up the federation; its population of more than 50,000 includes people living on nearby atolls. Because "free association" means that the islanders can easily immigrate to the United States, many have done just that. Today, a number of Chuuk communities are found in Hawaii and California.

Marshallese

Population: 66,400
Location: Marshall Islands
Language family: Malayo-Polynesian

The people of the Marshall Islands became best known to the wider world in 1946, when the United States began a series of nuclear tests on tiny Bikini Atoll. They are descendants of seafarers who came to these islands almost 2,000 years ago and speak a language that is most closely related to Pohnpeian in Micronesia.

Five small islands, 29 atolls, and hundreds of reefs form the islands' double chain, with those in the south having a more favorable, reliable climate. Gardening, fishing, and reef diving traditionally provided sustenance for the Marshallese. On northern atolls, arrowroot, pandanus, and coconut were cultivated, along with some breadfruit. Southern atolls enjoyed larger breadfruit yields, together with some taro.

Marshallese men always looked to the sea, where they fished from canoes that they had made. Women dominated activities on the land, making mats, baskets, and clothing from pandanus and coconut fronds. Marshallese women also prepared meals, cared for children, fished occasionally, and collected seafood from the reefs.

MARSHALLESE A boy proudly shows off a chicken. Marshallese grow food and raise poultry mainly for their own families.

On such small islands, land was inevitably a key factor in social life. On larger isles with more resources, the control of land permitted chiefs to exercise considerable power over those who depended on them for access. Links through females were basic to social organization—rights to land were inherited through women—but the paramount chiefs, who claimed descent from ancient gods, were male. Generally, social arrangements were flexible enough to provide people with what they needed for their livelihoods; residence and cooperation in socially valued activities could give individuals claim to resources that they might not have received through descent alone.

The Marshallese believed in many kinds of spiritual beings, both male and female. They associated major deities with constellations, and believed others inhabited specific local sites. Most people believed that ancestor spirits were able to exert a direct effect on their daily lives. The Marshallese were first missionized by American Protestants, and today, almost all adhere to some form of Christianity.

After living under different colonial powers and surviving the trials of World War II, the Marshallese were included in the U.S.-administered Trust Territory of the Pacific Islands in 1947. A Compact of Free Association with the United States became effective in 1986, creating the Republic of the Marshall Islands. The United States pays for the use of a missile testing base at Kwajalein. The United States also continues to compensate the Marshallese for the health and social problems that have resulted from past nuclear testing.

Palauans

Population: 14,100
Location: Palau
Language family: Malayo-Polynesian

Westernmost Micronesia is home to the Palauans, whose proximity to Asia has strongly influenced their culture over the years. The first Palauans are thought to have arrived in these islands more than 2,000 years ago; they were voyagers who perhaps came directly from Southeast Asia and chose to settle here rather than venture farther east. Today the people speak an Austronesian language that is more closely related to those of western Indonesia and the Philippines than to those of islands to the east. They have also marked out a political path separate from other Micronesians by establishing themselves as a self-governing nation: the Republic of Palau, or Belau.

In the prehistoric past, Palauans built inland villages on elaborate stone foundations. The village meetinghouse, or *bai,* was the center of the basic social unit, which not only controlled land for homes and taro patches but also had responsibilities regarding chiefly titles, certain valuables, and ceremonial functions. Members were related through maternal or, less often, paternal ties, and adoption was common within this extended kin group. Rank governed relations among siblings, among houses within a village, between titles in a political council, and among villages in the largest political grouping. Chiefly titles were ranked according to the social value of local landholdings, and titleholders assigned land to families according to their needs.

PALAUANS Palauan rowers on team System get a good workout in preparation for future races.

Fishing and taro cultivation provided the basis for traditional subsistence, with fishing symbolizing male virtue and taro cultivation serving as the emblem of female productivity. This division was paralleled by the control of valuables: Women exchanged hammered turtle-shell trays, while men used beads of foreign origin. Other items exchanged between districts depended on local resources and included canoe sails, pottery, and wooden tools.

Like other Micronesians, Palauans lived under a succession of colonial rulers: Spanish, Germans, Japanese, and finally Americans, who administered the Trust Territory of the Pacific Islands under the auspices of the United Nations. Because of its location, the Palauan archipelago was especially subject to Japanese influence during the period between the world wars. Japanese men sometimes married Palauan women, and even today, Japanese is spoken on Angaur Island, in addition to the local language. During World War II, Japan used the archipelago as a base from which to stage its attack on the Philippines.

Palau was the last of the four political units to emerge from what had been the Trust Territory of the Pacific Islands; the others were the Northern Marianas, the Federated States of Micronesia, and the Republic of the Marshall Islands. Negotiations were delayed because the Palauans, in 1981, adopted a constitution declaring their islands a nuclear-free zone—a sticking point for the United States, which hoped to maintain a strategic presence in the area. In 1994, the impasse was broken, and from that year on, Palauans assumed a new national identity as citizens of a republic.

Yapese

Population: 5,900
Location: Federated States of Micronesia
Language family: Malayo-Polynesian

The Yapese ranged widely as sailors in outrigger canoes from their base in western Micronesia, where they live on four islands surrounded by a coral reef. Despite historically close contact with a number of other groups, the Yapese differ from most of their Micronesian neighbors in both their language and social organization.

Yapese is an ancient Malayo-Polynesian language that some linguists tie to those spoken in the Melanesian Islands of Vanuatu far to the southwest. The Yapese today also speak English, the state's official language. Unlike most other Micronesians, the Yapese reckon descent patrilineally for the purposes of landownership, which traditionally has determined status and rank within and between Yapese village communities. Through their mother's line, though, Yapese belong to clans to which they can turn for emotional support and for refuge should they have a falling out with their patrilineal kin.

Where there is kindness,
there is life.

MARSHALLESE PROVERB

In other respects, the Yapese practiced the same subsistence activities organized by gender as other Micronesians, with the men responsible for providing fish and the women cultivating taro and yams. Like the nearby Palauns, the Yapese constructed their villages using large amounts of stone for foundations, pathways, seating areas, piers, and retaining walls. Yapese men also routinely sailed to Palau and Guam for the arduous and hazardous task of quarrying and transporting the large coral stone discs used as a kind of money in ritual exchanges on Yap, although they had no value for trade with other peoples.

The basic Yapese social unit is the family, or *tabinaw,* defined as a group who eats together. Typically this group includes a married couple and their children, and it can include the wives and children of married sons before they establish their own households. Traditionally, rank and status operated within the family and was manifest in the use of restricted cooking (and eating) pots for different family members, although this practice is fading.

The Yapese shared the area-wide experiences of Spanish and German colonization, Catholic missionizing, and Japanese and U.S. occupation in the 20th century, as well as U.S. bombing during World War II, culminating in the writing of a constitution for the Federated States of Micronesia in 1978, with sovereignty finalized in 1986. Many Yapese are now Roman Catholic, but the people persist in animistic beliefs linking female spirits to the sea and male spirits to the land, where they pose threats to the subsistence activities of humans of the opposite sex.

A number of changes in recent decades have combined to erode certain traditional structures of Yapese life. With about half of Yapese men working for wages, it becomes difficult for them to keep up with their contribution of fish, and they have had to resort to buying canned fish and meat for family meals, lowering their traditional status. Yapese youth spend less time learning cultural traditions from older men in the village men's house and more time with peers, sometimes with undesirable consequences. And young men and women are more likely to meet a future spouse on the school bus traveling to the central high school than they are within family-approved villages.

YAPESE Dressed for a traditional dance, a Yapese woman takes a quiet moment from daily chores.

ACROSS CULTURES

TO SAIL BY THE WISDOM OF THE STARS

Before global positioning systems, before sextants, even before compasses, native peoples navigated their ways across thousands of miles of Earth's waters.

Until A.D. 1500, many European scientists believed the world was flat—yet indigenous mariners of Oceania had been sailing the nearly 20 million square miles of Oceanic seas in double-hulled canoes for more than a thousand years.

How did they accomplish such a feat? Polynesian navigators had a sophisticated technology all their own. Their specialized science, now referred to as wayfinding, was a system of observation that synthesized patterns in nature and allowed seafaring cultures to navigate with keen accuracy. The shape and sequence of waves, the rising and setting of stars, even the flight patterns of birds served as instruments for the wayfinder. Intimately wedded to his environment, the navigator employed a highly developed facility for observation, memorization, and integration. These skills required a lifetime of study and commitment. In the Hawaiian system, the training was so rigorous that few apprentices ever achieved the rank of *kahuna ho'okele,* navigator-priest.

Wayfinding was a revered science and a priestly profession that linked the cosmos, sea, and land. The initiate had to be not only a superior seafarer but also a healer, leader, and interdisciplinary scholar in subjects as diverse as astronomy and botany. Once on the sea, a full-fledged navigator outranked even the highest chiefs of the land.

The ho'okele and his counterparts across Oceania had no paper charts, relying instead on conceptual constructs of the sky. One of these was "the path of the spider," the track of the sun and the planets around the Earth—the ecliptic, in Western terminology. The spider's web, like a knotted net covering a gourd, spreads out in star lines running north and south, east and west, and on the diagonal, much like a Westerner's star chart.

The Polynesian god of the ocean, Tangaroa (or Kanaloa), is represented by the octopus whose eight great arms are the sea highways between the far-flung islands. The head of the octopus is in the southern ocean, celestially represented by either the Southern Cross or Canopus, depending on one's culture. The tentacles of the octopus point north, much like eight lines of longitude, and are celestially represented by star lines.

The navigator also needed to know the horizon markers. "Stone canoes," like the carved rock located at Kukaniloko on Oahu, Hawaii, and one in Kiribati, taught the navigator the location of "star pits," the points on the horizon from which a star rose in the east

Polynesian men use stars to navigate a long sea journey by outrigger canoe between Hawaii and Tahiti.

For centuries, outrigger canoes have been carrying the seafarers of Oceania over millions of square miles of Pacific Ocean waters.

or into which a star fell in the west. The horizon markers served as compass points like that on a watch bezel. The navigator was required to know how far along the horizon he would need to look for the star pits from one sighting to the next.

The decan, or ten-day week common throughout Polynesia, was a navigational calendar based on the rotation of the horizon. The decan week, known in Hawaii as *anahulu,* was a system that worked in concert with the solar and lunar calendars as an invaluable system of temporal orientation for the navigator.

Stars as Guides

The navigators of Oceania also used the mariners' guide stars—Polaris, Sirius, and the Southern Cross, common to nearly all maritime cultures. Zenith stars, which pass directly overhead, served as another reference point to land. Satawalese voyagers from the Caroline Islands (now called Micronesia) referred to various storm stars throughout the year, including Canopus, Vega, and Arcturus, to predict inclement weather.

A navigator in training also learned to interpret the color of sunrises, sunsets, and clouds to determine weather conditions and proximity to land. The deep red, purple, and orange in a sunrise or sunset, coupled with certain shapes, types, and positions of clouds, forecast rain. The color of the cloud's underbelly suggested the type of land the mariner was approaching. A shining bright cloud, for example, indicated an atoll's sandy, shallow lagoon, while a greenish color reflected verdant land and grayish blue signified open sea. There were also myriad ways of referring to directions. A conceptual compass based on prevailing winds identified 16 directions. An astral compass marked the stars' rising and setting around the horizon.

The navigator's responsibility was so immense that he would not allow himself complete rest. Instead he would lie in the hull of the canoe, much like a child in a mother's womb, where he could keenly feel the direction, rhythm, and sequence of the waves. Such methodology might be lost on Western science, dismissed as intuition or mysticism. Yet the wayfinder was gathering explicit data. His facility to determine the dominant swell of a wave was as finely calibrated as any instrument. This was one of the navigator's key methods for maintaining the canoe's proper angle, particularly important during the day and on cloudy nights, when stars were not visible.

The patterns of living creatures also informed the navigator. In the morning, seabirds go in search of food, returning to their islands late in the day. Some birds fly far out to sea; some stay closer to land. By knowing the different birds and their behaviors, the navigator deduced how close he was to shore. Certain schools of fish indicated, among other things, sea lanes, similar to the messages we receive from highway signs. The color of the water and the smells carried on the wind provided signposts to the skilled wayfinder.

The navigator had to know what foods to bring and what to leave behind. Breadfruit and banana, for example, were stored in the form of fermented paste and fruit leather. Coconuts provided water and food. Some items had to be prepared long before the voyage began—dried fish, fermented foods, and dried fruits, for instance. Other foods, like chickens or pigs, could be carried live, but they too needed to be fed.

And this was just the beginning of what it meant to be a navigator. It wasn't enough for him to find his way. He was also responsible for the feeding and general well-being of those on board and, when necessary, was required to administer to the injured. Each culture had medicinal resources available from the environment. In Hawaii, for example, *'uhaloa (Waltheria americana)* was used for chest and throat congestion. *Ko'oko'olau (Bidens pilosa)* served as a general tonic, a treatment for kidney and bladder ailments, and a blood fortifier. *'Olena (Curcuma longa)* was used to treat the common cold and other ear, nose, and throat disorders—and,

The late Micronesian navigator Mau Piailug instructs his son and grandson on the intricacies of navigation using natural objects to represent stars, sea, and outriggers.

Master navigator Mau Piailug received a hero's tribute in Hawaii for reintroducing the art of navigating without instruments.

most important, was used and continues to be used for cleansing and purification rituals.

Navigators committed themselves to regimens of physical training similar in complexity to those of the martial arts. The *palu* of Micronesia, for example, practiced *pwaeng,* a physical discipline by which the navigator and his crew used their bodies to right a capsized canoe. During storms, palu mariners would also loosen the lashings on their vessels to accommodate conflicting stresses of winds, waves, and swells, reducing the likelihood of breakage. They had to secure themselves to the mast until the storm had passed, and then relash the canoe and continue on.

Without charts and maps on paper, ancient navigators repeated sailing chants, which taught the names of the stars, winds, rains, and key navigational reference points. Skylore and tribal mythology served as science texts among cultures where myth and legend were particularly developed.

As exceptional as these skills were, it was and still is the magic possessed by the priestly order of wayfinders that is most extraordinary. These seafaring sages rely on a multifaceted system of ritual, chant, myth, prayer, dreams, and visions. Polynesian cosmology takes as fundamental the interdependent relationships among humankind, the natural world, and deity. If the navigator prayed, he trusted—he knew—an answer would be revealed.

Navigating Our Future

Throughout the world—from Chowra, India, to Satawal, Micronesia—navigator-priests called the winds and calmed the seas by means of their magic. Such phenomena defied Western belief. Unfortunately, this mysticism was dismissed by classic science and demonized by Christianity. As native societies were conquered, decimated, and assimilated, these priests—Hawaiian *kahuna,* Micronesian palu, and many others—were forced to abandon or denounce their ancestral traditions or conceal their secrets. The judgments and consequent constraints imposed on native people almost destroyed their indigenous knowledge and practices. That period of history nearly sounded the death knell of native maritime science.

Fortunately, remnants of this ancient knowledge still exist within the vanishing clans such as the *palu* of Satawal, Micronesia. In March 2007, the late navigator Pius "Mau" Piailug, one of a handful of Satawalese sailing masters, conducted *pwo,* an initiatory ceremony, conferring 16 Micronesian and Polynesian men with the mantle of navigator—the first such ceremony in 50 years. Without Piailug's vision and dedication, wayfinding would be little more than a myth, traditional sciences shattered and forgotten, among the descendants of Satawal and other Pacific cultures.

Ancient seafarers were not unlike modern astronauts exploring new frontiers—their sea, our space; their canoes, our space capsules. We cannot overestimate the extent of their brilliance, their daring, and their achievements.

As we in the modern world teeter on the brink of environmental and social collapse, we are not unlike the ancient mariners who crossed uncharted waters, carrying finite resources along. Whether guided by stars, instinct, or global positioning systems, we are now the wayfinders. The success of our journey depends on everyone. As navigator Piailug said, "The canoe is our island; our island, the canoe."

The planet is our canoe; the canoe, our planet.

—Elizabeth Kapu'uwailani Lindsey

Chapter 5

EUROPE

EASTERN EUROPE

Kalmyks • Lithuanians • Roma
Romanians & Moldovans • Russians • Ukrainians

CENTRAL EUROPE

Bosnian Muslims • Croats • Czechs • Magyars
Poles • Serbs • Slovaks • Slovenes

SOUTHERN EUROPE

Albanians • Greeks • Italians

WESTERN EUROPE

Basques • Bretons • Catalans • Dutch & Frisians
French • Germans • Portuguese • Spanish

NORTHERN EUROPE

Cornish • Danes • English • Faroese
Finns • Icelanders • Irish • Norwegians
Sami • Scots • Swedes • Welsh

Europe has witnessed an astonishing florescence of settlement and cultural development over the course of many millennia. From an area that—including the European part of Russia—is less than half the size of North America, streams of influential ideas, technologies, political and legal structures, art, and other products of human endeavor have poured forth into the rest of the world.

Today, some geographers question whether Europe should even be considered a continent, attached as it is to Asia in the fashion of a large peninsula. Nevertheless, the traditional division that separates the two is a line that runs south along the Ural Mountains and the Ural River to the Caspian Sea before jogging west along the Caucasus Mountains to the Black Sea. More bodies of water provide other boundaries, and within them are a number of major islands, including Great Britain, Ireland, and Iceland.

EARLY EUROPE After Asia, Europe has the world's largest and densest population of the traditionally considered continents. Here, on just 7 percent of Earth's land, live more than three-quarters of a billion people, representing hundreds of distinct ethnic groups and speaking at least 164 languages.

On Krakow's Market Square, pigeons gather around a young Polish girl, hoping for a handout.

Europe's peopling by modern humans began only about 45,000 years ago, more than 200,000 years after the continent's occupation by Neanderthals. Neanderthals disappeared about 39,000 years ago, and all subsequent populations in Europe have belonged to the subspecies *Homo sapiens sapiens.*

During Europe's Ice Age, hunters of western Europe gathered regularly for ceremonial purposes, bringing an increasingly advanced tool technology and aesthetic sense. On the walls of their caves in northern Spain and southern France, they rendered dramatic and graceful images of horses, bulls, and other animals that figured prominently in their Stone Age culture.

The migrations that gave way to most European ethnic groups of today are believed to have begun some 5,000 years ago with Indo-European pastoralists from Central Asia. They brought the beginnings of the Indo-European language family, which in Europe survives in eight main branches, ranging from the Italic family—including modern Italian, French, Spanish, Portuguese, and Romanian—to the Thraco-Illyrian, with Albanian the lone modern representative.

Not all European languages are Indo-European, however. Finnish, Hungarian, and Estonian fall into the Uralic family, originating in the area of the Ural Mountains, while the Basque language of the Pyrenees is related to no other language known in the world and is believed to predate the Indo-Europeans.

Hunter-gatherers had already made the transition to agriculture when peoples from the Middle East and Egypt arrived on the island of Crete. Accomplished seafarers, they spread the influence of the Minoan civilization through the Mediterranean world. But by 1450 B.C., the balance of power had shifted to mainland Greece and the Mycenaeans. The ancient Greeks, especially those of the fifth century B.C., left a legacy of excellence in art, architecture, mathematics, science, philosophy, and political thought, bequeathing ideas and ideals that still shape European thinking.

CHRISTIANITY Romans helped filter this legacy through their widespread empire, which in the second century A.D. stretched from the British Isles to Iran. By the fourth century, the Roman people had adopted Christianity as their state religion, and the empire's decline did not end the faith's influence. As ethnic groups moved into areas that would become their homelands, leaders often embraced Christianity on behalf of their people. A split in the 11th century resulted in the distinction between Roman Catholicism and Eastern Orthodoxy that persists today

The effects of the 16th-century Reformation endure in the Protestantism of most northern European groups. Between the 12th and 20th centuries, Ottoman Turks conquered and controlled parts of Europe, and wholesale conversions to Islam took place; today, however, Islam survives on the continent mainly in Albania, among Bosnia's Muslims, and in the small European nub of Turkey. Followers of Judaism once formed significant populations throughout Europe, but their numbers were reduced tragically during the Holocaust.

NATIONALISM As Europe began to move from the feudal model of the Middle Ages to that of centralized authority in the second millennium, the concept of nationalism began to replace provincial allegiances. Nationalism became imbued with the notion of a common language and religion and thus helped create the modern states that formed toward the end of the millennium.

Nationalism also led to the spread of empires around the globe as rising powers emphasized their own cultures and promoted their own interests over those of other peoples. The dawn of the industrial revolution in the 18th century further aided Europe's domination of the world economy and turned Europeans into urban workers. In the 20th century, industrial might and nationalistic fervor fueled two world wars and other upheavals. Borders were redrawn, and millions of people were forced to relocate. Massive migrations occurred as people sought to avoid conflict, escape persecution, or find better economic opportunities.

Smeared with silica mud from Iceland's geothermal Blue Lagoon, two men enjoy spa time.

The current trend toward unity and cooperation, typified by the European Union formed in 1993, demonstrates the pragmatic nature of cultures and nationalities to work past historical divisions. Economic ties and threats to security also unite them, while intra–European Union migration alters the demographics of individual countries. At the same time, Europeans are indicating renewed interest in languages, music, and other aspects of cultural identity. Religion, seemingly less important in a secular world, still holds the ability to divide, as seen in the cases of Northern Ireland and the former Yugoslavia.

Eastern Europe

Although proximity to the sea characterizes most of Europe's territory, this eastern region contains some of the continent's landlocked nations. It is the region of contact between Europe and Asia, a boundary across which stretch the Ural Mountains and the central upland of Russia. Through this region flow legendary rivers: the Dniester and the Dnieper, the Don and the Volga, all emptying into the Black Sea, Earth's second largest inland body of water. Eastern Europe also includes the northerly reaches where lake-studded Finland meets Russia's northwestern corner.

Kalmyks

Population: 183,000
Location: Russia
Language family: Mongolic

The Kalmyks originally were a western Mongolian people known as the Oirat. Nomadic herders, they moved seasonally between camps on their traditional grazing lands in Mongolia, southern Siberia, and China while tending their flocks of sheep, goats, cattle, camels, donkeys, and horses. When it was time to shift to another location, they disassembled their large, circular felt tents called yurts, or *gers,* and moved on.

In the 13th century, the Oirat became involved in the infighting among Mongol tribes, starting out on the side of Genghis Khan and ending up affiliated with the khanate of his grandson, Kublai, after a stint with the opposition—Kublai's brother, Ariq Boke. It was at this juncture that they adopted the religious orientation of Tibetan Buddhism and became involved in Tibetan politics and the protection of the Dalai Lama.

In the mid-17th century, a large group of Oirats, perhaps as many as 50,000 households, continued to move west to the steppes between the Volga and Don Rivers on the northwestern shore of the Caspian Sea. Theirs was the last major westward migration of peoples from the steppes of Central Asia. Some time before this period, the name *Kalmyk,* which probably originated with a Turkish word meaning "left behind," became associated with these Oirats.

Now living in Europe, the Kalmyks continued their lives as nomadic herders. They maintained their language, a western branch of Mongolian, and their practice of Tibetan Buddhism, which made them Europe's only Buddhist people. They operated in the region as an autonomous khanate that was absorbed, as the Kalmyks lost their influence, into Russia's empire under Catherine the Great.

Then began the first of several attempts to Russianize the Kalmyks, getting them to abandon the nomadic lifestyle and settle into houses, and attempting to convert them to Russian Orthodox Christianity. Under communist rule, the Kalmyks resisted attempts to collectivize their efforts and deny their religion. About a third of them were slaughtered in the 1920s. Later, their religion was prohibited and their monasteries destroyed.

KALMYKS The Kalmyk National Ensemble performs in Elista, capital of the Kalmykia.

The situation worsened under Stalin: In one sweeping action, he deported all the Kalmyks to Siberia in 1943. Almost half died immediately. The survivors spent 14 years in exile, until 1957, when Nikita Khrushchev allowed them to come back to Kalmykia.

Returning Kalmyks found their land settled by Russians and Ukrainians, their houses lost, and their livelihoods stolen from them. Some resumed the nomadic way of living. But they retained few artifacts from their former lives, and their language came perilously close to extinction.

In the meantime, a group of 2,000 Kalmyks who had come under Allied control at the end of World War II had nowhere to go. Their kinsmen languished in Siberia, and returning to Kalmykia was tantamount to a death sentence. The United States took them in. Initial resettlement in New Mexico proved untenable, but the next stop, New Jersey, worked out. Eventually one of their spiritual leaders immigrated, giving the expatriate Kalmyks in the United States more of a sense of community. This population recently celebrated its 60th anniversary.

Some 183,000 Kalmyks live in their autonomous republic in the Russian Federation, facing the challenges of the post-Soviet economy. But now Kalmyk culture appears revitalized. Kalmyks of all ages are determined to keep their language and culture alive, whether learning ancient epic hero tales, traditional dancing, reviving Buddhist practice, or using modern technology like texting and social media.

Lithuanians

Population: 2.3 million
Location: Lithuania
Language family: Baltic

Lithuania, the largest and southernmost of the Baltic states, may have first been occupied by peoples who came from Asia some 4,000 to 10,000 years ago. The Lithuanians experienced a heyday from the 14th to 18th centuries, which included a period of union with Poland as Lithuania's power began to decline. Yet for much of their more recent existence, Lithuanians have tried to maintain their culture in the face of formidable challenges—German, tsarist, and Soviet control, and now a period of rebuilding in the post-Soviet era.

Today, about 80 percent of the country's population is ethnic Lithuanian; the remaining 20 percent is mostly Russian or Polish. Most speak Lithuanian, though, which along with Latvian belongs to the Eastern Baltic group of Indo-European languages. Lithuanian was an underground language during a period of Russian domination in the late 19th and early 20th centuries, when the language was banned and other attempts were made to thwart Lithuanian identity.

About 80 percent of Lithuanians today are Roman Catholic. Nearly all of Lithuania's Jews—some 240,000—perished during World War II. The strength of Lithuanian Catholicism often is credited with the culture's endurance. Yet this staunch Catholic faith was built on a pervasive animistic past. Lithuanians held sacred the forests and fires and worshiped gods who presided over the natural world. Ongoing regard for nature is seen in a number of popular names, such as Ausra (dawn), Giedrius (dew), and Gintaras (amber). It also persists in celebrations, such as the practice of dashing into the forest on Midsummer's Eve in search of a "fern blossom," which, though imaginary, brings good luck.

Lithuanians traditionally stretched the observance of Christmas and Easter to two days or more. Christmas Eve dinner, the most important meal of the year, featured 12 courses. Everyday meals included dairy products, especially sour cream and curds; potatoes; root vegetables; and meats such as pork and beef. Easter celebrations included roving groups of young men who went house to house, singing carols and begging for *marguciai,* or decorated Easter eggs.

In the secular realm, Lithuanians observe their independence in modern times, recalling post–World War I liberation from the Germans, the declaration of independence from the Soviets in 1990, and a standoff against Soviet tanks at the Vilnius television tower in January 1991, when 13 people died. A source of great pride is the music festival held in Vilnius every five years, which draws tens of thousands of singers who perform *dainos,* ancient folk songs often sung a cappella and in a minor key.

The Lithuanians came through the era of Soviet control with a strong sense of purpose but with a crumbling infrastructure and an economy that was limited by price controls and quotas. The outlook was rather bleak in the first years of independence, but the Lithuanians have made great strides toward establishing a strong market economy. Reward for their efforts came in 2004 with entrance into both the European Union and North Atlantic Treaty Organization.

Today, less than 10 percent of the people are involved in agriculture and about 20 percent work in industries such as machinery, textiles, clothing, and fertilizer; the rest are employed in services. While basketball is not an official export, the Lithuanians produce and send out a significant number of very skilled players, many of whom have ended up on college and professional teams in the United States.

LITHUANIANS Lithuanians relax in cafés on Town Hall Square in the capital city of Vilnius.

ROMA Roma musicians play on a street in Istanbul, a city that straddles Europe and Asia.

Roma

Population: 10–12 million
Location: Most European countries
Language family: Indo-Iranian

The Roma are known by a number of names, many of which—*Gitanes* in France, *Czigany* in Hungary, and *Zigeuner* in Germany—are variations of *Gypsy,* a term that erroneously ascribed their origins to Egypt. In fact, the Roma have been linguistically traced to northwest India. They speak Romany, an Indic language from which comes the word *roma,* meaning "man" or "people." One theory ties the group's 11th-century diaspora from India to their recruitment as troops who drove out Islamic invaders and pursued them for centuries into the Byzantine Empire and southeastern Europe.

Today, the Roma are a minority throughout Europe. Exact population figures are unknown for one reason: A long history of discrimination has made them reluctant to admit their heritage. Targeted by Nazis, more than half a million died in World War II. Communist regimes also suppressed their culture, but the collapse of communism did not stop the repression. A rise in nationalism led to more violence. In 1971, the first World Romani Congress established a much needed forum through which the people can address their struggle against racism and loss of cultural identity.

The Roma have long existed at the margins of other societies, partly to avoid discrimination and partly because of their migratory lifestyle. For centuries, families traveled in brightly painted, horse-drawn caravans, plying trades consistent with this kind of nomadism. They were horse traders, peddlers, metalsmiths, musicians, fortune-tellers, and carnival workers. In recent decades, some countries encouraged or forced permanent settlement, especially in Eastern Europe, where the Roma were put to work in

GENOGRAPHIC INSIGHT

AN ANCIENT TRYST

Genetic data are definitive in showing that modern humans emerged from Africa in the past 60,000 years. As we proceeded into the Middle East and Asia, following grasslands and water resources during the long march out of the tropics, we also encountered other creatures that were like us in many ways. Neanderthals and modern humans share ancestry in Africa more than 500,000 years ago, and the Neanderthals left shortly after this, while our ancestors stayed in Africa. Over hundreds of thousands of years, the Neanderthals proceeded down their own evolutionary path, acquiring the characteristics that we associate with them, heavy brow-ridges and burly bodies among them, as part of a suite of characteristics that were driven by adaptation to the colder climate of northern latitudes.

Neanderthal Genome

And then we appear on the scene suddenly, and within 20,000 years, the Neanderthals are extinct, swept aside by our success as we colonized Europe and Asia. After all, we don't see any Neanderthals walking around today . . . or do we? This was the prevailing wisdom until 2010, when a team led by Svante Pääbo of the Max Planck Center for Evolutionary Anthropology published the sequence of the Neanderthal genome. While obtaining a complete genome sequence from a long-extinct hominid was a technical tour de force and newsworthy in its own right, the comparison of the Neanderthal and modern human genomes held the real shocker: Modern humans from Europe and Asia appear to have a small amount of Neanderthal DNA (around 2 percent) in their genomes, while Africans don't. The implication was that as modern humans left Africa 60,000 years ago, they encountered Neanderthals and interbred with them, accounting for the small percentage of Neanderthal DNA carried by non-Africans today.

Denisovans Emerge

Later in 2010, Pääbo's team announced an even more startling finding. Based on a tooth and a pinky bone from the Denisova Cave in the Altai Mountains of south-central Siberia, the scientists were able to extract intact 40,000-year-old DNA and sequence the genome of the individual. The comparison to Neanderthal and modern human genome revealed that the sequenced individual belonged to a new species, distantly related to the Neanderthals, which the team dubbed "Denisovans." And perhaps most surprising, these Denisovans contributed as much as 6 percent to the genomes of modern Melanesians, while leaving virtually no trace in the modern inhabitants of Asia.

How does all of this fit together? The current thinking is that as the common ancestor of the Neanderthals and Denisovans left Africa around 500,000 years ago, one group stayed in the west, in Europe and the Middle East, while another group headed east into Central and East Asia shortly after this time. The former group gave rise to the Neanderthals, and the latter became the Denisovans. The mystery of why the Denisovans have such an unusual pattern of interbreeding with modern humans remains unsolved, but it is an excellent example of how ancient DNA is providing a previously unseen view of the complexities of human migration patterns.

Neanderthals looked different—but not that different—from the *Homo sapiens* with whom they interbred.

state-run industries. With the fall of socialism, many settled Roma have lost their livelihoods.

Over the years, the people have embraced a number of faiths, often depending on the dominant local tradition. But while they observe rituals of mainstream religion, they sometimes add their own practices. For example, followers of Roman Catholicism gather near France's Rhône Delta each May to honor Black Sarah, their patron saint, even though she is not recognized by the church. Recently, many Roma have been drawn to Pentecostalism.

The Roma are organized into clans that are usually based on occupation, each with a chief or head. Clan members, or large groups of Roma in general, travel long distances together, but now they frequently do so in mobile homes and RVs. Music and dance are integral parts of their culture and have influenced aspects of other European cultures. In recent years, Roma musical rhythms have enjoyed a surge of worldwide popularity.

Marime, the Roma code of ritual purity, is a remnant of the people's origins in India incorporating a number of taboos regarding social relationships and food. As a result, traditional Roma society is a closed one, shunning out-of-group marriages and other contacts with *gadjés,* non-Roma peoples who, as a group, are deemed unpure. Ideally, marriages were arranged and occurred at an early age, but such arrangements are becoming less common.

While birthrates for other European ethnic groups are declining, the Roma one is rising dramatically. Some demographers even speculate that Slovakia's Roma population will outnumber the country's Slovaks by the year 2060.

Romanians & Moldovans

Population: 19.9 million
Location: Romania, Moldova
Language family: Italic

When the Romans invaded what is now Romania in the first century A.D., they encountered the strong and prosperous Dacians, who had established a kingdom along the Danube, on the central plateau, and in the Carpathian Mountains. The Dacians fought

ROMANIANS These traditionally dressed Romanian dancers await their turn to perform at a wedding in Transylvania.

a good fight but succumbed to Roman troops and occupation. The introduction of Latin led to the development of the Romanian language and to influences that would last long after the Roman retreat in 271. For the next 1,600 years, the Romanians saw invaders come and go—barbarians, Ottoman Turks, and Hungarian Magyars among them. The last populated the western province of Transylvania, which has changed hands between Romania and Hungary a number of times. Communist control after World War II ended with the toppling of dictator Nicolae Ceausescu in 1989.

Today, ethnic Romanians make up about 92 percent of the country's population of 21.6 million, and about 70 percent of them follow the Romanian Orthodox faith, a branch of Eastern Orthodoxy. Roman and Uniate (Greek) Catholics and Protestants make up the remainder of the country's religious affiliation. Most of Romania's 400,000 Jews perished in the Holocaust, and many of those who survived immigrated to Israel.

Although religious expression was suppressed under communism, Romanians continued to practice their faith, which includes the veneration of icons in both churches and homes. These images, usually reverse paintings on glass, are believed to embody the physical presence of Christ, the angels, the saints, and other religious figures depicted on them. On entering a home, visitors acknowledge the family icon by bowing and making the sign of the cross before greeting their hosts. Each church holds numerous icons, which receive special veneration on feast days and by bridal couples during weddings.

Much of Romania's social life, especially in rural areas, traditionally centers around church activities such as parties and dances. The seasons leading to Christmas and Easter are elaborately observed, and at Christmas the Romanian love of music has given birth to an extensive repertoire of Christmas carols.

Pre-Christian practices survive in Romania as well, especially in rural areas. Belief in spirits in nature, witches, and the spirits of the undead still lead to precautions to thwart malevolent actions. The legend of Dracula is based on a number of elements in Romanian culture, including the historical figure of Vlad Tepes, a 15th-century local prince who fought ruthlessly against all who opposed him, earning the nickname Vlad the Impaler. Today, Romanians often reluctantly—but lucratively—promote Dracula tourism in Transylvania, which has visitors flocking to its mountains to listen for the cry of wolves as they tour ruins of Dracula's purported castle fortress.

There is no disputing a proverb, a fool, and the truth.

RUSSIAN PROVERB

The Romanians endured communist control for more than 40 years, until severe food and fuel shortages and the growing dissent in Eastern Europe led to the ouster and death of dictator Nicolae Ceausescu in 1989. Today, the country is still recovering from the privations of the communist regime and the mismanagement that followed. Most Romanians are poor by European standards, and inflation has cut sharply into their incomes. Housing is scarce, especially in cities, and accommodations are cramped. Roughly a third of Romanians work in agriculture, and in recent years they have been able to reclaim land taken from their families as far back as four generations, leading to increased agricultural production. Another third have jobs in industry, and almost as many work in the service sector.

Everyday Romanian cuisine features stews and soups, including borscht made from cabbage; pork and sausages; and *mamaliga,* a cornmeal mush considered the national dish. Lamb often appears on the table at holidays. Common beverages include local wines and a strong plum brandy called *tuica.*

Romanian identity extends to the Moldovans, residents of the republic to the east of Romania. A high percentage of the Moldovans consider themselves ethnic Romanians. The Moldovan language is a subdialect of Romanian, with many borrowings from Russian and Ukrainian. Like Romanians, most Moldovans are Orthodox Christians, but they follow Russian, not Romanian, Orthodoxy.

Many of the inhabitants of the eastern part of Moldova, called Bessarabia, have a more mixed ethnic

background. In the 1990s, the Moldovans considered unification with Romania but ultimately rejected it, choosing political independence over cultural unity.

Russians

Population: 166 million
Location: Russia
Language family: Slavic

The largest ethnic group in Europe, the Russians developed from ancient Slavic tribes who had settled in the area north of the Black Sea by the first millennium A.D. By the ninth century, when Vikings established a dynasty in the state of Kievan Rus', a number of tribes inhabited the area from the Black Sea to the Baltic.

Official conversion to Eastern Orthodox Christianity occurred in A.D. 988, linking Russians to the Byzantine Empire. Mongol occupation from the mid-13th to the late 15th centuries, however, sealed off the people from the cultural changes coming to Western Europe. These would not be felt until the early 18th century, when Tsar Peter the Great modernized and Westernized the culture and enlarged Russia's territory, a trend that continued under Catherine the Great.

As the tsars expanded Russia's boundaries, the original Slavic tribes assimilated groups with Finnish, Turkish, Siberian, and Baltic origins to form the present-day Russian people, who number about 166 million. In the post-Soviet era, ethnic strife has led to violence in some parts of Russia, notably in Chechnya, a region in the Caucasus Mountains.

The Russian language belongs to the East Slavic group of Indo-European tongues. An old form of Russian known as Old Church Slavonic survives in the Russian Orthodox Church liturgy. Up until the tenth century, Russian did not have a written form, but then disciples of the monks Cyril and Methodius constructed a hybrid alphabet from the Latin, Greek, and Hebrew alphabets. Today, a number of other languages also use the Cyrillic alphabet that the two monks devised.

Though officially condemned during the Soviet era, when perhaps up to 85 percent of churches were closed and their property confiscated, the Russian Orthodox

RUSSIANS Russian teenagers check their selfie in Moscow's Arbat district, known for its graffiti-covered Tsoi Wall.

Church survived and has been experiencing a resurgence of participation and elevation of status as a cherished cultural institution. Western proselytizers have also had significant influence in present-day Russia, as evidenced by the growing numbers of Baptists, Jehovah's Witnesses, and Seventh-day Adventists in the region.

Russians have traditionally been rural people. For centuries, many of them were serfs, but even after their emancipation in 1861, serfs saw little improvement in their condition. During the Soviet era from 1917 to 1991, the central government endeavored to hold all property and organized large collective and state farms. This did not deter food production on private plots, though, and people sold their surplus produce to alleviate shortages caused by a poor distribution system.

As the society and economy continue to adjust to the post-Soviet era, some situations have changed dramatically while at the same time others remain virtually the same. A warm, effusive people by nature, Russians of the Soviet era had become guarded in their public lives, and thus they tended to lead a dual existence, a situation that has since somewhat eased.

A slip of the foot is not nearly so dangerous as a slip of the tongue.

BULGARIAN PROVERB

Under President Vladimir Putin, Russia has emerged as an economic world power with considerable oil wealth, as a regional military power willing to meddle in neighboring states' affairs, and a high-profile host to an extravagant Winter Olympics at Sochi in 2014. Russians still struggle to build an open, democratic civil society in the shadow of an authoritarian regime that curtails freedoms of press and political organization. Though postcommunist Russia boasts many newly minted billionaires, the standard of living for most people is austere, and alcoholism and domestic abuse remain serious social issues.

Russians have a rich folk culture, often expressed in music and the decorative arts. *Byliny* are traditional epic songs, some dating back more than a thousand years. Songs are frequently sung to accompaniments played on the balalaika, a three-stringed, triangular guitar. Also popular among all levels of Russian society are classical music, painting, literature, and chess.

A typical Russian meal has four courses and often includes borscht, a red beet-and-meat soup served with sour cream. Among other traditional fare are *ikra* (caviar) and blini (small buckwheat pancakes). Sweetened hot tea, vodka, and brandy are typical beverages. Russian friends and family members congregate to celebrate the feast days of the Orthodox liturgical year as well as life-cycle rituals, including weddings and school graduations.

Ukrainians

Population: 34.5 million
Location: Ukraine
Language family: Slavic

Ukrainians are an amalgam of Slavic tribes who settled the vast grassy plains, or steppe, west of the Black Sea. By medieval times, the area around Kiev in the north had gained supremacy, and a union known as Kievan Rus' dominated the region between the 9th and 13th centuries. At that time, the people were known as Ruthens. A Ukrainian republic did not exist until the fall of tsarist Russia in 1918; it was short-lived, however, and another was not formed until the end of World War I, when Soviet leaders made Ukraine a socialist state. In 1991, Ukraine became an independent nation.

Today, approximately 34 million Ukrainians make up about three-fourths of the country's population. They speak Ukrainian, a language that belongs to the East Slavic group and is written in the Cyrillic alphabet. Since the end of the Soviet era, this language, along with other aspects of Ukrainians' ethnic identity, has flourished, while Russian influences and language use have declined. This resurgent Ukrainian identity and independence from Russia was evident in the 2000–2005 Orange Revolution and in the 2014 Euromaidan uprising.

Blessed with rich soils, Ukraine was regarded as the breadbasket of Europe until the 1986 nuclear meltdown at Chernobyl rendered some of the region's agricultural lands radioactive for many years to come. People continue to farm on lands that were unaffected,

UKRAINIANS A farmer oversees the work of combines on a Ukrainian wheat field.

growing grains, fruits, and vegetables and raising livestock. Ukrainian farmers live in villages that tend to be large, often holding more than a thousand residents. Until the 18th century, it was not uncommon for three generations to live together; since then, however, the nuclear family has been the norm in both rural and urban areas, which contain about two-thirds of the country's people. Industry, centered in the east, includes the mining of coal and mineral resources.

Most people follow a branch of the Eastern Orthodox faith. Because of their proximity to the Near East, Ukrainians became Christians early on, although the prince of Kievan Rus' did not officially adopt the religion on behalf of the state until A.D. 988. Of the many Christian holy days that Ukrainians celebrate, Easter remains foremost, surpassing Christmas in importance. Pre-Christian beliefs continue to influence observances of life passages, especially death, but weddings exhibit the fullest expression of Ukrainian life-cycle recognition.

Ukrainians are an effusive people who often greet others exuberantly with hugs, vigorous handshakes, or triple kisses. They entertain readily, frequently turning routine events into parties, at which toasting is a common form of recognition. In Ukrainian rituals, food plays an important role, as can be seen in the variety of breads used to commemorate different occasions. Traditionally a dinner is not considered complete unless borscht is served.

At least one form of Ukrainian folk art elicits worldwide recognition: *pysanky,* the Ukrainian name for an intricate hot-wax-removal technique for coloring Easter eggs, which results in delicate lines and brilliant colors. The art dates back to pre-Christian times but readily became infused with Christian symbols and meanings. Elaborate embroidery applied to clothing and everyday textiles is another typical folk art. Ukrainians are also enthusiastic dancers and singers, known especially for their unique New Year's carols.

Central Europe

Stretching from the Baltic Sea in the north to the Adriatic Sea in the south, central Europe represents a wide band of lands and cultures. Rivers begin in the Carpathian Mountains, curving through the heart of these lands, and flow in all directions. The greatest of the rivers in central Europe is the Danube, 1,770 miles long, which begins at two headwaters in Germany and flows through central Europe, forming several national boundaries, and finally emptying into the Black Sea.

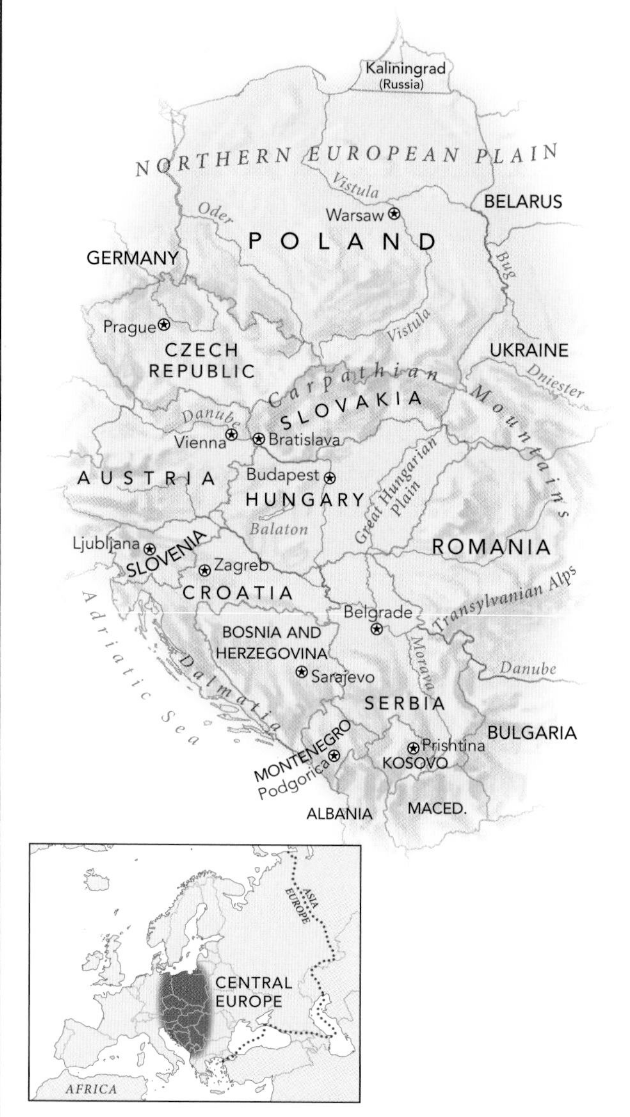

Bosnian Muslims

Population: 1.3 million
Location: Bosnia and Herzegovina
Language family: Slavic

Of all the countries that once formed Yugoslavia, Bosnia and Herzegovina (or simply Bosnia) is probably the most unusual. Distinctions there tend to be made on the basis of religion and culture, not ethnicity. About 40 percent of the total population are Muslims, 31 percent are Serbs who follow the Eastern Orthodox religion, and 15 percent are Roman Catholic Croats. All speak what was known before the breakup of Yugoslavia as a Serbo-Croatian language.

By the seventh century A.D., south Slavic peoples populated the area that is now Bosnia. During the long occupation by Ottoman Turks, from 1328 to 1878, many Bosnians converted to Islam from Christianity (in particular from a heretical sect called the Bogomils), because doing so was often a way to obtain economic or social advantage with Turkish officials. Bosnian Muslims were landowners and merchants then but tend to be urban dwellers now. The Turks strongly influenced other aspects of Bosnian culture, including architecture, cuisine, music, poetry, and dance.

Typically forgoing orthodoxy or fundamentalism, Bosnian Muslims (also known as Bosniaks) mostly follow the Sunni form of Islam, although Sufism, or Islamic mysticism, is also practiced. Muslims observe the month-long fast of Ramadan, which ends with a three-day feast known as Bajram (Id al-Fitr elsewhere). Bosnian Muslim women do not wear the all-enveloping coverings called *chadors,* but some wear raincoats and scarves for public occasions.

During the oil shortages of the 1970s, Yugoslav president Josip Broz Tito highlighted the religion of Bosnia's Muslims to obtain favorable terms from Saudi Arabia and other oil-producing countries; then, as fundamentalism took hold in the Middle East, Tito backed away from this approach, fearing its potential divisiveness in Yugoslavia.

With the breakup of the former Yugoslav federation in 1991, a great deal of religious and ethnic hatred has been released, much of it directed toward Bosnian Muslims. War broke out in 1992 as Serbs vigorously pursued a policy of "ethnic cleansing"—raping and

BOSNIAN MUSLIMS Keeping vigil for a relative, a Bosnian Muslim girl sits among the coffins of recently identified victims of the 1995 Srebrenica massacre.

murdering Muslims and forcing them from their towns—with the goal of forming a "greater Serbia."

Intervention by the UN and NATO, along with the Dayton Peace Accords signed in 1995, brought about some semblance of stability, but much of Bosnia struggles to rebuild. In the once beautiful city of Sarajevo, where many people were rendered homeless, residents are trying to rebuild their lives and their city. The Muslims have been effectively isolated, though, by a Croat alliance with Croatia and a Serb alliance with Serbia. At least 40 percent of the population is unemployed, and many young Bosnians of all faiths have left their war-torn homeland in pursuit of a better future. An estimated two million Bosnians have left, whether through expulsion, war, or voluntary migration, and live in diaspora in the United States, Canada, Germany, Sweden, and elsewhere.

Croats

Population: 4 million
Location: Croatia
Language family: Slavic

In the 7th century a south Slavic group known as the Croats settled in the Dalmatian region of the Adriatic coast, in the northern area of present-day Croatia, and in parts of Bosnia and Herzegovina. A Croatian state began to form in the 9th century and endured until the early 12th century. Strategically situated at a crossroads for Western and Eastern Europe and for Western and Eastern Christianity, the Croats sustained years of domination by Hungarians, Austrians, Italians, French, and the Turks; they were, however, more successful in containing the expansion of the Turks than others whose territories fell to the invaders. In the 19th and early 20th centuries, Croats resisted joining forces with other south Slavs, preferring their independence, and only reluctantly did they decide to become part of the newly created Yugoslavia following World War I—all the while remaining suspicious of Serbian expansion and control.

During the era of President Josip Broz Tito, Croatian nationalism was thwarted. But after Tito died in 1980 and communism began to wane, Croats voted for independence in 1991. They did so despite aggression from Serbs, who controlled the government and the army, even though they made up only about 12 percent of Croatia's population. After years of fighting, peace has been restored. Nevertheless, the country must contend with unemployment, now around 17 percent but as high as 23 percent in the past decade.

The Croatian language, which is spoken by the vast majority of people inside Croatia and by many outside as well, has three major dialects. It is written in the Roman alphabet and contains a number of loanwords, mainly from Latin and German.

Proverbs in conversation —torches in darkness.

BOSNIAN PROVERB

About 86 percent of Croatia's people are members of the Roman Catholic Church, and about 5 percent are members of the Eastern Orthodox Church, also adhering to the teachings of the pope. Catholicism has largely defined the country's culture, as it has in other parts of the former Yugoslavia, although the religion was suppressed during the communist era. While Carnival and Easter are important religious holidays,

CROATS In Zagreb, Croatian soccer fans anxiously await the outcome of a World Cup match between Croatia and Brazil.

Croatians also look forward to the celebration of Badnjak, or Christmas Eve, and their tradition of lighting the yule log.

The country of Croatia, which joined the European Union in 2013, is made up of 115 communes *(općina),* carryovers from socialism. Each commune comprises several villages and its own local government. A majority of the Croat people work in the service sector; about 27 percent work in industry; fewer than 5 percent work in agriculture. Croatians build their lives around family and a gregarious café-type culture, with much socializing in public places and homes.

Czechs

Population: 10.6 million
Location: Czech Republic
Language family: Slavic

In describing the modern Czech people, it is inevitable that they be compared with Slovaks, the ethnic group with whom they were politically united from 1918 to 1993 (except during World War II). Ethnically, the Czechs are an amalgam of Celtic, Germanic, and Slavic peoples who accepted Christianity in the 9th century. In 1085, they formed the kingdom of Bohemia, which was centered in Prague. By the 14th century, this kingdom oversaw the greatest expression of political and cultural development in central Europe. In 1526, a vacant Bohemian throne was filled by an Austrian Habsburg, whose house continued to control the kingdom until the end of World War I, when the Czechoslovak Republic was created.

Adolph Hitler annexed the republic's provinces of Bohemia and Moravia 20 years later, and Slovakia became a Nazi puppet state. After World War II, communists controlled a reunited Czechoslovakia until the "Velvet Revolution" under dissident playwright and political activist Václav Havel ushered in democracy. Leaders among the Czechs and the Slovaks disbanded the union in 1993, and since then Czechs have fared better than Slovaks at effecting economic and social progress.

In general, the Czechs have made a rapid transition to a market economy, with a long history of industrialization translating into highly regarded ceramics, glassware, and textiles. Agriculture is also heavily mechanized, and farms are very productive. The Plzen area, 50 miles southwest of Prague, is known worldwide for its beer production.

Over the years, Czechs have been far more urbanized than Slovaks, and today, many live in small towns and in the largest cities, Prague and Brno. People may also have *chaty*, or country homes, which they visit on weekends and holidays. About 10 percent of the population is Roman Catholic, and adherence to Catholic practice often tends to be quite relaxed. Evangelical Lutherans make up about 20 percent of the Czech population.

Most Czechs are part of small nuclear families, which may expand at some point as adult children assume the care of their aged parents. Many families make time in the middle of the day to enjoy their main meal, traditionally heavy in meats, game, and the characteristic *knedliky*, or dumplings. Current trends, however, seem to indicate that people are trying to consume lighter, healthier meals. Traditional folk costumes of embroidered shirts, pants, and vests for the men and embroidered blouses, skirts, and aprons for the women are worn mostly in the more rural southern areas.

CZECHS Czech racers participate in the 2015 Color Run, a fun race promoting peace and health, in Prague.

Youth is the gift of nature,
but age is a work of art.

POLISH PROVERB

The Czech language, like Slovak, is part of a cluster of Slavic tongues written in the Roman rather than the Cyrillic alphabet. Though Czech and Slovak are mutually intelligible, they are actually two distinct languages. Czech makes a distinction between formal or written communication and everyday speech.

According to a common saying, "Every Czech is a musician." While folk songs and classical music remain popular, rock has entrenched itself in the Czech culture, especially in Prague, and continues to be the medium of political protest.

Magyars

Population: 9.8 million
Location: Hungary
Language family: Ugric

Hungary's Magyars stand out amid the mostly Slavic peoples who live around them. As a result of population shifts occurring after World War II, Magyars now make up about 85 percent of the population of Hungary, establishing their country as one of the most ethnically homogeneous in the area.

Magyars first settled the Carpathian basin in the ninth century. They descended from a tribal people with origins in the forests between the Volga River and Ural Mountains. Their language, also called Magyar, is today spoken by some 12 million people worldwide; it belongs to the Ugric branch of the Finno-Ugric

MAGYARS Magyars make good use of the Széchenyi Baths, thermal mineral pools in Budapest's City Park.

family and is distantly related to Finnish and Estonian. Used interchangeably with *Magyar,* the name *Hungarian* probably derives from the Turkic words *on ogur,* or ten arrows, a reference perhaps to the number of Magyar tribes.

In A.D. 1000, pagan Magyars embraced Christianity under King Stephen I, who gained legitimacy for a Hungarian monarchy. About 37 percent of Hungary's people are now Roman Catholic and 25 percent are Protestant. Centuries of invasion and domination by the Turks, who conducted massive conversions to Islam, and the Austrian Habsburgs ended in the late 1800s, when Hungary entered a joint monarchy with Austria. World War I undid that alliance and greatly reduced Hungary's territory. The post–World War II era saw the development of a communist state, which in 1956 crushed a popular pro-democracy revolution quickly and harshly, causing hundreds of thousands of Magyars to leave the country. The communist regime ended with the declaration of a republic in 1989, although some socioeconomic changes had begun in the 1970s.

In the past 50 years, the Hungarian population has shifted from rural to urban, and even in areas where agriculture is predominant, many households derive income from both industry and farming. More than three-fourths of the women work outside their homes, and one in four workers of both sexes commutes a considerable distance. Holding down more than one job is not unusual here as people try to make ends meet or to afford a more comfortable lifestyle. Families tend to be smaller than in the past, and many Hungarian couples eventually divorce. About a sixth of all families live below the poverty line, and the suicide rate is one of the highest in the world.

Although Magyars are becoming more secular, they continue to celebrate significant days of the liturgical year as well as benchmarks in the agricultural calendar. As with other expressions of traditional culture, the distinctive Hungarian folk music, based on a pentatonic (five-note) scale, is in decline. Composers Béla Bartók and Zoltán Kodály preserved the best examples of this music form in the early 20th century. Hand kissing, a formal gesture still practiced by older men, has been replaced by the younger generation with the verbal greeting, "*Csókolom*—I kiss it."

Poles

Population: 36.6 million
Location: Poland
Language family: Slavic

Poland's physical geography, a low-lying plain open at both the eastern and western ends, has worked against its security for millennia: It is a region too easy for others to enter and exit. The earliest Slavic settlers most likely came through the eastern portal, although they left no evidence of when or from where they came. They have been an established presence in the region now called Poland since the eighth century. Embracing Christianity in the tenth century, the Polanie, or "people of the fields," eventually created a nation. In their lives and culture, Polish ethnicity and Roman Catholic Christianity have become inextricably linked. Poles prefer to consider themselves western Europeans, not eastern, and they do share a Western orientation.

Today, 98 percent of Poland's people are ethnic Poles; some 87 percent of them are Catholic. Poland's Jews, about 3 million strong before the Holocaust, now number only about 10,000. Other ethnic minorities

POLES Polish seniors participate in the annual Corpus Christi procession in Lowicz, Poland.

include Lithuanians, Belarusans, Ukrainians, Roma, and Germans—cultural groups found in Poland's neighboring countries.

The strong moral authority and political clout of the Catholic Church provided the support needed to carry the Poles through the difficult periods of their history. The Poles have undergone periods of Austrian, Prussian, and Russian domination from the 18th through early 20th centuries and then moved into an era of Soviet control following World War II.

A tree leans on a tree,
a man on a man.

SLOVENE PROVERB

Polish is a western Slavic language written in the Roman alphabet, with diacritical marks and characters used to represent unique sounds. There are a number of dialect groups, although education, migration, and the influence of the media have produced a standardized written and spoken Polish language.

Etiquette and polite formality are hallmarks of Polish social behavior. Close friends often greet each other with a triple kiss, and men may kiss women's hands at meeting and leave-taking. Such deference does not extend to home life, however. Although about half of all Polish women work outside the home, Polish men tend to avoid any involvement in housework or child care.

Farmers, gentry, workers, intelligentsia, and, recently, the upwardly mobile Soviet-era *nomenclatura,* or new class, form the broad categories of class among the Poles. In the late 20th century, the first four of these groups, along with the Catholic Church, came together to form the formidable union Solidarność (Solidarity). Outlawed through the 1980s, Solidarity nevertheless won a majority in the 1989 election. A year later Solidarity's leader, Lech Walesa, became president, and he initiated a plan to impose a sort of shock therapy to jolt Poland into a market economy. The plan failed, however, and resulted in inflation and high unemployment.

Today, industries employ about 30 percent of all Poles. Largely privatized, they are helping the Polish economy make new strides forward. During the era of Soviet control, farmers managed to avoid collectivization, and farms mostly remained in private hands. Holdings still tend to be small, and farms provide a livelihood for about 13 percent of the population.

The Poles' strong Catholic devotion inspires regular church attendance, at a rate estimated at about 75 percent of the total population—perhaps the highest rate of observance of any Catholic country in Europe. The Poles hold the Virgin Mary in special regard as their patron, honoring her with many feast days and making pilgrimages to Marian shrines such as that of the Black Madonna at Częstochowa.

The church calendar guides Polish holiday celebrations, with Christmas and Easter the most important of the year. Christmas Eve dinner is a sedate event featuring an odd number of meatless dishes as well as a thin white wafer, offered to each person with well-wishes and symbolizing reconciliation. The Easter meal often includes a festive roasted suckling pig served with a dyed red egg in its mouth.

Everyday Polish meals incorporate bread, meat, and plenty of potatoes—Poles consume an average of 300 pounds of potatoes per person each year. Potatoes also go into the manufacture of vodka, which Poles often claim as their own invention.

Some 13 million Poles live outside Poland, with large groups in the United States, Canada, and the United Kingdom. By comparison with the expatriates, Poles in Poland have a modest standard of living. Many still don't own automobiles, although public transportation services are extensive.

Serbs

Population: 5.9 million
Location: Serbia and Montenegro
Language family: Slavic

About 5.9 million Serbs form the ethnic majority in Serbia, one of the two republics that constitute the new, self-proclaimed Federal Republic of Yugoslavia; the other is Montenegro, also primarily a Serb country. In addition, there are Serb minorities in Croatia and in Bosnia and Herzegovina. Serbs speak Serbian, a south

DEEP ANCESTRY

THE SURVIVOR

Roberto Pasqualetti was born in Italy, and today lives in Switzerland. As far as he is concerned, he is completely Italian on both sides of his family—his father's family came from central Italy, while his mother's side hails from central and southern Italy. When he read about the Genographic Project in *National Geographic* magazine, he was curious about what it might reveal about his Italian roots.

DNA Results

His results were both reassuring and surprising. While his regional percentages are typically Italian, with a majority of the Mediterranean component, his mitochondrial DNA and Y-chromosome told a slightly different story. Roberto was especially intrigued by the spread of his lineages, including his mitochondrial haplogroup H1a1, into northern Europe. While both of his lineages are found at somewhat lower frequencies in Italy, the reason for their high frequency in northern Europe reveals an interesting chapter in European prehistory.

During the last glacial maximum (LGM), between 15,000 and 20,000 years ago, massive ice sheets covered northern Europe—Scandinavia and the British Isles (which were at that time joined to continental Europe) were all under ice. The European population at that time was largely confined to the southern periphery, near the Mediterranean, where the climate was more hospitable. As the Ice Age ended, 14,000 years ago, the ice sheets covering northern Europe began to melt. Game began to migrate northward, and with it came humans.

The genetic patterns in Europe reflect this migration, with western Europe connected to LGM populations living in southwestern France and Spain, the so-called Franco-Cantabrian refugium, and central/eastern European populations connected to LGM populations living near the Black Sea.

Roberto Pasqualetti, mitochondrial DNA haplogroup H1a1

Maternal H1a1: Once the glaciers retreated, central and northern Europe was populated from the south and east.

Widespread Lineage

Roberto's maternal lineage is part of one of the most successful and widespread mtDNA types in Europe, haplogroup H. His subtype, H1, seems to have been part of the repopulation of northwestern Europe from the Franco-Cantabrian refugium (see above). Roberto's specific lineage, H1a1, is widespread in western Europe and is also found in Italy. It's possible that his ancestors could have entered the Italian peninsula relatively recently, with the conquest by the Germanic Lombard tribe in the mid-first millennium A.D. Consistent with possible explanations, Roberto also has a Y-chromosome lineage that is more common in northern European populations. A Genographic Project study of genetic patterns in Italy published in 2014 found that the discontinuity between northern Italy—the Lombard-controlled region—and southern Italy is still visible in the genetic lineages of contemporary Italians.

—*Spencer Wells*

Slavic language mutually comprehensible with Croatian, but written in the Cyrillic alphabet.

Most people belong to the Serbian Orthodox Church, a self-governing body that has wielded a great deal of political power throughout the republic's history. In recent ethnic conflicts in the region, Serbian Orthodoxy has been drawn even further into the cause of Serbian nationalism.

During the sixth century A.D., Serbs and their flocks moved from the Carpathian Mountains into the Balkan Peninsula and occupied an area that stood to the east of the territory of the Croats, another South Slavic people. Within three centuries, the Serbs had established a state that would expand into a powerful kingdom by the mid-1300s. Internal struggles so weakened Serbia that Ottoman Turks were able to conquer the region and hold it for more than 400 years. Many Serbs fled west into the Dinaric Alps. A strong sense of nationalism prevailed, leading to several revolts by rival Serbian families and eventually forcing the Turks to grant autonomy to the Serbs in the early 19th century. Interfamily rivalries remained a theme of Serbian history into the 20th century.

Following World War I, the Serbs believed they had an opportunity to create a greater Serb kingdom, but what came about was a more inclusive state—the former Yugoslavia—uniting the south Slavs. The country became communist at the end of World War I, but under Josip Broz Tito, Yugoslavia withdrew from the Soviet alliance and forged a brand of communism that allowed more worker control and a greater degree of prosperity. President Tito's death in 1980 and communism's decline set the stage for the dissolution of the Yugoslav federation in the last decade of the 20th century.

Serbian nationalism and long-harbored animosities against other groups, such as the Croats and the Muslims, engendered aggression in Croatia and Bosnia. In Serbia, violence ensued against Albanian Muslims who formed the majority population in the autonomous province of Kosovo. The conflicts took tens of thousands of lives, led to large-scale movements of peoples, and brought worldwide sanctions against

SERBS The Serbian Orthodox Christmas, observed in January, provides an occasion for candle lighting in Belgrade's St. Sava Church.

the Serbs and their leader, Slobodan Milošević. His ouster in 2000 and the presence of UN and NATO overseers offered some hope of less conflict for the Serbs, their compatriots, and their neighbors.

Before World War II, most Serbs made a living through agriculture, but today many families have a mixed economic base involving farming and industry. As a result, some of the tasks once performed just by men have been taken over by women. The *zadruga*—extended family household composed of several generations in the male line—continues to survive in Serbia, though not in most other south Slavic communities; it is seen more often in rural areas than in urban ones.

Also important is the *vamilija,* or lineage traced from a common male ancestor. Vamilija members share a last name and venerate the same patron saint, whose annual feast day is one of the most important celebrations for the family. In this strongly male-dominated society, being heirless in the male line is considered a tragedy. A man without an heir can bring in a son-in-law to inherit his land and possessions, but such a step is almost always regarded as an unfortunate compromise.

Serbs have long been known for their grand tradition of oral epic poetry, in which bards retell heroic tales accompanied by a single-stringed instrument called a *gusle.* The people like to socialize outside the home, and in cities and towns they frequent sidewalk cafés for coffee and conversation. On a daily basis, their diet generally consists of bread and a hearty stew made with a lard base. Lamb appears on festive occasions such as Easter.

Slovaks

Population: 4.7 million
Location: Slovakia
Language family: Slavic

A Slavic people, Slovaks first inhabited the area between the Carpathian Mountains and the Danube River around the fifth century A.D. As residents of the former Czechoslovakia, they occupied the easternmost third of the country. The Slovak language—like Czech, a closely related tongue—belongs to the West Slavic group and is distinctive for its inflection and large number of words without vowels.

SLOVAKS A Slovak farmer uses horse power to harvest potatoes on a farm in eastern Slovakia.

After a period of hegemony by the Moravian Empire, named for the Morava River Valley, the Slovaks endured a series of invasions by the Magyars, or Hungarians, who controlled Slovakia from A.D. 907 to the end of World War I. The Magyars themselves were incorporated into the Austro-Hungarian Empire in the 16th century. Through a "Magyarization" program, the empire tried to suppress Slovak language and culture and, as a result, Slovak expatriates joined with their Czech counterparts at the end of World War I to press for the creation of Czechoslovakia. Except for the years from 1939 to 1945, this union persisted, albeit under communist control following World War II. In 1989, the so-called Velvet Revolution led by Václav Havel overthrew the communist regime, and about four years later, the Czech and Slovak republics became separate entities.

Slovaks traditionally have been a more rural people than the Czechs. Their hamlets and villages still have many of the simple one- and two-room homesteads that, slightly expanded, would often house three generations of a Slovak family. Rural families typically farmed, growing rye, wheat, corn, potatoes, sugar beets, and wine grapes. The steel, chemical, and aluminum industries established during the communist years have declined following the dismantling of the Soviet Union.

Much socializing takes place along gender lines, with men gathering in bars at night and women visiting at home. This is especially common in rural areas, as is adherence to pre-Christian beliefs and to superstitions about such things as the giving of flowers; cut flowers, for example, should be presented to one's host in an odd number; even-numbered bouquets are appropriate only at funerals. About 62 percent of Slovaks are Roman Catholics, who practice a much more observant form of Catholicism than their Czech neighbors, while about 8 percent are Protestants. A once sizable Jewish population was annihilated during the Holocaust.

Slovaks are a rather formal people, with ritualized greetings that include using a person's professional credentials in a form of address. Like Czechs, many are avid musicians who enjoy both vocal and instrumental forms of folk music. Folk dancing is a favorite activity as well. It is quite common to see traditional folk dress at festivals and on occasions such as marriage. For men, this means dark woolen suits, blouses, and knitted caps; for women, full skirts, aprons, and scarves. Celebrations are also good times for enjoying the traditional *bryndzové halusky,* potato dumplings served with sheep's cheese.

Slovenes

Population: 1.9 million
Location: Slovenia
Language family: Slavic

Slovenes are a Slavic people who have inhabited the area for more than 13 centuries. They speak Slovene, an old South Slavic language that has many dialects and subdialects. Together with the Croats, the Slovenes provided a catalyst for the independence movement leading to the 1991 dissolution of the former Yugoslavia. In many respects, the Slovenes have emerged from the country's breakup in a stronger position than the other ethnic groups who lived there. These people have long had a Western orientation and a higher standard of living, and that status has only been enhanced since independence.

Almost two million Slovenes inhabit the northwestern corner of the former Yugoslavia. Their new nation, Slovenia, is quite hilly and contains a sizable area of limestone karst; both geologic realities make the country unsuitable for large-scale agriculture and to some extent reliant on grain imports. Nevertheless, wheat, rye, oats, barley, and potatoes are grown and livestock is raised. Forestry, industry, and tourism are also major economic activities. Today, about 50 percent of the Slovenes live in urban areas. The largest and most populated of their cities is Ljubljana, the capital of Slovenia.

After joining a Slavic union in the seventh century, the Slovenes came under Frankish rule in the late eighth and early ninth centuries and were heavily converted to Christianity. For much of Slovenia's history, though, the people were under German control and influence, then Austrian and Austro-Hungarian rule until World War I, when Yugoslavia—"land of the southern Slavs"—was created. Communists took over after World War II, but Yugoslavia managed to forge significant regional autonomy.

During the communist era, the Roman Catholic Church fared better in Slovenia than it did in other areas of Yugoslavia. Today, its members make up more than half of the population, while a small proportion of the country's people are Orthodox or Protestant. For centuries, Catholicism has been inextricably intertwined with Slovenian culture. While regular attendance at Mass has dropped off, the church continues to influence cultural and life-cycle celebrations. Slovenes enthusiastically celebrate the pre–Lenten Carnival season, which they call *pust,* and combine Christian and pre-Christian elements in their festivities. Both children and adults don costumes at this time, and children go door-to-door in search of treats from their neighbors.

Music and dancing, both folk and modern, play a large role in Slovene culture. Choral music is quite popular and is supported by hundreds of different singing groups. For many Slovene people, entertainment involves family activities and visiting friends, and it is not unusual for city dwellers to spend weekends out in the country, helping relatives with their agricultural chores and relaxing together afterward. Many urban families also have small country cottages of their own.

Potatoes are frequently featured in Slovenian cuisine, as are breads and pastries. *Potica,* a festive bread made with either a sweet or salty filling, appears during holidays, especially at Christmas and Easter. So, too, do copious amounts of alcoholic beverages. In recent years, alcohol consumption has risen by 25 percent, and alcohol abuse is a recognized problem for many Slovenes, regardless of age or gender.

SLOVENES An orchard full of peach trees gets a careful pruning by this Slovene farmer.

Southern Europe

Southern Europe, considered the cradle of ancient Western civilizations, encompasses Italy, Greece, and Turkey, together with Albania, Macedonia, and Bulgaria. This is a region defined by water, its intricate coastlines outlining the Tyrrhenian, Adriatic, Ionian, and Aegean Seas, as well as the northern span of the Mediterranean. Thousands of islands populate these waters, from those with histories and cultures all their own—Malta, Sicily, and Crete—to tiny, uninhabited rock outcroppings. Greece alone contains nearly 10,000 square miles of islands, some—like Samothrace, Lesbos, and Ithaki—enjoying colorful histories as part of Greece's mythical past. Craggy mountains rise up from the shore of both landmasses, Italy and the Balkan Peninsula. Albania, Macedonia, and Bulgaria have distinct histories, even though their current boundaries have been recently drawn.

Albanians

Population: 3 million
Location: Albania
Language family: Albanian

For some ethnic groups, religious identity is paramount. Albanians, however, are fond of saying that they are, above all else, Albanian. Ethnically, they are a homogeneous group, probably descended from ancient Illyrians who controlled the Balkan area and parts of Greece in the 13th century B.C. Conquered by Romans a thousand years later, Albania was part of the Roman and Byzantine Empires before Ottoman Turks took control in the 14th century. Russia defeated the Turks in the late 19th century, and as the weakened empire came apart, Albania lost most of its territory to neighboring countries. Now, almost as many Albanians live in Bosnia, Macedonia, and Greece as in present-day Albania, whose population stands closer to three million.

The Albanian language, one of the nine original Indo-European languages, is written in the Latin alphabet. Although it was proscribed during Ottoman rule, it has made a strong comeback since the late 19th century. Its speakers make up two dialect groups—Ghegs in the north and Tosks in the south—with differences in pronunciation, but Tosk enjoys official status while Gheg remains mostly unwritten.

The religious lives of Albanians are difficult to discern at the present time. Before the communist takeover in 1946, about 70 percent of the people were Muslim (a reflection of Ottoman rule), 20 percent were Eastern Orthodox, and the rest were Catholic. From 1967 until independence in 1990, Albania was officially an atheist state. Today, many Albanian people are drawn to evangelical Christian groups; there is also the suggestion of a resurgence of Catholicism. Historically, Albanians have been tolerant of other religions, and in the era before communism, the Albanian clergy, in short supply nationwide, were respected and consulted by many individuals, no matter what their belief system. During World War II, Albania not only sheltered its own Jews but also took in those of neighboring

ALBANIANS In Saranda, three young construction workers relax during a lunch break.

countries, meaning that at war's end, its Jewish population was larger than the one it had at the beginning.

Albanians are a polite and formal people with elaborate habits of greeting and social interaction. Great respect is shown to elders, especially older men, to the point that boys and young men kiss their hands. A person's pledged word is his bond, and a pledge sealed with a handshake or embrace is the most binding of contracts. Traditionally, slights to a family's honor and transgressions such as theft and murder were answered with a blood feud. Since World War II and the communist era, women have gained status in this male-oriented country.

Women make the house, and then they destroy it.

GREEK PROVERB

About 35 percent of Albanians live in urban areas, 65 percent in rural. Despite valuable mineral resources and strong agriculture, Albania's transition to a market economy has been shaky, though not for lack of effort by its residents. Political repression continues to occur, even in the wake of democratic elections. Outside Albania, in the Kosovo region of Serbia, Muslim Albanians have been persecuted, killed, or forced out of their communities.

In cuisine, as in other aspects of Albanian life, the Turkish influence is strong. Bread is the main staple; lamb is the main meat. A Greek influence is seen in a multilayered stuffed pastry called *lakror.* The Mediterranean climate in the southern reaches of the country ensures an abundant supply of fruits and nuts. One of the most prized products is a three-star brandy named for George Kastrioti, a national folk hero from the 15th century, popularly known as Skanderbeg.

Greeks

Population: 10.7 million
Location: Greece, Cyprus
Language family: Greek

The more than ten million people of Greece occupy the mountainous southern tip of the Balkan Peninsula and many of the myriad islands off its convoluted coastline. There, their ancestors established the basis for Western civilization some 2,500 years ago, including the ideals of democracy. Conquest by the Macedonians in the fourth century B.C. ushered in a long era of foreign domination. Greece won official independence in 1832, although European royals still ruled. A bitter civil war between Greek royalists and communists occurred in the mid-20th century. Culturally, however, a unified Greek spirit prevails.

About 98 percent of Greece's population speaks Greek, an Indo-European language with more than 3,400 years of written records. Ancient Greek spread throughout Greece's colonies, even supplanting Latin as the language of the eastern Roman Empire. Although there are two major forms of Greek in general use—*demotic,* or the everyday language of conversation, and *katharevousa,* the language of government and media—the current trend favors the use of demotic in most contexts. A third form, *koine,* the Greek of the New Testament, is the language of church liturgy. Small numbers of Greeks speak other languages of the Balkans, as well as Turkish, and Greek children learn English beginning in the early grades.

GREEKS Evzoni, elite ceremonial guards, protect the Tomb of the Unknown Soldier in Athens.

The Greek Orthodox religion is a cornerstone of Greek culture, and it is the official religion of the nation. Services tend to be long, three hours or more, and it is not uncommon for people to wander in and out during a service. Maintaining Greece's many churches falls to parishioners, especially the women, who clean the buildings and often bake the bread to be served for communion.

The Greeks are a warm, effusive people who value relaxed socializing. They gather both privately and publicly for meals, coffee, and drinks. Conviviality often includes heated debate worthy of the ancient Greek philosophers as well as dancing. Ouzo, an anise-flavored beverage, and retsina, a resinated wine, may flow freely amid calls of "*Yiassas!*—Here's to health!"

Typical Greek foods include many lamb and seafood dishes, as well as dolmades—stuffed grape leaves—and moussaka, a casserole made with eggplant and lamb. The Greeks avidly consume many dairy products, including feta cheese and yogurt.

Annual celebrations follow the calendar of the Greek Orthodox Church, with Holy Week and Easter the most important. Greek Easter usually begins at midnight on Saturday with a candlelight vigil and service, followed by parties and a meal featuring roasted lamb. Instead of celebrating their own birthdays, Greeks mostly observe the day honoring the saint for whom they are named.

Greek social organization revolves around family, extended kin, and patronage relationships, with godparents and wedding sponsors playing important roles in an individual's life. Family inheritance passes to both males and females, and women traditionally receive their portion at marriage in the form of large and lavish dowries. A law passed in the 1980s tried, rather unsuccessfully, to limit the extravagances of this marriage practice. Newlyweds may live with the groom's family until they can afford their own home. Greeks consider owning a home, whether a hand-built, whitewashed house in a village or an apartment in Athens, the ultimate goal.

The Greek family will close ranks against other families if there is a dispute or if the family honor is violated. The concept of *philotino,* or love of honor, permeates Greek culture in both bold and subtle ways, guiding behavior, defining responsibilities, and contributing to an individual's or family's esteem or status. If breached, it can lead to long-lasting feuds and, on occasion, violence.

Many Greek businesses, large and small, are family enterprises. Fewer than 30 percent of Greeks farm, and they have moved from subsistence agriculture to growing cash crops, selling products such as grain and fruits. Greece remains relatively unindustrialized compared to other parts of Europe, although it supports a world-class shipping industry. Many Greeks have emigrated to northern Europe, North America, South Africa, and Australia in search of better prospects. In 2015, Greece was forced to adopt harsh austerity measures in an effort to repay its debts, and the economic outlook remains bleak.

Ethnic Greeks form the majority—some 80 percent—of the population on the island nation of Cyprus, which lies in the Mediterranean south of Turkey. The rest of the population are mostly Muslim Turkish Cypriots, who have claimed an unrecognized independence for the northern third of the island. As a result of conflicting claims, Cyprus has been a hotbed of civil strife and violence in recent times. Greek Cypriots share the language, religion, and other aspects of mainland Greek culture. They are mostly agriculturalists, and most live inland. Cyprus has few coastal settlements due to poor fishing and a history of pirate attacks.

Italians

Population: 61 million
Location: Italy
Language family: Italic

The Italians inhabit one of the most recognizable landforms on the planet, a boot-shaped peninsula that juts into the Mediterranean Sea. Etruscans, Germanic peoples, Greeks, North Africans, and Phoenicians all contributed to the Italian heritage. A topography consisting of 80 percent mountains or hills encouraged the development of regional cultures. These eventually became the city-states ruled by powerful individuals or families that would unite under the impetus of 19th-century nationalism to become the Kingdom of Italy in 1861.

Italy's population numbers more than 60 million; nearly all of them speak Italian, a Romance language derived from Latin, albeit in several hundred regional dialects. Over time, the Tuscan dialect has become the national standard. Along the borders with France, Austria, and Slovenia live populations that speak French, German, and Slovene, as well as pockets of Albanian speakers in the south.

One who loves does not smell the odor of garlic.

SICILIAN PROVERB

Italians tend to be warm and engaging, fixing the steadiness of their gaze on conversation partners and punctuating their phrases with touches and gestures. Family provides the anchor of their social life. Although kinship is reckoned equally on both sides of the family, there is commonly more attachment with the mother's kin, and newlyweds may live for a time with the bride's family. In recent decades, birthrates have fallen sharply, leading to predictions of a dramatic drop in population—perhaps by a fifth—by 2050.

In general, northern Italy is more prosperous and has a longer history of industrialization than the south. The north boasts the fertile Po Valley and also the industrial triangle of Milan, Turin, and Genoa, which, in addition to being Italy's major shipbuilding and financial and commercial center, turns out textiles, clothing, iron, steel, furniture, cars, airplanes, and motorcycles. Italian farms produce wheat, vegetables, olives, and grapes; Italy's wine production ranks near the top worldwide.

With Rome as the epicenter of world Catholicism, it is to be expected that some 80 percent of Italians would be Roman Catholic. What may seem surprising is that about 2 percent of Italians are Jews, a population that has endured since the days of the Roman republic. There are also small numbers of Muslims and Orthodox Christians.

Most Italians boast of the superiority of their regional cuisine, more specifically of their hometown's, and above all of their own mother's, cooking. While pizza and pasta have entered the international vernacular as Italian foods, these items do not begin to represent the culture's everyday cuisine, although in most parts of Italy, some kind of pasta dish appears at the day's main meal, which may also include soup, bread, and fish or meat. A family's status can be determined in part by how much meat they can afford.

Italian holidays call for festive meals and special food traditions. In addition to the numerous preparations for Christmas and Easter, holiday traditions include special bread for the feast of St. Joseph and various foods shaped like eyes in honor of the martyr St. Lucy. Family life revolves around these special occasions, although any visitor to an Italian home will be offered the hospitality of food and drink.

Italians live among some of the most splendid works of art and architecture of three millennia, starting with the frescoes of the Etruscans, continuing through the glories of Rome—exemplified by the Pantheon and the Colosseum—and by no means ending with the Renaissance and baroque wonders of Michelangelo, Leonardo da Vinci, and Gian Lorenzo Bernini.

Italy's population includes the five million inhabitants of Sicily, the Mediterranean's largest and most populous island. The Sicilians are an amalgamation of the island's invaders: Greeks, Romans, Carthaginians, Byzantine Greeks, North African Muslims, Normans, Angevin French, and Spanish. Some linguists classify the Sicilian language as a dialect of southern Italy, although it is not mutually intelligible with standard Italian. Most young Sicilians speak Italian as well.

Sicilians of the interior are mainly farmers, while those on the coast fish and also work in industry and services. A notorious fact of Sicilian life is the existence and power of the Mafia, known locally as Cosa Nostra, "our thing." Rising from lawless conditions centuries ago, families of organized crime are involved in the international drug trade and bogus government contracts and extortion. It remains difficult to convict Mafiosi because their members observe a strict code of silence.

ITALIANS Neapolitan pizza, famous around the world, is readied for the wood-burning oven in a Naples restaurant.

Western Europe

Powerhouse nations through history—Portugal, Spain, France, Germany, Austria, Switzerland, Belgium, and the Netherlands—join as members of the region of western Europe. An intricate coastline wraps around the region, edging the Mediterranean Sea, the Atlantic Ocean, and the North Sea. Mountain ranges punctuate the landscape and form natural boundary lines. France and Spain meet in the Pyrenees; France, Germany, Switzerland, and Italy share the Alps. Great rivers run north, south, east, and west, notably Germany's Rhine; France's Seine, Loire, and Rhône; and Italy's Po. Geophysically, Europe is an extension of the landmass of Asia, and it is as much because of its history and economic significance as because of its geographic character that it is called a continent.

Basques

Population: 2 million
Location: Spain, France
Language family: Language isolate

In their Pyrenees stronghold straddling the border between France and Spain, the Basques developed one of the most enigmatic cultures in Europe. Many scholars consider them the oldest ethnic group on the continent, saying that they lived in the area before Indo-European peoples arrived more than 3,000 years ago. Some suggest they are descendants of Upper Paleolithic Cro-Magnons who occupied the region about 30,000 years ago.

The Basque language, Euskera, occupies its own family and is not related to any other language spoken in Europe; scholars have looked for its roots in the Russian Caucasus and in North Africa as well. It is an extremely complex tongue, and a Basque tradition holds that the devil spent seven years trying to learn it but finally gave up. Nevertheless, Euskera has lent itself to a vast oral tradition kept alive by storytellers and by spontaneous versifiers called *bertsolariak*. Most Basques are bilingual, speaking French or Spanish and sometimes both.

Although the overwhelming majority of Basque people are Roman Catholics, a millennium has passed since the Basques inhabited a unified community. Today, they occupy three provinces in France and four in Spain—a homeland that includes the Pyrenees Mountains, the foothills of the Pyrenees, and the coastal plain along the Bay of Biscay.

An intensely independent people, the Basques have long managed to maintain some autonomy even when dominated by foreign powers. Under the 1939–1975 regime of Spain's Francisco Franco, however, Basque language and culture were vigorously suppressed, giving rise to the ETA—Euskadi ta Askatasuna, Basque Fatherland and Freedom—a strong,

BASQUES Dressed in traditional caps and kerchiefs, Basque men celebrate a religious festival in Vitoria-Gasteiz, Spain.

sometimes violent independence movement that has endorsed terrorist acts.

Traditionally, Basques worked as fishers, farmers, and shepherds. Noted mariners, some even crewed for Columbus and Magellan. Many Basque people now work in the area's large industrial sector, and while only 20 percent of Basques continue to farm, the *baserria,* or farmstead, remains a fixture in the rural areas inhabited by the Basque people.

Large three-story stone dwellings provide room for livestock on the bottom floor, living quarters for an extended family on the middle level, and storage for hay on the top.

One child within each baserria is designated the heir, the one who will receive the landholding at the time of his or her marriage. Males are favored, but females may also inherit. In the past, other siblings typically left the farm, sometimes settling for a religious life, and many also emigrated to other parts of Europe and beyond. A Basque village comprises 10 or 12 farmsteads, a church, a school, taverns, a town hall, and a handball court, often an extension of the church wall. A number of fast-paced ball games, including jai alai, originated in Basque country.

Bretons

Population: 3.1 million
Location: France
Language family: Celtic

A look at a map showing Brittany, a peninsula jutting into the English Channel and the Atlantic Ocean, gives a hint that this region's population stands apart from other ethnic groups in France.

The Bretons are a Celtic people, with ties to the British Isles dating back to migrations that occurred between the third and fifth centuries A.D. Over the years, numerous invasions thwarted their attempts to remain independent, and in the early 16th century, their land was annexed to France. Provisions for some autonomy, however, and distance from the state administration in Paris served to isolate and insulate the Bretons, preventing their assimilation into the French nation until well into the 19th century.

With some 2,100 miles of coastline, including islands, Brittany has a distinct maritime orientation and a regional cuisine in which various seafoods figure prominently. The economy has long been based on fish and crustaceans and, until recently, algae gathering. Brittany also relies on its strong agricultural base. For generations, Bretons have worked the lands of the interior, planting vegetable crops and raising livestock on farmsteads marked by dense hedges. Today, mechanization of farming has made Brittany the premier agricultural region of France, complemented by large food processing and agricultural machinery industries.

The traditional rural scene of Brittany was composed of scattered homesteads—rectangular granite buildings with roofs of thatch or slate and chimneys at their gabled ends—as well as small villages and larger settlements called *plous.* The largest plous were subdivided into *treviou̇.* Vestiges of this system survive in the many place names beginning with *Plou-* or *Tre-*, especially in northwestern Brittany.

The Breton tongue, today spoken by just over 200,000 people, belongs to the Brythonic branch of Celtic languages, making it closely related to Welsh and Cornish. Despite the decline in their spoken language, Bretons avidly foster their Celtic heritage and enjoy its many traditions, particularly folk music—which features a small bagpipe called a *biniou*—and folk dance. They routinely invite Celts from the British Isles to participate in their festivals.

Most Bretons are Roman Catholics, with their regional celebrations tending to be religious in nature. They show much devotion to hundreds of local saints, make pilgrimages to the region's nine cathedrals, and participate in festivals called *pardons.* These feature large processions, as well as traditional music and dancing, and they take on a secular atmosphere once religious devotions have been completed.

BRETONS Costumed Bretons dance in the town streets of Quimper, the cultural heart of Brittany.

Catalans

Population: 3.7 million
Location: Spain
Language family: Italic

Like their Basque neighbors to the west, Catalans retain a strong sense of identity and autonomy despite their longstanding incorporation into the national structures of France and Spain. These people reside in the Països Catalans, the "Catalan Countries," which historically included the Catalonia region of northeastern Spain, as well as Valencia, the Balearic Islands, the independent state of Andorra, and the French province of Pyrénées-Orientales. These people speak Catalan, a Romance language with ties to the Provençal language of southern France.

Spanish Catalonia has been an area of strategic importance for a long time. Union with Aragon in the 12th century and with Castile and León in the 15th century widened the region's political sphere but subjugated Catalan interests. The people later experienced repression of their language and culture under Francisco Franco's dictatorship, lasting from 1939 to 1975. Finally, in 1979, regional autonomy for Catalonia was established and the Catalan language was allowed to flourish again, fueling the cultural pride that has led to a renaissance of Catalan music and literature, especially poetry.

Catalans are overwhelmingly Roman Catholic, with the cycles of the church year defining cultural observances. Each village and town honors its patron saint at an annual festival, during which processions typically feature huge papier-mâché representations of ritual figures: *capgrosses* (bigheads) and giants, some 15 feet tall. All of Catalonia reveres Sant Jordi (St. George). On his feast day in April, which is also the official opening of the book publishing season, Catalans distribute his symbols of roses and books to their loved ones. Another feature of these Catalan celebrations is the *sardana,* the traditional dance whose stately cadences reflect the highly regarded concept of *seny,* a kind of refined good sense and self-realization at the heart of the Catalan self-image.

Economically, the region depends on a strong base of industry and tourism—seen at such places as Costa Brava, where hundreds of thousands of Europeans, including Catalan families, vacation each year. Despite a pastoral past, only 10 percent of the approximately seven million Spanish Catalans currently farm, while 25 percent reside and work in cosmopolitan Barcelona, the Catalonian capital.

CATALANS Catalan men build a traditional human tower during a festival in Tarragona, in northeastern Spain.

Dutch & Frisians

Population: 15.7 million
Location: Netherlands
Language family: Germanic

Some 15.7 million Dutch inhabit the Netherlands, or "lowlands," at the southeastern edge of the North Sea. These people are descendants of German, Frisian, and Frankish tribes who entered the area in pre-Roman times. Long tied to the fortunes of the Frankish and Habsburg empires, the Netherlands came into its own with the development of cities and subsequent involvement in commerce, trade, and colonial

DUTCH Strolling Dutch pass under the elegant Dom Tower of St. Martin's Cathedral in Utrecht's pedestrian center.

expansion. In the 17th century, the Netherlands entered a golden age of wealth and culture splendidly documented in the paintings of Rembrandt, Vermeer, and Frans Hals.

The Dutch language belongs to the Germanic family of Indo-European languages and is therefore related to English. Although regional dialects abound, they are mutually understood, their differences tempered by media influence. Dutch schoolchildren learn English as well, and many people, especially young urbanites, speak it fluently.

The Dutch are vigilant stewards of the land, much of which has been reclaimed from a former North Sea inlet, the Zuider Zee (now the IJsselmeer). Despite their renown for growing tulips and for other kinds of horticulture and agriculture, only 4 percent of the Dutch still farm. Many now work in services and in industries such as food processing, petrochemicals, and electronics.

Modest and reserved in public, the Dutch place much emphasis on a *gezellig,* or cozy, family life. The nuclear family *(gezin)* has long been the ideal, with families tending to be small and children generally receiving much care and attention. People take fierce pride in their homes and gardens, and an evening stroll reveals the old practice of *schmeren:* Household curtains are open to let passersby get a lingering glimpse of the respectable gezellig gezin within. The ubiquitous bicycle serves many transportation needs in a country that boasts 6,200 miles of bike trails.

Though noted for their racial and ethnic tolerance and for neutrality in war, the Dutch traditionally divide along lines of religion. Currently more than a third of the people are Roman Catholic, and about a quarter are Protestants, spread among six groups, the largest being the Dutch Reformed Church.

Religious affiliation forms the basis of *verzuiling,* a "pillarization" or segmenting of Dutch society into separate groups, all with their own schools, newspapers, political parties, labor unions, and other organizations. While divisions have weakened—and were always moderated by the Dutch emphasis on mutual respect—verzuiling still significantly influences rural communities.

He who hunts after bargains will scratch his head.

CATALAN PROVERB

The Frisians, contributors to the original mixture of populations who became the Dutch, also maintained their own culture in Friesland, a small region on the coast of the upper North Sea that includes several barrier islands.

Since the early days of settlement there, around 400 B.C., the Frisians have battled the sea, constructing and continuously maintaining the "golden hoop," a system of dikes designed for protection against frequent storms and flooding. Laws specific to Frisian life were codified by Charlemagne into the Lex Frisionum, recorded in the early ninth century A.D. Among them

GENOGRAPHIC INSIGHT

WAVES OF REPLACEMENT

When DNA from participants of the Genographic Project is analyzed in the laboratory, the goal is to place each lineage on a specific branch of the human family tree. Generally this is done through the analysis of genetic mutations, or markers, which when shared by people unite them into an ancestral clan called a "haplogroup." By analyzing both the geographic distribution and genetic diversity of markers within each haplogroup, geneticists are able to trace back to the group's most recent common ancestor, allowing us to pinpoint where their predecessors came from, and when.

The European Chapter

The emergence of modern humans in western Europe marked the appearance and spread of what archaeologists call the Aurignacian, a culture distinguished by significant innovations in tool manufacturing such as end-scrapers for preparing animal skins and tools for woodworking. These inventive early humans, named the Cro-Magnon after the cave where the first specimens were found in southwest France, also used bone, ivory, antler, and shells as part of their tool kit. The large number of archaeological sites found in Europe from around 40,000 years ago indicates that they were experiencing a population expansion, and jewelry found at several of the sites—often an indication of status—suggests a more complex social organization was beginning to develop.

The Cro-Magnon are responsible for the famous cave paintings found in southern France that depict animals like bison, deer, and horses in colors far more intricate than anything seen prior to this period, providing evidence for a sudden blossoming of artistic skills as they moved into Europe. Their emergence heralded an end to the era of the Neanderthals, a hominid species that inhabited Europe and parts of western Asia from about 250,000 to 40,000 years ago, and it is thought that the better communication skills, weapons, and resourcefulness of the Cro-Magnon probably allowed them to outcompete Neanderthals for scarce resources.

A Dutch woman wears patriotic orange shades during the celebration of King Willem-Alexander's birthday.

Ice Effects

But these early Europeans were hardly in the clear, and around 20,000 years ago, the climate window that had allowed their rapid diffusion from Central Asia slammed shut, expanding ice sheets forcing people to move south. During the glacial maximum, all of northern Europe, almost all of Canada, and the northern half of the West Siberian Plain were covered by huge ice sheets, and for a period of about 2,000 years, humans were restricted to isolated refuge regions in southern Iberia and the Balkans, areas close to the more temperate Mediterranean Sea.

Through these bitterly cold times, human populations dwindled. The reduction in population size also reduced the group's genetic diversity. Once temperatures warmed and the ice retreated, beginning about 12,000 years ago, people emerging from the southern regions had a better shot at passing their DNA on to subsequent generations. Migrations northward allowed people to recolonize areas that had been vacated during the Ice Age.

What can we say about the genetic lineages associated with these migrations? Until recently, the best interpretation was that the genetic patterns of the original settlers survived into the present relatively unchanged. However, ancient DNA has revealed a more complex pattern; in fact, genetic replacements have occurred throughout the past 10,000 years of European history, associated with shifts in the archaeological record. The Y-chromosome and mitochondrial DNA lineages we see in Europe today are a product not only of the earliest Paleolithic settlers but also of the spread of agriculture out of the Middle East over the past 8,000 years, as well as the migrations of Bronze Age peoples who are thought to have carried with them the Indo-European languages widespread in Europe today.

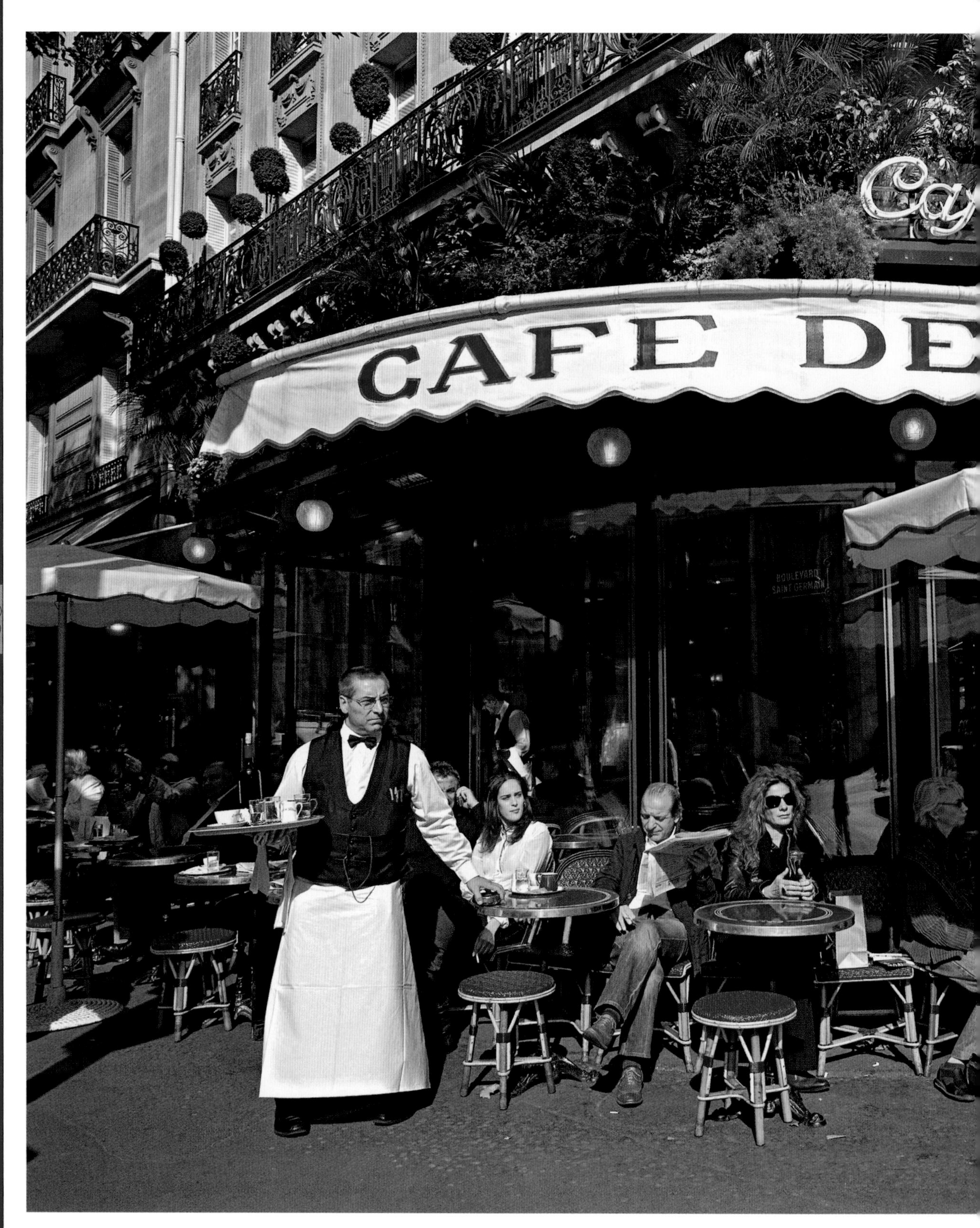
CAFE DE
BOULEVARD SAINT GERMAIN

is *bourrereplicht,* or "neighbor's duty," requiring that neighbors assist one another in a crisis.

Historically, trade was important in Friesland. Today, many Frisians participate in agriculture by growing crops or dairying, as well as by breeding their prized milk cows. The typical "head-neck-body" house, a fixture in the Frisian countryside, consists of living quarters attached to a large barn by a passage that contains a kitchen, milk cellar, and churning area.

The Frisian language is related to Dutch, and the region's 600,000 Frisians speak it mostly at home. With its 11 cities, Friesland has remained fairly independent of the rest of the Netherlands; among the Frisians, a modern independence movement persists.

French

Population: 60 million
Location: France
Language family: Italic

Descendants of Celts, Romans, and Franks, the French inhabit Western Europe's largest country, one of its oldest, having become a unified kingdom in the 15th century. The French language, which derives from Latin, originated in and around Paris. It now is spoken by virtually the entire population of France, as well as by hundreds of millions of people worldwide, in countries as diverse as Senegal, Canada, and India.

Shortly after the French Revolution, regional languages and dialects were suppressed and a mandate given that standard French be taught in all schools. Today, the Académie Française guards the purity of the language against outside influences, especially those of English. Nevertheless, the French do speak of *le parking* and *le weekend,* and the language of the Internet and social media further transforms standard French.

Although about 65 percent of the French are Roman Catholic, attendance at Mass is quite low on any given Sunday. About 2 percent of the French are Protestant, mainly Lutheran and Calvinist.

Following World War II, large numbers of immigrants from former North African colonies began to

FRENCH Café de Flore on the Boulevard Saint-Germain exudes the ambience of Parisian café culture.

arrive in France. Although these immigrants spoke French, their Muslim religious affiliation and their distinctions in dress and other customs have caused some conflicts in a larger culture that generally prizes conformity. Today, about two million French people are Muslim, more than half a million are Jewish, and a significant number are Buddhist, primarily immigrants from Vietnam.

Though regional and individual differences certainly exist, the French tend to be a formal people. They have a strong sense of correct behavior and procedures that they expect others to share. They also maintain distinctions of class based on occupation and education. Carrying formality into their home lives, they entertain guests in their public rooms while all other rooms remain off-limits, behind closed doors.

One should not let the grass grow on friendship's road.

FRENCH PROVERB

Since World War II, French families have become significantly smaller. To encourage population growth, the government offers subsidies for pregnant women and for children from birth until young adulthood, regardless of a family's financial worth. French towns and cities offer preschools—*écoles maternelles*—and nearly all children attend full time from age two on.

Much of French culture is geared toward food—its purchase, preparation, and consumption. French cuisine, with all of its regional nuances, benefits from France's fertile farmland, still abundant forests, and wide access to the bounties of the sea. Daily meals unfold in a ritualized way that usually includes soup or another starter, meat and vegetables, salad, cheese, and dessert, along with wine. On weekends, the main meal usually occurs at midday and can last for hours.

Many holidays, such as Christmas and Easter, are marked with a further elaboration of foods, wines, and other spirits. The French national holiday, Le quatorze juillet (the Fourteenth of July), also known as Bastille Day, commemorates the storming of the Bastille in Paris during the French Revolution and the overthrow of the monarchy. Bastille Day is celebrated with parades, dances, fireworks, and other festivities.

The French symbolize their republic as an idealized woman, called Marianne. In recent times she has been depicted in statuary modeled after famous actresses like Brigitte Bardot and Catherine Deneuve.

Although France is self-sufficient agriculturally, only 3 percent of the people are involved in farming; this corresponds to a trend that now finds four-fifths of the population living in cities and suburbs. But that 3 percent produce some 60 varieties of grapes, mostly used in winemaking, and 400 types of cheese.

About one-third of the French work in industry, and the majority—more than 70 percent—work in the service sector. The French workday tends to be long, due in part to a lengthy lunch break, and so weekends tend to become very important. Many urban French escape as often as they can to second homes that they own or rent in the countryside.

The French revere education. The whole country closely follows the yearly back-to-school transition, called *La rentrée,* which is chronicled on the nightly news with a countdown to opening day. The national government sets school curricula as well as the questions for the month-long *baccalauréat,* the nation's secondary school exit exams. Results of this stressful ordeal determine eligibility for a university education, which is free for French students. Higher scores qualify students for enrollment at that nation's more prestigious universities.

Germans

Population: 69.8 million
Location: Germany
Language family: Germanic

The reunion of East and West Germany following the dismantling of the Berlin Wall in 1989 brought all 82 million Germans together in a single nation for only the second time in almost 120 years. German-speaking peoples had occupied the land east of the Rhine River from the first century B.C., eventually becoming incorporated into the Holy Roman Empire. In 1871, amid strong stirrings of nationalism, Chancellor Otto von Bismarck of Prussia united

GERMANS Beer conjures romance for Germans attending the International Beer Festival in Berlin.

independent German cities and provinces in a nation-state. Following the German defeat in World War II, the Federal Republic of Germany was formed, and the eastern part of the country became the Soviet-controlled German Democratic Republic. In the years that followed, East Germans fled to West Germany by the thousands, a diaspora that was thwarted by a wall erected in the divided city of Berlin, beginning in 1961.

The German language is closely related to English and Dutch. Dialects of German fall into two broad groups: the High German spoken in the southern German highlands and the Low German spoken in the northern lowlands. Religious affiliation follows a similar geographic trend, with a mainly Protestant north and a largely Catholic south and Rhineland. Germany's Jewish population, about half a million before the genocide of the Holocaust, now numbers significantly fewer than 100,000.

The Christian religious calendar sets many annual celebrations, including Karneval or Fastnacht before Lent, as well as Easter and Christmas. Many towns sponsor a yearly Weihnachtsmarkt, a Christmas market, at which people can buy seasonal foods and holiday trimmings. A secular holiday on October 3 commemorates German reunification.

German cuisine incorporates a wide variety of starches, including bread, potatoes, and noodles, along with root vegetables. The favored meat is pork, often made into tasty sausages. World-famous German beers have been made according to the same laws of purity since the 16th century, which allow the use of only grain, hops, yeast, and water. Today Germans can choose their libations from more than 1,200 breweries in their country alone.

Germans typically prepare themselves for a career by seeking extensive education and training, whether the goal is to become a physicist or a plumber. A highly skilled labor force, respect for precision, and attention to detail contribute to German success in science and engineering, represented by the highly valued automobiles that roll out of German factories. The German industrial force contributes to the world's second largest export economy.

Although less than 3 percent of the population is engaged in farming, almost half of Germany's land is under cultivation, and the nation produces some 85 percent of its own food. Overall, Germans are well compensated for their work and enjoy a high standard of living that includes ample vacation time. *Gastarbeiter,* or guest workers, coming from Turkey, the Balkans, and other countries supplement the German workforce.

For centuries, the bucolic and impeccably maintained German countryside has stood in contrast to the country's highly organized and bustling cities, where commerce and trade flourished. Today, the central areas of many cities have been transformed into pedestrian-only zones where people can stroll, shop, and perhaps take in a concert in the city square. Germans also put a great deal of care and attention into their homes, and they relish family time, especially on the weekends. Many of them also join clubs and other interest-based groups in their local communities, and on a larger scale, they join unions and other advocacy organizations

Many Germans remain self-conscious about the horrifying events that occurred during the Nazi period of the 1930s and 1940s. As a nation and a people, Germans have spent the intervening decades trying to come to terms with the factors that allowed the Third Reich to take form and eliminating aspects of the regretted past. Display of the *Hakenkreuz,* or swastika, is now forbidden, and measures are in place to ensure that the media cannot be co-opted for propaganda purposes.

German reunification has come with its own set of concerns. There is still economic disparity, with higher unemployment and housing shortages in the east, and there are social implications as well. Germans still frequently differentiate between *Wessi* (westerners) and *Ossi* (easterners) in conversation, and some maintain that their diverse experiences on either side of the physical wall contribute to different mind-sets, a result of the *Mauer in den Kopfen,* or wall in the mind.

Portuguese

Population: 10 million
Location: Portugal, Brazil
Language family: Italic

The Portuguese occupy about a sixth of the Iberian Peninsula, clustered mainly in towns and cities and in rural settlements in river valleys and along the coast. They also inhabit the Azores and Madeira, islands in the Atlantic off the coast of western Africa.

The Portuguese people have descended from Iberian groups who entered the area in the third millennium B.C. and, later, Celtic peoples who entered the area after 900 B.C. The establishment of boundaries as they exist today occurred in the 13th century and solidified the Portuguese ethnic identity. These people displayed their maritime prowess several centuries later as explorers and colonizers of an empire that formed rapidly.

A fairly strong north-south distinction characterizes many aspects of Portuguese life. Northerners tend to be more traditional in outlook than the people living in the south, a region known as the Algarve. Foods grown in the north include corn, potatoes, wine grapes, and vegetables. Southern cash crops include wheat, olives, and cork. People along the coast fish, and *bacalhau,* or salt cod, is the national dish. Portugal also maintains a large fish canning industry.

Virtually the entire population speaks Portuguese, a Romance language based on Latin. On the whole,

PORTUGUESE Colorful trams provide popular transport for Portuguese in Lisbon.

DEEP ANCESTRY

THE METALLURGIST

I have, of course, tested my own DNA through the Genographic Project. I belong to one of the most common male lineages in Western Europe, known as R1b, which is consistent with my father's English ancestry. R1b is found throughout western Eurasia in varying flavors, depending on the precise set of markers. One sublineage is limited to Central Asia, while another is found in the Middle East and North Africa.

DNA Results

My own, defined by the marker U106, is one of the most common R1b sublineages in northern Europe, reaching frequencies in excess of 25 percent in England and the Netherlands.

R1b has undergone a radical reevaluation in recent years. When the Genographic Project was launched in 2005, it seemed likely that R1b had been in Europe since the Upper Paleolithic, at least 25,000 years ago, and its distribution was a result of the same postglacial recolonization of northern Europe as mitochondrial DNA H1a1 (see page 215). More recent work has suggested a younger age for R1b, and ancient DNA has provided even more insights: R1b is simply not found in ancient European remains prior to the Bronze Age, circa 5,500 years ago.

The latest research suggests that R1b expanded recently, within the past 4,000 years, along the routes of migration define by Bronze and early Iron Age cultures, from central Europe, northward and westward. The reason for its high frequency in the populations of western Europe is explained by the dominance that metalworking gave to the people undertaking these migrations—people such as the Unetice Culture of central Europe. It is likely that these same cultures spread Indo-European languages throughout western Europe, where languages belonging to a different family were widely spoken before the expansion of Indo-European. Today, Indo-European languages—including Romance languages such as Spanish and French, Celtic languages such as Irish Gaelic, Germanic languages such as English and Swedish, as well as the Slavic languages of eastern Europe—are spoken throughout the region. The only remaining non-Indo-European languages are Basque and the Finno-Ugric languages of Hungary and northeastern Europe.

Spencer Wells, Y-chromosome haplogroup R1b

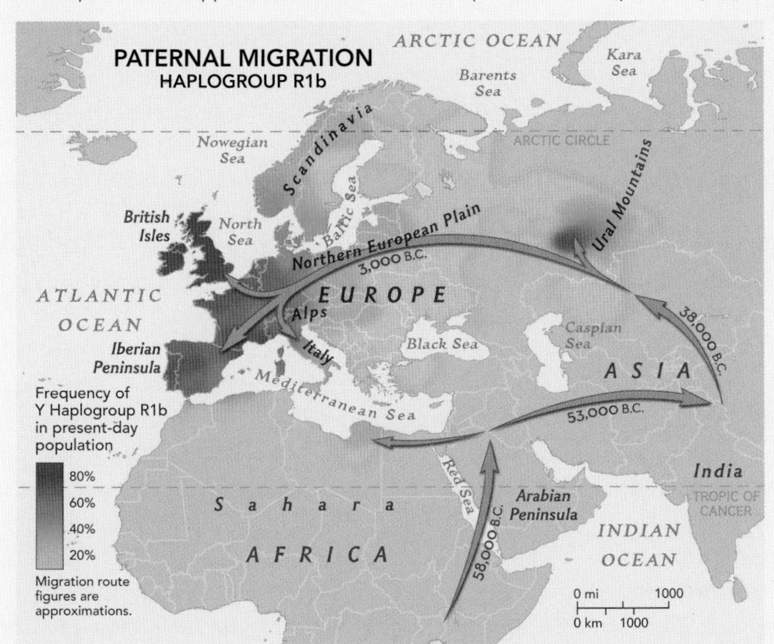

Paternal R1b: Many European men trace their deep ancestry to Central Asia some 40,000 years ago.

Importance of Migration

This rapid linguistic replacement, tied in with the expansion of the R1b lineage, demonstrates the importance of migration over cultural diffusion in European prehistory. While today's Europeans carry genetic markers that connect them with the earliest Europeans of about 40,000 years ago, they also carry the signals of later migrations, ranging from Middle Eastern farmers to Bronze Age warriors. European genetic patterns, as with the rest of the world, are a result of the complex interactions of many peoples—and their DNA—over tens of thousands of years.

—Spencer Wells

Portuguese is spoken by more than 200 million people, including those living in regions in Africa, Asia, and South America formerly colonized by Portugal. At home, the Portuguese respond with polite dismay when visitors assume that Spanish can be spoken there as a lingua franca.

Portuguese society is less formal and hierarchical than it was in the past, though status still permeates social relationships as well as businesses and other organizations. A great deal of time is spent cultivating appropriate relationships, as loyalties are forged between individuals, not companies.

About 80 percent of the Portuguese are Roman Catholic. A tradition of *romarias,* pilgrimages to sites associated with particular saints, combines with *festas,* local festival days. Portugal has gone through several periods of anticlericalism in which the status of the clergy was challenged to the point of shutting down convents and monasteries. Currently, priests receive tempered respect, acknowledged for both their spiritual leadership and their human fallibility.

God is a good worker but loves to be helped.

BASQUE PROVERB

Catholicism coexists with strong beliefs in magic, sorcery, and witchcraft. The Portuguese also fear the malevolent effects of *inveja,* or envy, and practice measures to avoid the evil eye. Christian and pre-Christian beliefs surround death. Brotherhoods called *confrerias* help arrange funerals, assist with funeral expenses, and provide Masses of remembrance, which take place for years after the death of a loved one. Mourning practices may continue for years; widows usually wear black clothing for the rest of their lives.

Like most other Europeans, the Portuguese are avid soccer fans. Their traditions in the sport of bullfighting differ from those in neighboring Spain in that much of the contest takes place with *cavaleiros* on horseback, and the bull is worked over but not killed in public at the end.

Portuguese musical forms include the complex and stylized vocal tradition known as *fado* (from the word for "fate") performed by men or women, but more often female *fadistas.* Accompanied by two guitars, a fado vocalist performs poignant narratives that embody the quintessential Portuguese concept of *saudade,* a pervasive and ineffable longing.

Spanish

Population: 38.4 million
Location: Spain
Language family: Italic

The Spanish people inhabit the bulk of the Iberian Peninsula as well as the Canary and Balearic Islands, making Western Europe's second largest country one of its less densely populated. Roman conquest of indigenous tribes in 200 B.C. brought Latin to the peninsula, giving rise to Spanish as well as the Catalan and Galician languages, but not supplanting Basque, a pre-Roman language of the Pyrenees. In the fifth century, the Visigoths invaded, followed by the Moors three centuries later. The Moors finally were driven out in 1492, when their last ruler fled from his stronghold in Granada. In the modern era, a bloody civil war from 1936 to 1939 set the scene for the long dictatorship of Francisco Franco. At his death in 1975, a constitutional monarchy was reinstated.

Geographic diversity characterizes the Spanish landscape. Mountains, rolling hills, valleys, coastal plains, and the vast tableland that forms Spain's interior promote regional variations in culture, although regional dialects of Spanish are mutually intelligible with Castilian, the country's official language.

Spain is 94 percent Catholic, and although the church wields far less influence than in the past, the religion permeates the culture and the landscape. Beautiful churches and cathedrals draw visitors even as weekly attendance at Mass declines. People celebrate the feast days of saints, especially that of Santiago (St. James), the country's patron saint. Other holidays include October 12, the Dia de Hispanidad, commemorating colonization of the New World.

Spanish families value togetherness, often organizing their workdays around the afternoon meal. Many businesses, shops, and schools close between 1 and

SPANISH Regional Spanish folk dances include these traditional steps in Santiago de Compostela.

2 p.m. so that people can return home for what most consider the main meal of the day. It may consist of a stew or perhaps paella, saffron-infused rice with vegetables and seafood or meat that is associated with the Valencia region and has become a national dish. A siesta may follow the meal, and people return to work and school by 4 or 5 p.m. The evening meal, usually less elaborate, may be eaten as late as 10 p.m. Weekend meals expand in social importance, offering opportunities for relatives or friends (known as *cuadrillas* or *peñas*) to share a leisurely repast.

Spanish homes usually are built to maximize a family's privacy, often by means of an interior patio, while public spaces let people see and be seen. Housing tends to be tightly clustered, and people spend a great deal of time outside the home, going for *paseos,* or walks, in the street or plazas, visiting friends, shopping, or stopping for a beverage and snacks called tapas.

The Spanish live mostly in nuclear families. At marriage, a couple usually establishes a household. In rural areas, the child expecting to take over the family farm may stay on at the farmstead after marriage. Class differences matter less now than in the past, and now more people can achieve higher status through education or professional achievement.

Until the mid-20th century, the Spanish worked predominantly in agriculture. Typical crops are wheat, rice, grapes, citrus fruits, and vegetables, along with cured hams, Spanish wines, and sherry, a fortified wine named for its origins in Jerez de la Frontera. Industrialization came later to Spain than to more northern countries, but it is now well established, the major industries being arms and munitions, machinery, and automobiles. Today, tourism and related services provide the largest number of jobs.

Although a controversial entertainment in these times, the bullfight remains a Spanish national symbol. The event unfolds in three highly ritualized stages, ending with the killing of the bull by the matador. The Spanish also identify with the iconic figure of Don Quixote, from the 1605 novel by Cervantes, whose ingenuous optimism endears readers worldwide.

Northern Europe

The region of Northern Europe comprises the peninsulas and island nations in the northwestern stretch of the European continent. The British Isles—Ireland and Northern Ireland, Scotland, Wales, and England—while forming a closely knit geographical unit, have been joined together and split asunder politically in a variety of combinations over the centuries. Demanding landscapes and rugged lifestyles of the nations of Scandinavia—Denmark, Norway, Sweden, and Finland—give grounds for a common regional identity. Out on its own in the Norwegian Sea, the volcanic island of Iceland stakes a claim in world literary history.

Cornish

Population: 532,000
Location: United Kingdom
Language family: Celtic

Descendants of one of three Brythonic kingdoms (the others were Wales and Strathclyde, now southern Scotland), the Cornish reside in Cornwall, at the southwestern tip of Great Britain. For generations, they spoke a language that was closely related to Welsh and a cousin to Irish and other Gaelic languages. In recent years, Cornish has been brought back from extinction, and people are once again speaking, singing, and writing in it as a symbol of cultural identity.

About 532,000 people live in Cornwall today. In 2014, the United Kingdom finally granted the Cornish official minority status with the kinds of protections already afforded to the United Kingdom's other Celtic peoples: the Welsh, Scottish, and Irish.

For several decades, a solidarity movement has pushed the identity issue forward, and many people are now asserting the uniqueness of their culture and history. Many hope that Cornwall will eventually see the establishment of its own regional government—somewhat like Scotland and Wales, each of which have independent parliaments and yet are part of Great Britain.

Cornwall has long been known not only for the skill of its miners but also for a ubiquitous meat-stuffed pastry dubbed the Cornish pasty (pronounced PASS-tee). In fact, the histories of the two are deeply intertwined. Pasties were convenient food for miners to carry to work with them, and the miners are said to have offered corners of the pasty to the "knockers," mine spirits who could prove troublesome. Cornish miners emigrating to the American West brought this tradition with them, and around mines from Colorado to Montana, it is still common to hear stories about the "tommyknockers."

CORNISH A Cornish couple carries the day's catch along the quay in Coverack, Cornwall.

The last Cornish tin mine closed in 1998, yet stories and traditions associated with mining persist.

Danes

Population: 5.3 million
Location: Denmark
Language family: Germanic

The Danish people originated in southern Sweden, crossing the narrow sound to the Jutland Peninsula—the northern extension of the West European Plain—and the several hundred islands that surround it. For centuries, their rural agricultural communities practiced a communalism of open fields and pooled livestock herds, a tradition that has carried over into the agricultural cooperatives in Denmark today. Aversion to conflict and confrontation are Danish traits fostered by the communal spirit. Children are taught to suppress aggression and work differences out through dialogue and compromise.

Danes universally speak Danish, a Germanic language known for dialectal variations and related to the other Scandinavian languages and to English, German, and Dutch. Danes also typically speak a second language, either German or English—and sometimes both.

At birth, a Dane automatically is considered a member of the Evangelical Lutheran state church, although actual church affiliation includes other denominations such as Danish Baptist and Roman Catholic. Lutheran confirmation remains an important rite of passage, but modern Danes otherwise spend little time in church, except perhaps at holidays. For the pre-Lenten observance of *fastelavn,* costumed children visit door-to-door, singing songs and asking for treats. Bonfires light the night sky on Saint Hans Day at the summer solstice, when an effigy of a witch is burned in the blaze.

Holiday meals tend to be festive and include a wider range of foods than the traditional daily fare of open-faced sandwiches, boiled potatoes and vegetables, and fried meats with gravy. Special seasonal Danish beers complement holiday foods.

There is an economy of scale to almost every physical feature in Denmark, natural and built. The land itself has little elevation, all from gently rolling hills. Houses, both old-fashioned and newly built, are low, rarely rising more than a single story. Few commercial buildings are taller than four or five stories.

The Danes excel at design and have inspired several generations of manufacturers worldwide with their sleek, functional, and easily assembled furniture. The small, interlocking blocks called Lego also have Danish roots, and visitors to the Legoland theme park in Billund can see their possibilities carried to the most fanciful extremes.

About two-thirds of the Danish landscape is cultivated, and Danish farmers grow much more food than the population needs. Cooperatives produce dairy products such as milk, butter, and cheese. The yields of commercial fishing and livestock raising are consumed locally as well as exported. Industries and services now employ many more Danes than agriculture, and a very high proportion of women—some 80 percent—are found in the workforce.

Danish family life is based on a privacy that usually admits only family and close friends. Currently, though, nuclear families consisting of married parents and children make up fewer than half of all households. Elderly parents once were automatically cared for by the family of the eldest son, but now extensive Danish social services, which include universal health care and free higher education, provide elder care as well.

DANES A young Dane displays enthusiasm at the queen's birthday celebration in Copenhagen.

English

Population: 55.6 million
Location: United Kingdom, United States, Canada
Language family: Germanic

England had been settled for thousands of years by Celtic tribes when the Romans occupied and unified it, beginning in the middle of the first century B.C. The Germanic tribes of Angles, Saxons, and Jutes who followed brought the basis for the English language (the Angles supplying the root of the name), with further additions from the Vikings and especially from the Norman French.

Today, English is spoken by more than 55 million English and by hundreds of millions of others worldwide. In England, the language exhibits strong regional and class differences. Standard English is based on pronunciation in southeastern England and is the version spoken by most media announcers. It is often called "BBC English," after the national broadcasting network.

Although churchgoing has declined dramatically, the Church of England still serves as a defining institution of English culture. Its Book of Common Prayer, first published in 1549, has served as a rich source of expression in the English language, together with the plays of Shakespeare. Today, about half the English are Anglican; the rest of England's Protestants are mostly Methodists and Baptists. About 10 percent of the English are Roman Catholic, and the country includes sizable numbers of Jews, Hindus, Sikhs, and Muslims, most of whom represent England's large immigrant populations.

Criticized by some as anachronistic, the monarchy remains an enduring English symbol. The escapades of younger royals may fill the tabloids, but the English still expect to see their queen, Elizabeth II, in familiar roles and places: opening Parliament, inspecting her troops, and broadcasting a Christmas Day message.

The English value reserve, civility, and a sense of fair play. The habit of "queuing," or lining up in an orderly fashion at bus stops and other venues, expresses all of these qualities. They also highly value the varied landscapes of their small country, including the built environment, and work to preserve them. Extensive walking trails throughout the island nation offer a chance to explore the countryside, and by law, it is the duty of private landowners to allow access to the paths that run through their property.

English cuisine long endured a worldwide reputation of being bland, based on traditional fare of well-cooked meats, boiled vegetables, and fried foods such as fish and chips. Tea is the national drink (although the recent proliferation of chain coffee shops makes coffee a viable alternative), and the English consume about a third of the world's production. Afternoon tea—the beverage plus dainty sandwiches and cakes—traditionally was an upperclass habit. Working people, who often ate their main meal at lunchtime, took "high tea" of more hardy fare in the evening. Workers and business-people alike often stop in at their local pub (short for "public house") for a pint of ale or lager after work. In recent decades, the English have widely broadened their culinary horizons, embracing international cuisines.

Avid sports fans, the English enthusiastically follow the fortunes of their favorite teams, whether the game is football (soccer), rugby, or cricket—all English inventions. It is legal to gamble on sports in England, and many people frequent betting shops to play the odds, as they also do on horse racing.

ENGLISH Shearing time means a sheep roundup for this English shearer in Gloucestershire, England.

A falsehood is the best traveler.

WELSH PROVERB

For centuries, gardening has been a favorite English pastime. An English home almost invariably includes a garden, and even a city window box is likely to receive considerable attention. DIY (do-it-yourself), promoted by home improvement programs on television, now rivals gardening's popularity.

Most English—about 90 percent—live in urban or suburban areas. Despite the popular notion of bucolic rural England, few people actually farm, and as the industrial north of England has declined, the number of people employed in the service economy, especially in the southeast, has risen dramatically.

In the 17th and 18th centuries, thousands of English sailed to the New World and established settlements there. They became a foundation of the nonnative populations of Canada and the United States. Canada remained in England's fold as a member of the Commonwealth, and its population of English descent remains close in many ways to the country of origin. Although the American Revolution broke the political connection, England's language and the legal system reveal America's deep roots.

Faroese

Population: 66,000
Location: Faroe Islands
Language family: Germanic

The Faroese way of life centers on small villages spread out among the North Atlantic's 18 Faroe Islands, which are administered by Denmark. Intensely proud of their beautiful homeland, the people are primarily fisherfolk who exploit rich Gulf Stream waters surrounding their archipelago. The Faroese have enormous boating skills, and these are frequently exhibited at festivals where wooden boats are raced. The events recall a time when large groups worked together to hunt whales from such vessels. But the sea is not the only focus here: Sheepherding is also a feature of Faroese life, with twice as many sheep as people occupying the islands.

The Faroese are scattered across the islands, dwelling in and among towns that may have as few as 500 inhabitants. People often gather for the religious holidays observed by the Lutheran Church, which continues to be an active force in the culture. Chain dances where people form circles, hold hands, and sing catchy, rhyming songs were first known in the Middle Ages and are still popular here, even among residents who drive modern vehicles and carry cell phones. Because of their geographic isolation and small numbers, the people are untroubled by most urban problems, but the collapse of the fishing industry in the 1990s created a recession and encouraged emigration. Though the Faroese have recovered, their dependence on fishing leaves them vulnerable.

With a history much like that of Iceland—these islands were settled in the ninth century by Norwegian Vikings and people of Celtic origin—the Faroe Islanders speak a language very similar to Icelandic. But unlike Iceland, the Faroe Islands are part of the Danish kingdom. A movement to break free of Denmark continues to grow in popularity, mostly because of differing attitudes toward European unity; Denmark is a member of the European Union, while the islands are not. The Faroese, however, already enjoy a high degree of autonomy, and they would find it difficult to survive without financial subsidies from Denmark.

FAROESE A soccer fan shows his stripes—literally—at a World Cup qualifying match between Germany and the Faroe Islands.

Finns

Population: 5.1 million
Location: Finland
Language family: Finnic

Until 1809, Finland was part of the kingdom of Sweden and had been for more than 500 years, causing some people to think of it as a Scandinavian country (it is not). Furthermore, about 20 percent of the population in southern Finland, especially in the Åland Islands, is of Swedish descent. But while Finns may resemble Scandinavians, with light skin, hair, and eyes, their culture is very distinct from that of Scandinavia.

The spender will get what he will spend and the saver won't get what he will save.

HIGHLAND SCOTS PROVERB

The people speak a Finno-Ugric language, which has more in common with Hungarian than with Swedish, suggesting that they may have originated in eastern Europe or Siberia. The *Kalevala,* an epic poem about a mythical hero who not only brings music and prosperity to Finland but also rescues the sun and moon, is considered the national story. It has motifs, such as the Earth hatching from the egg of a duck, that are thought to be more than 3,000 years old. Composed in Finnish, the poem is sung and performed on national holidays as a way of celebrating the distinct Finnish identity. It was based on folktales collected in Karelia, an eastern part of Finland where the tradition of singing complex poems lasted well into the 19th century. A large section of this region now lies within Russia's borders, and its people follow the Russian Orthodox religion.

Politically, Finland was annexed by tsarist Russia in 1809, and for many years it was influenced by Russian culture, architecture, and political thought. Even so, it became an independent republic in 1917 and later observed strict neutrality during the Cold War. Today's Finns see themselves bridging the cultures of western and eastern Europe; some adhere to Eastern Orthodoxy,

FINNS A furry and friendly Siberian husky proves quite a handful for a young Finn.

but more than 90 percent practice evangelical Lutheranism, the dominant faith in Scandinavia.

Though Finns are typically known for their reserved personal style, they are a highly technologized and active people. Their main recreational activities include mambo dancing, winter sports, *pesäpallo* (a form of baseball), and relaxing in saunas, which they invented. Several Finnish designers and architects enjoy world renown, and the design style often associated with Scandinavia actually finds its origin in Finnish aesthetics of austerity, beauty, and practicality.

Icelanders

Population: 300,000
Location: Iceland
Language family: Germanic

Little more than a quarter of a million people live on the volcanic island of Iceland, first settled by Norwegian Vikings in A.D. 874 and included in the Scandinavian kingdoms until its people declared independence in 1944.

Because of that long connection, Icelanders are similar to Scandinavians in more ways than just physical type: They enjoy a high standard of living, have socialized health care, eat the same kinds of food, and are Lutheran. A number of early settlers, however, were Celtic slaves who had been captured in Scotland and Ireland, and their contribution to the gene pool means Icelanders are as likely to be brown-haired as blond. After a thousand years of intermarriage, Scandinavians and Celts have become one people—a tightly knit culture in which almost everyone is related to everyone else by blood, if only distantly

ICELANDERS A small and placid Icelandic horse does not seem to mind a gentle petting at mealtime.

He that hath found the handle, hath found also the blade.

WELSH PROVERB

The language of Iceland, related to archaic Norwegian, is very different from the ones of modern Scandinavia. It is considered a main distinguishing ethnic marker, with the government pursuing an aggressive policy of language purity and even maintaining a list of approved names for newborns. Iceland's people are also distinguished by a love of writing and storytelling that led to their creation of the Icelandic sagas, a medieval body of literature that continues to be read and cherished. One of the most literate peoples in the world, Icelanders most often give books as birthday and Christmas presents.

While rooted in the past, the nation of Iceland is also very modern. More than half of the Icelandic population resides in Reykjavík, the capital city, where advanced technology, tourism, music, and fashion have overtaken fishing and farming as the most common occupations. Living on a geologically active volcanic island, where natural hot springs abut huge glaciers, Icelanders have pioneered the use of geothermal energy to produce electricity for industry and to provide heat and light for their homes.

Although they have high-quality public education through the university level, Icelanders often go abroad for specialized education, which also contributes to the international influences on the country's population. Wherever they go, though, Icelanders remain attached

GENOGRAPHIC INSIGHT

LOST SONGLINES

Northern Europe encompasses a geographic range where local populations speak languages belonging to completely different families, practice ways of life from reindeer herding to whale hunting to farming, and have cultural histories that range in age from 40,000 years old to as recent as a few hundred. And yet, beginning with the European age of exploration and continuing with the rapid periods of industrialization and globalization that have characterized these past two centuries, its people and their culture have now arrived on every continent, and they have forever changed our planet's genetic landscape.

At the turn of the 20th century, Europeans arrived by the scores at Ellis Island, the main entry facility in New York for immigrants entering the United States. Hoping to make a fresh start, they adopted new names and moved to towns they had never heard of before, breaking with their cultural past in hopes of a better life. More than 12 million people, many from northern Europe, entered their names in Ellis Island's official ledger over a 60-year period, and today more than 100 million Americans can trace their ancestry to those people, the first immigrants to walk through its golden door.

Hunting and Gathering

This pattern of recent migration has been mirrored around the world, and so today, people are for the first time in history confronted with the difficult reality of not knowing exactly where their own ancestors come from, what cultural traditions they practiced, and which languages their people spoke. Because many people now live in places where everybody comes from somewhere else, we no longer have the oral histories or ancient songlines to educate us about our distant past.

Surprisingly, genetic research is now allowing people to reconstruct the birthplace of those long-lost migrations. DNA markers take a long time to become historically informative, which is why they are traditionally used to help answer questions of anthropological interest. While mutations occur in every generation, it typically requires hundreds, if not thousands, of years before they can be used as windows back into the distant past, signposts on the human trail that tell us something about the people who have come before us. But if we accumulate enough genetic information from enough people around the world, our own genetic sequences may reveal details of the smaller, more recent branches of that trail. It may be difficult to say anything about the history of these subgroups, but they can reveal the distribution of other people alive today who are more closely related to us, helping to fill in events that have occurred as recently as the past few hundred years. It is a useful way to help bridge the anthropology of population genetics with the popular study of genealogy, to which we are accustomed.

A Northern Irish bartender pulls pints in a Belfast pub.

Ancestral Gene Pools

Public interest in genealogy has become a global phenomenon, and people are now turning to DNA to answer key genealogical questions. As the genetic genealogy databases grow and people from around the world sign up to have their DNA tested, scientists are beginning to recognize a real value in numbers.

A recent survey for Scandinavian ancestry in one such database revealed that some lineages characteristic of this northern European region were showing up in people from the British Isles. When the team looked more closely, they found that the villages where these people lived were predominantly sites of known Viking influence, suggesting an entire scenario of travel and cultural interaction that happened many centuries ago.

These realizations provide some of the first evidence for the impact these seafaring traders and warriors had on the gene pool of places they had raided. Research that was initially intended to address questions of great antiquity is now helping to solve the genetic riddles still puzzling the past few dozen human generations.

to their little nation. Today, Iceland's cultural influences come primarily from Europe and the Americas, befitting a nation that sits astride the Mid-Atlantic Ridge, where the European and North American tectonic plates meet.

Irish

Population: 4.1 million (in the Republic of Ireland); 1.8 million (in Northern Ireland)
Location: Ireland
Language family: Germanic (English), Celtic (Irish Gaelic)

In the Republic of Ireland, the Irish people number a little over four million, but it is impossible to know just how many Irish live and work throughout the United Kingdom and farther abroad. In terms of global Irish culture, it is probably equally impossible to quantify how many millions of people worldwide claim some degree of Irish ancestry.

In the modern era, after a long period as a colony of England, Ireland asserted its independence with the Easter Rebellion of 1916. This movement eventually led to the creation of the Irish Free State in 1921. Ratification of the Free State meant an independent Ireland, but it also ensured British sovereignty over Northern Ireland. In 1949, the Republic of Ireland that we know today was born, while Northern Ireland remained a province of the United Kingdom.

Following the tragic potato famine of 1846–1851, some two million Irish emigrated to England and America to seek a better life. Even in more recent years, Irish workers have continued to emigrate and pursue their fortunes abroad, but many expatriates came home to participate in the country's wide-ranging economic revival. Growth in computers, telecommunications, and other technology industries increased the opportunities in Ireland for professional employment, although recent recessions brought serious setbacks. In addition, Ireland's expanding infrastructure, spurred by its inclusion in the European Union, has brought about a marked increase in work for skilled tradesmen.

IRISH A County Mayo farmer rests with his dogs on his lush, green land.

Ireland's cultural traditions remain firmly ensconced in its people, more than 85 percent of whom are Roman Catholic, with minority populations of various Protestant denominations. Northern Ireland shows a different proportion, with about 48 percent Protestant and 45 percent Catholic. The Irish, and the rest of the world, revere the legacy of intellectual centers such as Trinity College in Dublin, where the most famous medieval Gospel book, the *Book of Kells,* resides. Literature, music, and theater are staples of Irish life, along with soccer, Gaelic football, and hurling, an exuberant relative of field hockey.

Like their English neighbors, the Irish have an appreciation for life in the public house. At the pub, friends gather for snooker, darts, and other games while listening to music or debating James Joyce or the finer points of Irish politics—all, of course, accompanied by a pint of Guinness or Murphy's stout.

The principal language of Ireland is English, but Irish is still important in the contemporary culture. Irish, or Irish Gaelic, is spoken by a large part of the population, especially in the north and west, and cities, towns, and roads bear Irish names, as do many organizations and businesses. Learned either traditionally or in schools, the Irish Gaelic language is used professionally, informally, and artistically.

Norwegians

Population: 4.6 million
Location: Norway
Language family: Germanic

With only four and a half million of them in a more than thousand-mile-long country, the Norwegians find themselves with plenty of room to stretch out; in fact, they have the lowest population density in northern Europe. Mountains and water are defining features of their land, in the form of a long chain that runs Norway's length and numerous rivers, lakes, and fjords, as well as a long, rugged coastline.

Norwegians descend from the Vikings who unified the territory in A.D. 900 under Harald Fairhair, whose moniker still describes a great many of the people. They became Christians a century later, and as the Reformation became entrenched in northern Europe, they adopted evangelical Lutheranism as the state religion in 1814, upon the dissolution of a 400-year union under Danish kings. A merger with Sweden ended this brief period of independence.

NORWEGIANS Fiddle music is a prominent component of the Norwegian folk repertoire, often accompanying dancing.

By 1905, Norway was free again, but without a monarch, so a Danish prince was appropriated and made Håkon VII, King of Norway. By this point, the Norwegians had begun in earnest to stress their Norwegian heritage to distinguish themselves from their neighbors and create links to their Viking past—a process of reinvention that continues to this day.

The Norwegian language is closely related to other Scandinavian languages and exists in many spoken dialects and two literary forms: *Bokmål,* book language, a Danish-oriented version from eastern Norway, and *Nynorsk,* new Norwegian, consciously forged in the 19th century from peasant dialects and tied directly to Old Norse. Norwegians today use both or may simply write their own dialect as spoken.

More than a third of Norway stretches beyond the Arctic Circle. The country's extreme northern location means long days in summer. On the longest, Midsummer's Eve at the summer solstice in June, Norwegians celebrate all night with bonfires, feasting, and dancing. They also emphasize Christmas, which they treat as a season lasting into January, and observe with candles, greens, carols, and traditional foods including lutefisk, dried cod soaked in lye.

The main secular feast is Constitution Day, on May 17, and is celebrated with parades and public ceremonies followed by gatherings of family and friends at home. In keeping with their reverence for the rural Norwegian

SAMI Sami camp in a *lávut,* or traditional herder's tent, in northern Finland.

past, many individuals own complete folk costumes that they wear at celebrations or on formal occasions.

The discovery of offshore oil and natural gas in the 1960s has turned Norway into a prosperous country and one of the world's top exporters of these fuels. Norway's farming and fishing base involves only about 5 percent of the people, industry employs some 23 percent, and services the rest. Well paid and well covered by socialized health care and other state-run programs, Norwegians on the whole enjoy a very high standard of living.

Although they are known as a nation of readers—supporting more daily newspapers than many larger populations—Norwegians spend much of their leisure time out of doors, often gathering together in little huts or cabins in the mountains for weekends and in the summer. Skiing in all its forms is the national sport, and even for residents of Oslo the ski trails are only a tram ride away.

Sami

Population: 227,000
Location: Norway, Sweden, Finland
Language family: Finnic

Most likely those who have lived the longest on the Scandinavian peninsula, the Sami are a non-Germanic ethnic group living in northern Norway and Sweden, as well as in northern Finland and northwestern Russia (the Kola Peninsula, in particular). The Sami were traditionally nomadic pastoralists and occasional hunters, heavily dependent on their reindeer herds, which are

DEEP ANCESTRY

THE FARMER

Holly decided to participate in the Genographic Project after hearing about it from a friend who worked at the National Geographic Society. She felt that she knew her family history fairly well—a descendant of Samuel F. B. Morse, inventor of the Morse code, on her father's side, and her mother's side traced back to Germany. "I'm European—I'm not expecting any surprises!" she said when she swabbed her cheek. Her results were consistent with that European ancestry, but they revealed a deeper story as well.

DNA Results

Holly's results were typical for people from northern Europe, but her mitochondrial lineage was more interesting. Her haplogroup is T2b, which is found at low frequencies in many European countries. T2b has been called the "Royal T"; it is common in many members of the ruling families of Europe, including Nicholas II of Russia, Olav V of Norway, and George III of England. These European nobles all trace their female ancestry to Barbara of Celje, wife of Sigismund, Holy Roman Emperor in the 14th century. Belonging to this lineage doesn't mean that you are a descendant of a European noble family, but you do share maternal ancestry with them; you are genetic cousins on your maternal side.

Haplogroup T, the ancestors who gave rise to T2b, arrived in Europe from the Middle East in the past 8,000 years. This genetic marker was introduced by people from the Fertile Crescent region who brought with them their farming culture—one of the most dramatic lifestyle shifts in human history. Prior to 10,000 years ago, and going back millions of years along the hominin line, humans had lived as hunter-gatherers. The advent of farming changed their lifestyles forever, and genetic data show that it spread via the migration of people and their DNA, not simply the spread of culture.

Holly Morse, mtDNA haplogroup T2b

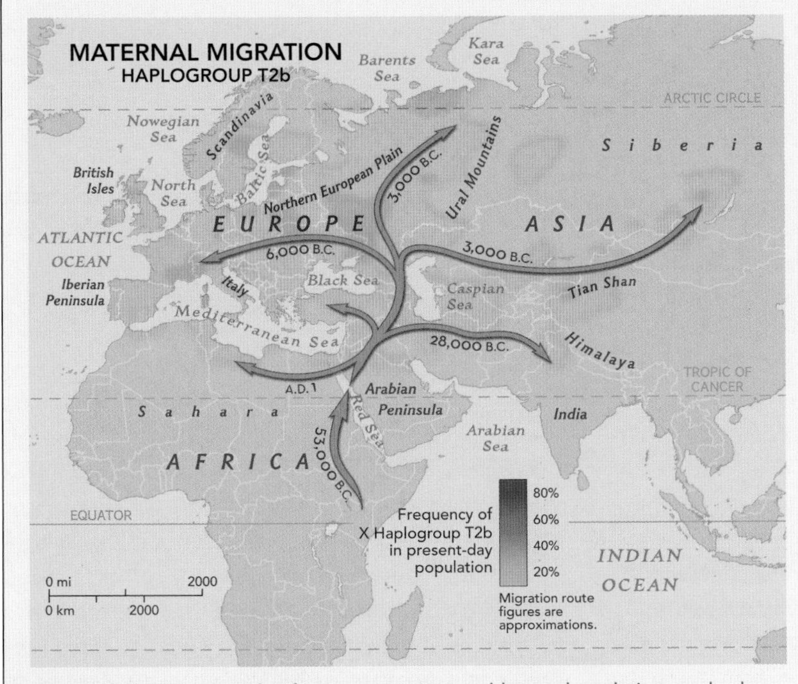

Maternal T2b: Once outside Africa, our ancestors could spread, exploring new lands and resources.

First Farmers

It is perhaps fitting that Holly herself is an avid gardener. One of her fellow T2bs was one of the most famous gardeners of all time, George III of England (1738–1820). He was known sarcastically by his subjects and others as "Farmer George" because he preferred to spend his time at his country house in Kew rather than in courtly London. His "farm" formed the basis for what eventually became Kew Gardens, one of the first and most important modern botanical gardens. It is fitting that Holly, George, and the gardens at Kew all share this connection to the first farmers in the Middle East more than 10,000 years ago.

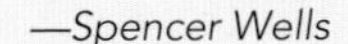

—Spencer Wells

likely the inspiration for the reindeer that accompany the cultural icon Santa Claus. The Sami practiced a shamanistic religion and saw power in natural features such as rocks and mountains. But because of their nomadic way of life, they were discriminated against by Scandinavians, who called them Lapps, a derogatory name.

You cannot catch fish until you wet your feet.

SAMI PROVERB

The Sami are well adapted for living in the north. To get from place to place, for example, they rely on skis, which they invented as a means for accompanying their sleds. Traditionally, they used every part of a reindeer for survival; animal skins were made into clothing or draped around poles to create a transportable dwelling similar to a tipi. Their brightly colored ceremonial costumes, worn with upturned boots, include jackets, skirts, and hats that are ornately decorated with strips of colored material.

Although they follow a way of life that shares customs with Arctic groups, the Sami are genetically different. Some researchers place their origin in the Alps, while others say ancient Siberia. In any case, they most likely came to northern Europe thousands of years ago, following reindeer herds as the polar ice cap receded. The people speak a Finno-Ugric language like Finnish, but the two tongues are not mutually intelligible. In fact, there are three main dialects of Sami, and these developed probably through the isolation of the different bands in the hard-to-cross terrain of the northern reaches of Scandinavia. Today, Sami is a vibrant, living language that continues to be taught in schools, and it is a distinguishing ethnic marker of the people who speak it. The vocabulary reflects a particular way of life, and at least one word, *tundra,* is used worldwide.

More than a thousand years of close cultural ties with Scandinavians and other Europeans have created considerable variation within the Sami. A good half of today's Sami people still herd reindeer, and many subsist as farmers and fishers or live in villages, working as miners and loggers. But changes in their traditional lifestyle have not weakened the Samis' sense of who they are, and this group is reasserting its unique identity today.

Sweden, Norway, and Finland once tried to limit claims of Sami identity to the people who still practice reindeer herding. But all of the people who speak the Sami language or who had Sami parents and grandparents, regardless of occupation, consider themselves to be Sami. They also are demanding political autonomy for northern areas of the Scandinavian peninsula that have traditionally been their home. To secure their rights and ensure the continuance of their culture, the Sami have organized a representative body, the Sami Parliament, in Norway, Sweden, and Finland.

Scots

Population: 5.2 million
Location: United Kingdom
Language family: Germanic (English), Celtic (Scots Gaelic)

Scotland's population is largely concentrated in major urban centers such as Edinburgh, Glasgow, Dundee, and Aberdeen. Scots are also found beyond the Scottish border, living and working elsewhere in the United Kingdom and in many other parts of the world. Most people are members of the Church of Scotland, which is Presbyterian.

For much of its history, Scotland has been viewed by its English neighbors as a rugged, untamed land filled with rugged, untamed people. And like widely held views of the American West, this conception has some measure of truth and a goodly portion of fancy. The latter stems from English prejudice against Scotland's warring clans. These groups wore tartans, woolen clothing of variously colored plaid patterns that served the political function of identifying one's clan affiliation. Worn less and less frequently, though still preferred for formal wear, kilts continue to be symbols of Scottish ethnicity and the time of independence before 1707, when England and Scotland were united under one king.

Known as the birthplace of golf, Scotland is perhaps equally famous for its poets, its military heroes, its whisky, and its national dish, haggis—a pudding made from a mixture of organ meats and oatmeal, stuffed and boiled within a sheep's stomach. Haggis originated in Scotland's agrarian past, when the herding of sheep was a mainstay of the economy.

While many Scots continue to ranch and farm, the modern economy increasingly revolves around such high-technology endeavors as semiconductor production, software design, genetics, and environmental science. The birth of the world's first cloned sheep, called Dolly, at the Roslin Institute in Edinburgh bears ample testimony to Scotland's ability to blend cutting-edge science with traditional occupations. New industries coexist with longstanding work in petroleum production, shipbuilding, and the making of malt whisky, known to the world as Scotch.

But if Scotch is a drink, the name "Scots" means a people and their language. Scots English, known from the 18th century as Broad Scots, is a Germanic dialect akin to English. Although it was the language of the great Scottish poet Robert Burns (1759–1796), it was largely abandoned in the latter part of the 19th century—banned in schools and discouraged everywhere else—so that Scots could compete in the dominant English economy. Scots English enjoys only limited use today and is principally observed in dialects such as Glasgow Patter and the Broad Buchan of the northeast coastal region.

Scots Gaelic, a Celtic language that had nearly disappeared by 1970, has recently experienced a resurgence. There are now Scots Gaelic development programs, theater groups, and television programs, and the language is taught in some schools. The new popularity of bagpipe music, played by an instrument perfected in 18th-century Scotland, is another sign of the survival of Scottish ethnicity.

Scotland's greatest heroes—William Wallace, Robert the Bruce, and Bonnie Prince Charlie—all fought for independence. Today, after the 1997 referendum to convene a new parliament, a measure of that independence has been realized. In September 2014, some 85 percent of the Scottish electorate voted in a historical referendum on Scottish independence from the United Kingdom. The result was a no vote from 55 percent of voters.

SCOTS A worker tastes a "wee dram" among the aging casks of whisky in a Scottish distillery.

Swedes

Population: 8.8 million
Location: Sweden
Language family: Germanic

The Roman historian Tacitus wrote in the first century A.D. of the Sviones, a people "mighty in ships and arms," indicating that they traded with the Roman Empire. Some seven centuries later, Viking longboats were setting out for trading voyages to western Europe and Russia, and the Swedes were soon to become a unified—and Christian—kingdom. They also began to struggle politically with their Scandinavian neighbors: Sweden controlled Finland for almost 800 years and was united with Norway for almost a century, until 1905.

Today, some nine million Swedes live in a country the size of California, mostly clustered around three major cities in the south. Their nation boasts 100,000 lakes, and about 70 percent of the land is forested. Climate varies dramatically, from the cold and snowy north above the Arctic Circle to milder, relatively snowless regions in the south, which experience little more than one snowy month each year.

SWEDES A Swedish mother and children stroll among the cobbles of the Haga district of Göteborg.

All Swedes speak Swedish, a Germanic language related to Danish and Norwegian. This is also true of the Sami people, who live in the far north, and also of Sweden's Finnish population. Until recently, Swedes automatically became members of the state Lutheran church at birth, but the 21st century brought a separation of church and state. Still, some 87 percent of Swedes are Lutherans, with the rest mainly Jews, Roman Catholics, Muslims, and Greek Orthodox.

In modern times the Swedes have chosen to keep a neutral position in international conflicts. During World War II, Sweden took in large numbers of refugees and gave many political asylum, as they have in recent years, along with many immigrant workers. Sweden also awards the annual Nobel Prizes, bequeathed by Swedish dynamite inventor Alfred Nobel before his death in 1896. These prizes honor significant advances in the sciences and literature, and they also include the all-important award given to a person or persons who have made noteworthy contributions toward achieving world peace.

The Swedes enthusiastically spend a great deal of time outdoors in all seasons. Many of them make frequent use of vacation homes—often abandoned farmhouses, of which there are more than a half million in the countryside. Boating, skiing, bicycling, and tennis are popular sports. Many of their holidays celebrate aspects of nature and ties to Sweden's rural past. The coming of spring is celebrated, as is the summer solstice.

Like other Scandinavians, Swedes relish open-faced sandwiches and smoked, marinated, and cured fish, especially salmon, whitefish, herring, and eel. Hot meals may include small meatballs, or *köttbullar,* served with lingonberry jam.

Swedes enjoy one of the world's highest standards of living. High taxes mean comprehensive social services, including child care, parental leave, and other assistance, allowing more than 80 percent of Swedish women to work outside the home, often in upper-level positions.

Within the past few decades, the number of marriages in Sweden has declined considerably, and more couples are opting to live together. In Swedish, this situation is called *sambo,* and legislation enacted in 1988 entitles the partners to a number of legal rights. Many sambo couples end up marrying, before or after children arrive.

The majority of Swedes are employed in services and industry; only 2 percent farm. About 85 percent of Swedish workers belong to unions. The Swedish system stresses dialogue and negotiation between workers and employers, and the government, since a general strike in 1909, has generally been able to keep strikes at bay. A strong desire to prevent the escalation of disputes led the Swedish to develop the concept of the ombudsman—a representative who investigates complaints against a corporation, government, or other organization—which has spread to many parts of the world.

The Swedes were latecomers to the European Union, joining in 1995, although they opted out of using the euro as currency and have declined membership in NATO. In 1980, the Swedes voted to get rid of all nuclear energy plants and are following a plan for systematic decommissioning.

Sweden's immigrant population has grown to about 15 percent of the total, dramatically changing its demographic landscape. As a result, some Swedes no longer support the country's welcoming and inclusive immigration policies.

Welsh

Population: 3.1 million
Location: United Kingdom
Language family: Germanic (English), Celtic (Welsh)

The population in Wales currently stands at about three million, but countless Welsh and people of Welsh descent live throughout the world. Ever since the decline in the coal and steel industries after World War II, the majority of citizens in Wales have lived in or near urban centers such as Cardiff, the capital. Nearly all are Protestants, with Methodism foremost.

Although many people now enjoy city ways, life in smaller villages remains an important touchstone to cultural identity. Nowhere is this more apparent than in the annual *eisteddfodau,* or "chairing" festivals, where towns choose poetry champions who are crowned and raised aloft on chairs or thrones. These festivals have expanded in recent years to cover the breadth of Welsh cultural life and now include music in all forms, as well as dancing, drama, arts, and crafts. Winners hope to compete in the Royal National Eisteddfod of Wales, Europe's largest folk festival, and no other language but Welsh may be spoken on the poetry stage.

WELSH Sheep greatly outnumber the people of Wales, while providing bucolic landscapes.

Currently, Welsh, a Celtic language, is spoken by at least half a million people in Wales alone. A living tongue, Welsh is learned at home and in schools, and it is used in any circumstance, whether professional, artistic, or informal. *Wales* itself is not a Welsh word; in fact, it comes from the Anglo-Saxon *wœl,* meaning "foreigner" or "slave." The country's name in Welsh is Cymru (KUM-ree), from which Cambria, another name for Wales, survives in English use today. For most Welsh people, English is the language of everyday use.

Issues of language use have become more prominent since 1998, when the Government of Wales Act was passed by the British Parliament. This act established the National Assembly for Wales, which exercises administrative authority over such important concerns as transportation, education, housing, and the environment. Wales still retains representation in the British Parliament as well.

The advent of the National Assembly has accompanied a gradual transformation in the Welsh economy. While Wales maintains an agrarian base—it is Europe's most important sheep-raising center—its modern economy revolves around many high-technology enterprises. These include software design, precision manufacturing in aerospace and hydrodynamics, and the development of sustainable environmental strategies, all of which contribute to a Wales at ease with its past and its future.

ACROSS CULTURES

NOTES ON THE WORLD'S SCALES

In something as simple as a sequence of notes to form a musical scale, a culture expresses its unique identity and its relation to others around the world.

The neck of a guitar provides opportunities to make a limitless number of notes. Place your finger anywhere along the string, pluck the string, and you hear a pitch. But the pitch you find along the string might not sound like any note you've ever heard before. Beginning guitar players learn quickly that the right notes lie only in a few places on the instrument. But what about all the space—and all the possible notes—in between?

Musical traditions around the world vary in many ways: their melodies, instruments, rhythms, and even what notes they use. Western music uses only a certain set of notes, which splits an octave into 12 half steps. Our word *octave* is based on the Latin root *octo,* "eight," because of the seven whole steps within an octave and the next note up, the eighth, that finishes the scale. The word *octave* can, however, be used to talk about scale systems with any number of notes.

The Pentatonic Scale

One of the most common ways of splitting an octave into a scale is to use only five notes in each octave. Pentatonic (five-note) scales are used in traditions from folk music in the United States to percussion ensembles in Indonesia. Playing the five adjacent black keys on a piano produces one pentatonic scale, and there are many other varieties.

The sound of a pentatonic scale often makes Western listeners think of Asia, and for good reason. Musical traditions throughout Asia rely on pentatonic scales: the Indonesian gamelan, for example. A gamelan is a musical ensemble based primarily on a set of metal and wood percussion instruments. Percussion instruments are the core of the ensemble: The word *gamelan* comes from the Javanese word for "striking with a hammer," and the melodic instruments of the gamelan are metallophones, made of metal bars laid in a single row and struck with hammers. Gamelan music may also include voices, strings, and wind instruments, and it is often used in religious ceremonies or to accompany dance or shadow-puppet performances.

Gamelan music uses a few different scale systems. Each percussion instrument is tuned when it is made and can play only the notes of a particular scale. Instruments from one ensemble, then, cannot necessarily play with instruments of a different group, and even within one gamelan, some instruments will never be played at the same time.

One of the two most common gamelan-tuning systems is *slendro,* a pentatonic scale. The five notes in slendro are spaced roughly evenly in the octave; this means that the notes do not follow the whole and half steps used in Western music. The other common scale system is called *pelog,* a system of seven notes, only five of which are used in any given piece of music.

Pelog splits the notes of one octave into seven approximately even intervals. Any five-note subset of pelog, then, will sound different from the five-note scale of slendro, because the pitches of the notes in the two systems are different. The two tuning systems are associated with different moods or emotions in gamelan music, with slendro used for more melancholy pieces than pelog.

While guitarists have to determine the intonation of the notes they play, by tuning their strings and then putting their left-hand fingers down in the proper place, gamelan players are more like pianists, with the instrument determining what pitches they produce. Unlike pianists, though, gamelan musicians never play alone.

A Spanish guitarist plays the now global music of electrified rock in Barcelona.

A note produced by only one instrument is considered incomplete in gamelan music. To play a complete note, two performers must play the same note on a pair of nearly identical instruments. Gamelan instruments are crafted in pairs, with the two instruments of a pair tuned to slightly different notes. Played simultaneously, the two instruments produce nearly identical sound waves, which interfere with each other to create a characteristic shimmery sound.

Microtones and Bent Notes

The notes that gamelan musicians can use are limited by the instruments they play; once the instruments of an ensemble are made, the pitches are set. Other musical traditions value the ability to adjust pitches delicately. The basic structure of composition in Indian classical music, for example, uses microtones (pitches more closely spaced than half steps, the smallest interval in Western music) and bent notes (notes with contoured pitch).

A distinction is often made in Indian classical music between Carnatic, southern Indian, and Hindustani, or northern Indian. The traditions are very different, but they share some basic concepts, including the role of the scale system in compositions. Like Western music, Indian classical music uses seven named notes, or *swara*. Certain swara can be sharp or flat, to create a 12-tone scale like the Western scale.

Indian classical music is based on a raga, which has a function somewhere between a scale and a melody. The raga specifies what notes will be used in a piece, to some extent what order they should be used in, and the intonation of the notes. One particular raga can call for an extra-flat or an extra-sharp note, or it can call for a note that begins at one pitch, veers down, and then comes back to the starting pitch.

These precise intonation requirements mean that an instrument with fixed pitches, like an Indonesian metallophone or a piano, is not ideal for playing Indian classical music. Traditionally, an Indian ensemble has three musicians: one playing a tabla drum or other percussion instruments, one playing a drone, and one playing or singing a melody. The percussion maintains an underlying rhythmic structure, while the drone establishes the pitch to which all the notes of the melody relate.

The melody instrument or vocalist must be able to produce carefully controlled pitches. A traditional Indian melody instrument is the sitar, a lute with a very long neck. Violins, introduced by Westerners, are quite common and accepted melody instruments, and more recently Indian musicians have adapted slide guitars to play traditional music.

Creating a National Instrument

Like all other parts of musical traditions, the notes and scales of a musical style change over time. The music of Zimbabwe illustrates this. The Shona are the largest ethnic group of Zimbabwe; their folk music is centered on the *mbira,* or thumb piano, which is tuned differently according to each performer and each situation.

The mbira is made of 22 to 28 metal keys mounted on a wooden sounding board. To amplify the instrument, mbira players place the instrument inside a gourd; to create a buzzing sound, they attach metal beads or shells to the bottom of the sounding board. In modern times, the metal keys have been made of sofa springs, bicycle spokes, or any other source of steel, and the buzzing beads and shells have been replaced by bottle caps. A performer uses the thumbs of both hands and the right forefinger to pluck the metal keys of the mbira. Tuning is a matter of personal preference, in both the pitch of any note and the intervals between notes. The mbira is rarely played on its own; more commonly it is played in pairs or with other instruments, so performers playing together must agree on tuning.

Another major instrument in Zimbabwe is the marimba, a wooden xylophone, which is tuned with equally spaced intervals, using the same pitches as Western music. The mbira, found throughout southern Africa, is closely identified with tribal cultures, while the marimba has no traditional roots in Zimbabwe—which is precisely why it is now the national instrument there.

A highland Bolivian man plays his region's traditional panpipes, fashioned from reeds.

In 1960, the area that would become Zimbabwe was still under the rule of the British government, and the black majority of the country wanted white-minority rule to end. One political initiative was to create a national instrument, taking inspiration from the marimbas from Central America and other parts of Africa and adding buzzer membranes, like the beads and shells of the mbira. The creators of this instrument wanted to make an instrument that could play both African and Western music, and therefore they tuned it to a Western C-major scale, with the addition of an F-sharp, which is not part of the traditional Western scale. The Zimbabwean marimba has been fully accepted as a national instrument, despite its foreign tuning system and untraditional birth.

Hearing the World in Music

The range of world music is astounding. Cultures around the world have developed their own instruments from local materials, their own melodies and lyrics, and their own ways of using music in daily and ritual life. The scale systems of the world are extremely diverse and are closely linked to the notes available on traditional instruments as well as to the cultural importance of music. In South America, tuning systems range from the paired panpipes of Aymara speakers in Peru, which must be played together to include all the notes of a melody, to the practice among the Suyá of Brazil, who listen for the amount a pitch rises rather than any fixed pitch. The Otomí of Mexico similarly assign specific healing qualities to intervals between pitches but place little importance on individual notes.

In contrast to the uniformity of its classical music, the folk music of Europe is extremely varied. The folk songs of Georgia, for example, are almost always performed by an ensemble singing in harmony, with a number of complex tuning systems including the whole-tone pentatonic scale (which you hear when you play the adjacent black keys on a piano). Norwegian traditional music centers around the hardanger fiddle, which plays intervals somewhere between a half step and a whole step. Music played on this instrument involves unusual harmonic effects created by using all the strings as drones throughout a piece and by having extra sympathetic strings, which are never played but resonate with the notes of the played melody. Traditional music in the United States includes various tunings, such as the blues scale, a pentatonic scale with the additional "blue note"—a very flat diminished fifth above the first note of the scale (the first interval in the song "Maria" from *West Side Story* is a diminished fifth).

Today, we can witness the interaction of unrelated musical traditions: performers like Krakatau, an Indonesian ensemble that combines elements of pop and jazz with traditional gamelan music, or the Jaipur Kawa Brass Band, which plays jazz inspired by classical Indian music. Despite diversity, there are constants. Musical traditions worldwide include the concept of a scale that determines the relationships of pitches. The notes used in scale systems are the music of this sphere.

—*Miranda Weinberg*

Stringed instruments accompany a Mongol girl as she dances the traditional "wild horse" on a Mongolian steppe.

Chapter 6

NORTH AMERICA

THE AMERICAN ARCTIC
Alaskan Eskimos • Aleut • Inuit

NORTHWEST COAST
Kwakiutl • Tlingit

SOUTHWEST & FAR WEST
Apache • Havasupai • Hopi
Navajo • Pomo • Tohono O'odham

PLAINS
Blackfeet • Cheyenne
Crow • Dakota • Mandan

PLATEAU & GREAT BASIN
Klamath • Nez Percé • Northern Paiute

NORTHEAST
Americans • Amish • Canadians
Chippewa • Cree • Iroquois

SOUTHEAST
Cajuns • Cherokee
Creek & Seminole • Lumbee • Gullah & Geechee

In 1491, the year before Christopher Columbus set foot in the Western Hemisphere, millions of people of great cultural, linguistic, and ethnic diversity were already living in North America. They spoke hundreds of languages and had a range of political systems, family structures, art forms, spiritual beliefs, and adaptations to local environments. The marked variation of physical traits bespoke many peoples, many ethnicities.

In the southeastern and midwestern regions of what is now the United States, streams teemed with fish, forests held abundant game, and the climate provided the long growing season needed for agriculture. Out of these favorable conditions arose the most highly developed Native American civilization north of Mexico: the prehistoric Mississippian tradition. This complex, class-structured society featured populous towns with temple mounds where elaborate ceremonies were held.

To the north, Iroquoian speakers of the Saint Lawrence lowlands dwelled in fortified villages supported by intensive horticulture, fishing, and hunting. The more severe climate of northern Algonquian speakers generally precluded farming—they hunted, gathered, and fished—but it was perfect for white birch trees, from which people built wigwams and highly maneuverable canoes.

Positively patriotic: A young South Carolina girl goes all in for Independence Day.

EARLY MIGRATIONS In the vast boreal forest of the Alaskan and Canadian subarctic lived a number of related tribes who spoke languages of the Northern Athabaskan family. They were probably descendants of Paleo-Indians who crossed the Bering land bridge and began spreading south and east perhaps 20,000 years ago. Unlike their Inuit and Eskimo neighbors, groups such as the Koyukon and Kutchin, or Gwich'in, saw their extreme environment as a dark and dangerous place, so they developed taboos and rituals to appease numerous forest spirits. Northern Athapaskans were also very adaptable, and their cultures easily incorporated useful customs from other groups. Thus, the Sarsi and Sekani absorbed the horse culture of the Great Plains and became almost indistinguishable from the neighboring Blackfeet, while the Eyak took on the appearance of a Northwest coast tribe. Still other groups wandered thousands of miles into the American Southwest.

Athapaskans migrated from Canada sometime before European arrival, adapted to a new landscape and new neighbors, and became the Navajo and six Apache tribes. The Kiowa-Apache and other groups mingled with Plains tribes, living in tipis and sharing their customs, while the Navajo and Western Apache incorporated farming and ceremonies from the Pueblo. Named by the Spanish, for their compact villages of multistoried houses, Puebloan peoples have been living in the Southwest since 500 B.C. On the Great Plains, drought and population pressures caused significant population shifts long before white settlers began displacing tribes. Around A.D. 1300, the agricultural Pawnee moved into Nebraska from east Texas; by 1400, the Mandan had left eastern prairies for the upper Missouri Valley. Both groups lived in villages made possible by agriculture near rivers, while groups such as the Siouan tribes were bison-hunting nomads who carried their belongings on dog-pulled travois. Horses would transform their lives. The Lakota and other Sioux got their first horses between 1750 and 1770, after which they reached the height of their powers, riding freely across vast grasslands.

Far from the grasslands are the lush cedar forests and salmon-streaked waters of the continent's northwest coast, a region whose rich resource base enabled people to develop complex societies and elaborate ceremonies. California, south along the coast, offered a benign climate and a plentiful supply of wild foods. Home to the greatest concentration of divergent and unrelated languages, California was probably settled over a long period by peoples who spoke different languages and who, over generations, developed a wide array of dialects that distinguished their home villages and ethnicity.

PRESERVING THEIR HERITAGE By the late 19th century, most tribes in the United States had been isolated on reservations. Children were often removed from communities to hasten assimilation into white culture; forced to attend boarding schools, they were forbidden to speak their own languages, a situation that led to the loss of most Native American tongues. Canadian Indians, designated "First Nations," did not experience such wholesale removals and great reductions in tribal land because competition for resources was not as intense at the time. Established in 1876, the Canadian reserve system—so called because Indian land was "reserved" as hunting grounds—continues to administer Indian affairs, but today the First Nations are confronting economic pressures to exploit the oil and mineral resources found on their lands.

A Navajo rider surveys the exquisite formations of Monument Valley, located on Navajo lands.

In 2000, more than half the Native Americans in the United States were living in urban areas. The government has recognized 566 Indian tribes, and more than 200 groups are petitioning for federal recognition. Contemporary Native Americans are not only preserving their cultures but also reshaping them in ways that make them more meaningful to their peoples. They are still under considerable pressure to assimilate, yet they continue to celebrate individual tribal identities and their pan-tribal ethnic identity as American Indians.

Many immigrants or their progeny also hold fast to beliefs and traditions brought from other continents. Among the millions of arrivals over the years were Amish from Switzerland, France, and Germany; Hutterites from Moravia; and Old Believers from Russia—groups who sought separate ethnic identities based on religious beliefs. Descendants of these pacifistic, agrarian peoples still favor plain garb and reject "worldly" practices, eschewing electric lights and mechanized vehicles. During the slave trade, more than nine million Africans were brought to the New World against their will; many of their descendants are rediscovering their ancestral cultures. Between 1820 and 1924, many Asians came to work on railroads and farms in the West, followed later by large groups from the Philippines, India, and Korea. Recent decades have seen the arrival of millions more from Asia, Eastern Europe, and elsewhere in the Americas, almost all of them drawn to North America, a land that is still home to many peoples, many ethnicities.

The American Arctic

Roughly the northern one-third of the state of Alaska lies above the Arctic Circle—north of the 66° 30' latitude line—together with Canada's Arctic Archipelago, a group of about 50 large islands in Nunavut and the Northwest Territories. Tundra and permanent ice characterize this mountainous region. Its landforms include the Brooks Range, which runs inland along the northern coast of Alaska, giving way via the North Slope to the Beaufort Sea and the Arctic Ocean. On the west side of the Brooks Range, the De Long Mountains and the Baird Mountains form the Noatak National Preserve; to the east, the Endicott Mountains are the main feature of the Gates of the Arctic National Park and Preserve. The Dalton Highway, the only major road through the American Arctic, runs straight north to Prudhoe Bay on the Beaufort Sea.

Alaskan Eskimos

Population: 67,200
Location: United States, Russia
Language family: Eskimo-Aleut

Of all the Eskimo populations today, the Alaskans are the most numerous and most diverse, dwelling in a variety of environments from the Arctic tundra of the North Slope to the forested seacoasts of the south. Several thousand years ago, Alaska's Eskimos split into the northern Inupiat and the southern Yupiat, or Yupik, peoples.

In the north, the interior Eskimos were mainly caribou hunters, while the coastal groups had diverse economies alternating between sea-mammal hunting, fishing, and inland hunting. Because of these seasonal subsistence activities, social organization was flexible, often bilateral, with ties relating individuals to the families of both their mothers and fathers, maximizing territorial connections. The nuclear family—father, mother, children, and perhaps an older relative—was the basic social unit during most of the year. The larger, extended family band met and lived together during times of abundance in the spring or fall.

Non-kin extensions of social ties, such as trading partnerships solidified by wife exchange, were also important. Belying the notions of popular fiction, wife exchange was considered an economic necessity: The tasks of both men and women were essential for survival, and men often had to be away hunting for weeks at a time. Partnerships established between interior and coastal villages were beneficial in that they assured fairness in trading, access to each other's territory in times of need, and availability of products not found in one's own home environment.

Infanticide (abandonment of newborns, especially females) and senilicide (voluntary suicide by the elderly) were practiced by all Eskimo groups in times of extreme hardship; such acts were never common, however, and they were always occasions of great sadness. The elderly were much valued for their wealth of life experiences, and children were cherished, as shown by the frequency of adoption and the complete lack of social stigma attached to an adopted child.

As culturally diverse as the Inupiat in the north, the Yupik-speaking Eskimos have long lived south of

The wind is not a river;
at some point, it stops.

ALEUT PROVERB

Norton Sound, scattered along the coasts of the Bering Sea and Pacific Ocean; some dwell on Siberian shores, including those of Russia's Kamchatka Peninsula, west of the Bering Strait. Because they have had access to an abundance of sea mammals and fish, Eskimo settlements and populations on Kodiak Island and elsewhere in southern Alaska have been among the largest in the circumpolar Arctic.

Today, Alaskan Eskimos are not entirely dependent on hunting and fishing. Most lead relatively settled lives in villages, where they operate their own businesses or work for wages, and if they go hunting, they may take snowmobiles instead of dogsleds, and they may carry rifles rather than harpoons. Sometimes they supplement their incomes with sales of local art, such as soapstone and fossil-ivory carvings of animals;

ALASKAN ESKIMOS Amid a substantial haul, an Alaskan Eskimo woman thanks a fish for giving sustenance.

the Nunamiut at Anaktuvuk Pass are particularly famous for their skin masks.

In 1971, the Alaska Native Claims Settlement Act authorized a grant of 44 million acres and a payment of $962 million to Alaska natives. Village and regional corporations were established, and these entities have functioned more or less successfully to encourage economic growth.

Aleut

Population: 19,200
Location: United States
Language family: Eskimo-Aleut

The hardy, seafaring Aleut originally occupied the 1,100-mile-long Aleutian archipelago, more than 70 volcanic islands stretching between North America and the Asian mainland, with the Bering Sea on the north and the North Pacific to the south. Until two centuries ago, the Aleut maintained a population of perhaps 20,000, but because of exploitation and introduced diseases, the population since then has been steadily declining. Often relocated by the U.S. government, the people live in scattered villages, sometimes not even in the Aleutians.

Compared with northern Alaska, the Aleutian Islands enjoy a relatively temperate climate, made less harsh by the warming influence of the Pacific's Kuroshio (Japan Current) flowing from the west. Even so, the islands receive their fair share of extreme weather. They are often lashed by violent winter storms, soaked by summer squalls, and blanketed by heavy fog. The windy, wet climate has had a profound effect on the Aleut way of life. People settled in fairly permanent villages with large multifamily houses. Because staying dry was as important as staying warm, the Aleut developed highly sophisticated rain parkas made of bird skin, fish skin, and sea-mammal intestine.

ALEUT A trio of Aleut seniors chat in the Commander Islands, Russian territory, in the western Aleutians.

GENOGRAPHIC INSIGHT

INTO THE AMERICAS

Nowhere else on Earth is the human capacity to adapt more evident than inside the Arctic Circle. In an area where the annual average temperature is below 20°F, it is no surprise that a region comprising nearly 10 percent of the planet's landmass harbors only 0.2 percent of its human population. To put this into perspective, South Asia packs roughly a thousand people into every square mile, the Arctic not even one. The major reason is that humans are a hairless species of tropical primate. Our morphology developed on the African savannas, where it tended toward taller stature and longer extremities, both favorable for animals needing to dissipate heat in an equatorial climate. But head to the Far North—even with warm clothing—and increased surface area is the last thing we want. The cold weather makes larger body mass and a more compact stature—think of the polar bear—the best configuration for generating and maintaining body heat. In short, humans were not built to live here.

Adapting

And yet a few select thrive. Around 35,000 years ago, migrations originating in and around the fertile Altai Mountains of southern Siberia brought people to the inhospitable reaches of the Arctic. These intrepid people were the end result of a long, difficult struggle for survival, a tug-of-war between resilient humans and an unrelenting Mother Nature, and over time our biology responded to the new environment. Shorter, stockier frames better equipped for the frigid weather became the norm and are retained in today's Arctic populations, many of whom have average heights of around five feet four inches.

But biology is not the only mechanism of adaptation, and an important innovation tens of thousands of years ago finally made life here feasible. The earliest groups into the region relied heavily on reindeer, bison, and mammoth herds, following them from Central Asia into the cold beyond. They hunted with complex group coordination honed on the African savannas of yesteryear, but as their hunts led them onward, they encountered cold their African ancestors could never have imagined. It's debated exactly when and where people began exploiting the skins of their prey for clothing and shelter, and because the materials used by these primitive tailors were organic and therefore quickly degraded, there remains no direct evidence for this important cultural adaptation prior to the last 10,000 years. Regardless, the advent of footwear and other warm clothing meant that humans could follow reindeer across the Arctic tundra. In doing so, they walked into a world never before seen by humans.

Long years in a harsh environment accentuate the age of an Alaskan Eskimo.

Linguistic Clues

The timing, number, and location of migrations into the Americas remain among the most hotly disputed topics in anthropology. Linguistic work in recent years on the structure of languages spoken by populations from both sides of the migration route has played a pivotal role in moving the debate forward.

Languages indigenous to the Americas, linked by descent from their common proto-language, fall into one of three primary language families: Eskimo-Aleut, which is widespread in the Arctic; Na-Dene, spoken throughout Alaska and in isolated parts of southwestern North America; and Amerind, spoken everywhere else in the Americas.

In 1986, linguist Joseph Greenberg and his colleagues presented a seminal, though controversial, paper showing that each of these language clusters more closely with a set of languages spoken in Eurasia than they do with each other, suggesting that they could not have come as one homogeneous migration.

The earliest of these groups likely came across the Bering land bridge that had been created during the last glacial maximum, around 16,000 years ago. Then, around 14,000 years ago, the ice sheets parted briefly and allowed small groups of hunter-gatherers, adapted to life inside the Arctic Circle, to cross through the narrow, passable corridor and enter the rest of the Americas—a new era in a new world.

The Bering Strait region contains Earth's greatest concentration of animal protein, but the sources of all this protein are usually deep in the water or under the ice. To get at them, the Aleut became some of the world's most capable hunters of seals, sea lions, walrus, and whales. They excelled as builders of seaworthy skin-covered sea kayaks (the Russians called them *baidarka*) and as seamen and navigators. It was not uncommon for hunters to paddle for days on end in search of sea mammals, covering hundreds of miles in often turbulent waters.

When the tide is low,
the table is set.

TLINGIT PROVERB

The Aleut used nearly every part of the animals they caught—skin, bones, organs, sinews, even whiskers—for food, clothing, fuel, and tools. Excellent weavers, they made baskets from tundra grasses and whale baleen. They also were very knowledgeable about animal and human anatomy and the principles of physiology, even practicing mummification; they wrapped corpses in woven mats and placed them in caves facing the sea.

The Aleuts' extraordinary seamanship, combined with a wealth of sealskin and furs in the region, made their first contact with Russian explorers and traders, in 1741, an unmitigated disaster. In the 18th and 19th centuries, the area was plundered by Russian hunters, adventurers, thieves, and princes, some of whom commissioned Aleuts to build two-, three-, and four-hole baidarka to transport missionaries and traders. The Russians wiped out many villages, and diseases further decimated the population.

During World War II, key battles protecting American territory from Japanese aggression disrupted the dwindling Aleut population, now concentrated in only a few seaside towns. One such town is Unalaska. Its population of 150 is about three-quarters Aleut, and although the people no longer hunt sea mammals, they have recently renewed their tradition of crafting boats made from animal skins.

Inuit

Population: 80,000
Location: Greenland, Canada
Language family: Eskimo-Aleut

The earliest recognizable Eskimo peoples came into the Bering Strait area around 6,000 years ago and over the next 2,000 years moved east toward Greenland. After about A.D. 1000, these earliest cultures began to be absorbed or replaced by groups rapidly migrating into the region from the west—sea-mammal hunters who were the ancestors of today's Inuit. The term *Inuit* encompasses not only the Greenlanders but also the Polar Eskimos and the Central Eskimos (Caribou, Iglulik, Netsilik, and Copper Eskimos).

Around A.D. 1500, the climate began to change, initiating what some scientists call the Little Ice Age and hastening the Norse abandonment of Greenland. It also forced the Central Eskimos of what is now northern Canada to give up semipermanent coastal villages for a new lifestyle: In summer, they fished and hunted caribou; in winter, they hunted sea mammals out on the ice. For winter encampments, they built snow houses, while their Greenland cousins, able to maintain the stable coastal subsistence of former times, still lived in partly subterranean houses constructed of sod, driftwood, and stone. The Greenlanders also continued to hunt from their single-cockpit skin kayaks and larger open *umiaks.*

Most Eskimo and Siberian groups used skin boats. The Polar Eskimos remained an exception. This small, isolated Inuit group, whose culture inspired the popular, legendary image of Eskimo life, inhabited the northernmost reaches of western Greenland and Ellesmere Island, where the polar seas were frozen all year, rendering boats useless. Almost every group also used dogsleds, with the animals harnessed in a fan shape by the Inuit and in tandem by the Alaskans. The Aleut were an exception and did not have dogsleds because the warmer Aleutian climate made the sleds impractical.

Inuit cosmology was similar to that of other Eskimo societies. It was directly linked to hunting and fishing, with the most common religious practices related to taboos about subsistence. Religion was animistic; that is, all things—human, animal, and inanimate—had souls, and all were related to each other. Fantastic ivory carvings dating from A.D. 1000 to 1300 attest to the rich spiritual life.

A shaman, or *angakok,* served as a direct link to animal spirits, and he could prophesy and cure illness by sucking out objects that had intruded into souls. In Alaska, the shaman's job also included retrieving lost souls in an elaborate villagewide ritual. Although successful shamans could accrue a good deal of power, community leaders were chosen because of their abilities as providers. Social control was generally informal; leaders were usually followed by consensus.

The Inuit finally were accorded full Canadian citizenship around 1960. Their political clout increased in the 1970s when ten assembly seats were allocated to the Northwest Territories for Inuit territory. In 1979, the first Inuit was elected to the national House of Commons. The people of the Northwest Territories later voted in 1982 to create a new territory, called Nunavut, and 11 years later, the Nunavut Act was passed. The legislation set aside nearly 750,000 square miles for the Inuit people and authorized the government to pay them $1.1 billion between 1993 and 2007. In a historic step toward self-determination, the election for the First Nunavut Legislative Assembly was held in 1999.

The Inuit story is different in Greenland, which became part of Denmark in 1953. Complete home rule was ultimately achieved in 1979, but administrative control has been slower in coming, although a 2008 referendum increased it

Today, hunting and fishing remain important activities, but they are now abetted by modern equipment such as rifles and snowmobiles. A number of Inuit communities—Baffin Island's Cape Dorset, for example—have established arts cooperatives that attract much international interest.

INUIT On Canada's Baffin Island, an armed Inuit man keeps a lookout for polar bears.

Northwest Coast

The mountainous, lush swath of coastline running from the Yukon down the west coasts of British Columbia and Washington State is dominated by the Coast Mountains through British Columbia and the Coast and Cascade Ranges in Washington State. The Rocky Mountains form the eastern boundary of this heavily forested region, where more than 80 inches of rain fall each year. Though there are some major urban areas such as Vancouver and Seattle, some areas of this region are vast, uninterrupted woodland and mountains. The Coast Mountains are sparsely populated, with fewer than nine people on average living in each square mile. Fishing, forestry, and manufacturing are the major industries in this region.

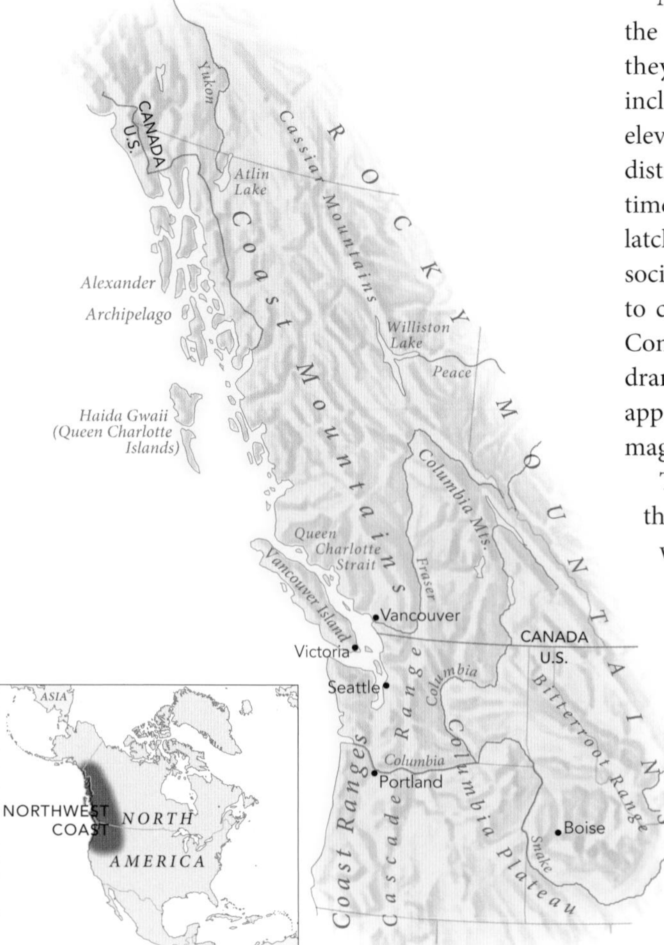

Kwakiutl

Population: 7,000
Location: Canada
Language family: Wakashan

The 30 or so Kwakiutl tribes occupying the area around the Queen Charlotte Strait in British Columbia exploited a shore environment so rich in marine life that they were able to follow a settled village lifestyle that only agriculturalists usually enjoyed.

Residing in their permanent villages, the Kwakiutl developed an extensive ceremonial and artistic life that included the carving of totem poles to represent clan figures. The poles had become a well-established art form long before Europeans arrived on the Pacific coast, but they were much easier to carve—and began to increase in number—after the Kwakiutl acquired the iron and steel blades introduced by the newcomers.

Members of the Kwakiutl aristocracy demonstrated the degree of their greatness by the amount of wealth they gave away at potlatches. These events, which included much singing, speech giving, and dancing, elevated an aristocratic family's social status as they distributed gifts to vast numbers of guests, and sometimes destroyed gifts they had received. Because potlatches were required for membership in secret religious societies, wealthy people were also initiates who learned to conduct the societies' enthralling dance-dramas. Commoners composed most of the audiences for these dramas and were usually spellbound by the performers' apparent ability to influence supernatural powers; such magic reinforced the status of the aristocrats.

The most terrifying ceremonial performance was the Hamatsa, or Cannibal Dance. For the Kwakiutl, who avoided contact with the dead, the concept of eating human flesh was both disgusting and awe inspiring. The bodies of the dancers supposedly were inhabited by cannibal spirits, and as the performers flung themselves around in a frenzy, they screamed and pretended to eat human flesh. Hidden attendants made strange howling sounds just offstage, and carved bird monsters swooped over the anxious audience. Eventually certain rites pacified the cannibal spirits, and the dancers vanished in puffs of smoke.

KWAKIUTL A Kwakiutl man swings a colorful, fringed traditional shawl along the shore in British Columbia.

The Kwakiutl once built specialized canoes for whaling, sealing, fishing, freight transport, river travel, racing, and war; by 1900, however, canoemaking and seagoing travel were all but forgotten. Then, in the 1980s, a Haida Indian carved a 50-foot-long canoe. His craft inspired the Kwakiutl to make similar canoes and renew paddling events, such as the 1989 Paddle to Seattle, with participants from several groups along the northwest coast. Bringing back the great canoes led to the revival of potlatches and the songs, dances, and prayers associated with them, instilling a sense of cultural pride.

Tlingit

Population: 20,000
Location: United States, Canada
Language family: Athabaskan-Eyak-Tlingit

Sharing customs and a common language, various Tlingit tribes dwell on reservations and reserves in British Columbia, the Yukon Territory, and Alaska's southeastern panhandle. Here, mountains rise almost directly from the sea, isolating the mist-shrouded land. The 80-mile-wide archipelago off the fjord-carved coast is also mountainous, and its islands, like the mainland, are covered by dense forests.

In the past, when the fishing season had ended and stores of salmon and other food had been set aside for winter, families moved from their summer fishing camps to permanent winter villages. Plentiful resources from the ocean, rivers, and land—sea mammals, shellfish, fish, deer, bears, sheep, and mountain goats—enabled people to obtain enough food to last through winter, the time for potlatching. If a person had amassed enough wealth to validate an inherited clan leader position, he or she hosted a potlatch, a massive feast and giveaway for another clan. The size of the gathering and the value of the gifts were thought to reflect the clan leader's status. By accepting gifts, guests confirmed the host's right to the inherited privileges and tacitly agreed to reciprocate. Potlatches not only served to circulate material wealth but also helped to establish alliances between clans.

The Tlingit developed complex ceremonies within their class-conscious society that was headed by high-ranking individuals. Commoners made up the majority of the people, while slaves, usually women and children captured from other tribes, were forced to perform menial labor for their masters. Each clan had several crests that members displayed on totem poles, canoes, dishes, hats, and blankets, named for the river, worn by high-ranking individuals.

Today, Tlingit women still weave Chilkat blankets from cedar bark and mountain goat's wool, with designs based on clan crests that men painted on wooden boards. Recently, weavers resurrected the technique for making raven's tail robes, an ancient art involving bold black and white geometric forms that had been replaced by the curvilinear designs of Chilkat weaving. Wood carving has also experienced a revitalization in the form of bas-reliefs, transformation masks, portrait masks, house posts, and totem poles. Although artists sell much of their work, many of them dedicate a significant part of their time to carving items for ceremonial use in their villages.

TLINGIT Tlingit women wear colorful woven hats and Chilkat button-blanket capes, often given as gifts at ceremonies.

Southwest & Far West

From the sunny west coast of southern California to the remote, desertlike regions of western Texas, the Southwest and Far West of North America form a dry and arid landscape. The temperate year-round climate of southern California's heavily irrigated coastal region turns searing east of the Coast Ranges. Temperatures in Death Valley soar consistently into the hundreds. Western Texas's dry plains give way northward to the cool peaks of the Rocky Mountains and Rio Grande in New Mexico. In northwest Arizona, one of North America's most remarkable natural formations, the Grand Canyon, has been cut through the landscape by the erosive action of the Colorado River over thousands of years. The region has a number of striking natural formations, such as mesas and buttes.

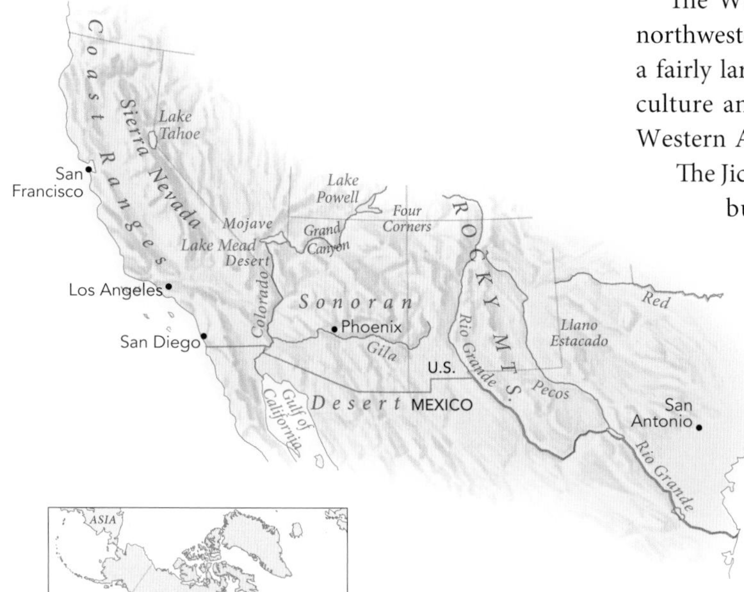

Apache

Population: 111,800
Location: United States
Language family: Athabaskan

The six main tribes known as Apache have their origins among the Athabaskan-speaking peoples of Canada, who also are the ancestors of the Navajo. Migrating southward, the Apache arrived in the Southwest around 1400 and fanned out into varied environments there. The kinds of cultural and economic pursuits they developed depended on their surroundings and on relationships with other peoples in the area such as the Navajo, the settled farmers of the Pueblos, or the bison-hunting Plains Indians.

Regardless of their economic orientation, most Apache took up raiding when they acquired horses. They would descend on other Indian groups—including their trading partners—and take food and whatever material objects they wanted. This activity differed significantly from warfare, which was intended to avenge the deaths of individuals.

The Western Apache covered nearly all but the northwesternmost quarter of Arizona and farmed on a fairly large scale. They shared many similarities in culture and language with the Navajo, making the Western Apache a likely late offshoot of that group. The Jicarilla of northern New Mexico also farmed, but culturally were similar to the Plains Indians and hunted bison. The Jicarilla probably gave rise to the wide-ranging Lipan of eastern New Mexico and western Texas, who were driven down into Mexico by the Comanche. The Mescalero of south-central New Mexico displayed some Plains traits and mostly hunted and gathered.

The Chiricahua also were hunters and gatherers and lived in southern Arizona, western New Mexico, and northern Mexico. They hold the distinction of being the last Indians in the United States to lay down their arms. They surrendered in 1886, and for the next 27 years they were treated as prisoners of war and shuffled between prison sites in Florida, Alabama, and Oklahoma. When they finally were freed in 1913,

APACHE Apache, like this man in Arizona, work in a number of fields, including the tourism industry.

most chose to join the Mescalero and Lipan Apache on their reservation in south-central New Mexico. They are represented by the Chiricahua Apache Prisoners-of-War Descendants today, whose ancestors include Cochise, Geronimo, and warriors who rode with the famous chiefs.

The Kiowa-Apache, also known as the Plains Apache, followed a trajectory different from the five other tribes. When those groups were pushing farther into the Southwest, the Kiowa-Apache stayed on the plains and became associated with the Kiowa, a typical Plains tribe, eventually functioning as a Kiowa band, although they retained their Apache language. Together with the Kiowa, they obtained horses early on and used them to hunt bison. Today they are known as superb musicians, dancers, and artists. Other Apache tribes currently experience a wide range of cultural and economic circumstances on their reservations. Apache horsemanship has turned some descendants of fierce raiders into skilled cowboys.

Many Apache ritual traditions continue to the present day. Wearing painted wooden-slat headdresses and holding wooden swords aloft, Crown Dancers embody mountain spirits known as *ga'an*. Along with a medicine man, singers, and drummers, they evoke supernatural power to cure and protect. Some Apache also perform the Sunrise Ceremony for girls reaching puberty, infusing them with the White-Painted Woman's powers to ensure long life, vitality, prosperity, and a good disposition. During the elaborate four-day ritual, singers recount tribal history from the creation of the universe to the present.

Havasupai

Population: 1,000
Location: United States
Language family: Cochimi-Yuman

For nearly a millennium, the Havasupai have dwelt amid one of the planet's most spectacular landscapes: the Grand Canyon. Unusual for a postcontact Indian tribe, they continue to inhabit the same land where their ancestors lived. The Havasupai are an Upland Yuman people, related to the Yavapai and also the Walapai (Hualapi), who probably were their ancestors and with whom they still intermarry. The language of the Havasupai, whose name means "blue-green water people," is mutually intelligible with other Upland Yuman languages and remains strong, spoken by young and old. A written language was created toward the end of the 20th century. Most Havasupai also speak and are literate in English.

HAVASUPAI A Havasupai elder performs a ceremony to reopen the Bright Angel Trailhead at Grand Canyon National Park.

The Havasupai and other groups were part of the southwestern Desert Culture of hunters and gatherers who lived lightly on the land and developed a simple material and ritual culture. They constructed dome-shaped huts of branches covered with bark or thatch. Contact with the pueblo-dwelling Hopi inspired them to enrich their economic and cultural pursuits.

Under Hopi tutelage, the Havasupai planted corn, beans, and squash, as well as tobacco and cotton, on the floor of Cataract Canyon, a side branch of the Grand Canyon about 3,000 feet below the rim, and learned to irrigate the crops with water from Cataract Creek. As novice agriculturalists, the Havasupai also adopted Hopi rituals aimed at ensuring success, such as prayers at the time of corn planting and the Rain Dance.

After the fall harvest, the Havasupai would ascend the narrow trails to the canyon rim and resume hunting and gathering on the plateau. The women collected pine nuts, mesquite pods, berries, and honey, while the men hunted rabbits, deer, pronghorn, and other game. Then they returned to the canyon floor to start the agricultural cycle.

The Havasupai led a calm and peaceful existence in their out-of-the-way canyon and were recognized by other tribes for their noncombative nature. They lived in family groups under the leadership of a head chief. Disputes within the group often were worked out by the men during sessions in the sweat lodge.

In the 1880s, the Havasupai were stripped of most of their land and granted only 518 acres within the canyon as the Havasu Creek Reservation. They also were prevented from hunting and gathering on the plateau, and so they began more intense cultivation in the canyon, adding peaches to their crops and a peach festival to their annual celebrations. In 1912, they began to raise cattle.

Since then, agriculture has been mostly replaced with wage labor and federal employment tied to tourism. The Havasupai act as mule guides for tourists who want to visit the canyon floor, maintaining the mule herds as well as operating a lodge and other services and selling crafts such as baskets. In 1975, the government restored 185,000 acres to the tribe.

Havasupai children attend the local school until eighth grade, when many are sent to boarding school in California. But they remain part of the community.

Hopi

Population: 18,300
Location: United States
Language family: Uto-Aztecan

For the Hopi people occupying 13 villages on and around three mesas in northern Arizona, each year centers around the arrival and departure of the *katsinam,* spirit beings who carry prayers to Hopi deities. The people believe that katsinam live among them from roughly January to June and then for the next six months dwell in the spirit world atop San Francisco Mountain. These beings are not worshiped but are instead respected as powerful spirits who bring rain and other blessings.

The word *katsina* refers to the incorporeal spirit and to the dancer who brings this spirit to life. As a baby, a Hopi girl receives the first of many *katsintithu* carvings representing spirits as real beings who will bestow on her the gift of reproduction. Such carvings are sacred and made exclusively for the Hopi—as opposed to the carved kachina doll figures that are on sale to the public.

The Hopi are also known for their distinctive basketry, jewelry, and pottery. The art of pottery making was revived around 1900 when a woman named Nampeyo, a Hopi-Tewa, began to copy fine prehistoric ware that had been found at nearby Sikyatki, an abandoned Anasazi pueblo. Developing a style based on the ancient pottery as well as Hopi-Tewa ceramic traditions, Nampeyo painted stylized birds and katsina beings in black, orange, and white designs on cream, white, or dark red backgrounds.

HOPI The family of Lori Ann Piestewa, a Hopi and the first female soldier to die in the Iraq war, mourns their loss.

In the Hopi world, corn is life and is tended with almost the same depth of care that people give to their children. To grow this crop in such an arid environment (it receives just a few inches of rain or snow each year) is an act of faith requiring deep commitment—and the help of katsinam. These beings are considered masters of the art of raising corn and other plants, and songs and ceremonies with katsina dancers evoke the reverential attitude that the Hopi believe is necessary for growing corn. The rituals renew the people's faith, as well as their aspiration and commitment to the ideals of compassion and cooperation.

Today, the Hopi continue their ceremonies, including the well-known Snake Dance. Some ceremonies are closed to visitors. Others are open, as the Hopi believe every person can contribute spiritual energy and heartfelt prayers. Nevertheless, they are discouraged by the visitors who push aside members of the Hopi audience, disregard signs that prohibit photography, and dress inappropriately

NAVAJO An elderly Navajo woman weaves on a loom in a hogan in the Monument Valley Tribal Park.

Navajo

Population: 332,100
Location: United States
Language family: Athabaskan

Sprawling across parts of Arizona, New Mexico, and Utah, the Navajo reservation is home to a group of people who speak an Athabaskan language, like their Apache cousins and distant relatives in northwestern Canada. By A.D. 1400, small bands of Athapaskans had begun to arrive in the Southwest, and sometime later, the Navajo separated from the rest, moving into the Four Corners area and settling on land that could not support the agricultural towns of their Pueblo neighbors.

The Navajo thrived, however, and by the early 1800s, they counted their wealth in horses and flocks of sheep. As pastoralists and farmers, they lived in isolated family groups surrounded by grazing land, and they built earth-covered, domed dwellings known as hogans. Most Navajo live in houses today, but some still dwell in hogans, which are also settings for ceremonies.

Navajo ceremonies often focus on the physical, mental, and spiritual healing of persons who become ill after failing to maintain proper relations with supernatural powers. The many songs, prayers, rites, and sandpaintings in a ceremony must be precisely replicated for it to be effective. Each chant, or ceremony, is a ritual drama that tells the story of a person who overcame great obstacles to receive healing. For some ceremonies, the chanter and his helpers make a sandpainting from 12 to 20 feet in diameter on a hogan floor. After completion, the painting is blessed, and the patient sits on it to absorb healing power. The power of the Holy People, who created the ceremony with its sandpainting designs, penetrates the person's body and restores a state of harmony and health. Unlike the sandpaintings that are applied to a board and sold as art, ceremonial works are sacred and ephemeral and can include crushed flowers, pollen, and cornmeal.

Today, a major part of Navajo ethnic identity centers on the Long Walk, an event people still talk about as if it had happened yesterday. In 1863, after burning Navajo fields and property, the U.S. military forced nearly everyone to walk some 300 miles to Bosque Redondo, New Mexico, where many died of starvation and disease. Five years passed before the Navajo signed a treaty that established a reservation and allowed them to go home.

Long before trading posts were built in the late 1800s, Navajo women had created a market far beyond the Southwest for their magnificent blankets. Then, in an effort to increase the commercial value of Navajo weaving, traders introduced non-Navajo designs, such as geometric motifs from Plains beadwork and hooked elements from Persian rugs. Today, weaving is a vital part of Navajo culture and women practice the art.

During World War II and after, Navajo recruits assisted the U.S. armed forces by delivering and deciphering coded messages based on the complex Navajo language over military telephone and radio. Their work was critical to U.S. success at Iwo Jima.

In 1969, the Navajo opened the first tribal college in the United States. Now called Diné College (the Navajo name for themselves, *Diné* means "the people"), it integrates typical college classes with Navajo culture and language courses.

Pomo

Population: 10,300
Location: United States
Language family: Pomoan

When the Spanish came to California in 1769, some 300,000 native people were already living there and speaking scores of mutually unintelligible languages that differentiated into an unknown number of dialects. Although there are other language hotspots in the world where many languages are spoken within a small geographical area, California is one of only a few places in North America where this level of diversity is known to have occurred.

As different ethnic groups moved into the area, they discovered an abundance of natural resources and a relatively mild climate. Each group found its own territory, and with time, isolation from each other led to variations in speech and dialect. The Pomo, for example, speak seven distinct and mutually unintelligible languages of the Hokan language stock—more appropriately called the Pomoan family of languages. The most divergent of these differ more from one another than English does from Icelandic.

POMO The Pomo people excel at basket weaving. Here, a woman fashions a gambling tray, used in a traditional dice game.

The Northern Pomo lived north of San Francisco in the Russian River Valley area where their territory had 22 miles of coastline. Like other California Indians, they formed tribelets, small autonomous groups occupying a territory that was recognized by neighboring communities. The rocky coast provided crustaceans, sea mammals, and fish, while streams held plenty of trout and salmon. In summer and fall, the Pomo relied on temporary shelters when they were gathering wild plants and hunting game, but the dependability of their food supply allowed them to live in permanent camps with substantial houses for the rest of the year. Acorns, which they ground into flour to make gruel or bread, were a staple of their diet.

Known for weaving baskets that are considered to be among the finest in the world, California Indians have revitalized this once dying art as an essential part of their tribal identities. Learning to weave also means learning the songs and cultural knowledge related to the making and use of baskets. Traditionally, Pomo women made feathered, twined, and coiled baskets—some of them as small as pearls—while men wove bird traps, fishing weirs, and the baby baskets that were begun immediately after the birth of a child. Basket making requires dedication, the same trait California Indians are using to revitalize their cultures.

Tohono O'odham

Population: 23,500
Location: United States
Language family: Uto-Aztecan

In the Sonoran Desert of southern Arizona and northern Mexico, where drought is a constant threat, the Tohono O'odham once divided the year between summer villages near their fields and winter villages near permanent springs in mountain foothills. Their reliance on mesquite beans led the Spanish to call them Papago (Bean People), but in 1986, they became officially known as the Tohono O'odham (Desert People) Nation, their name for themselves. Today, they dwell on three southern Arizona reservations.

In the intense summer heat, these people harvest the crimson fruit of the giant saguaro cactus, such a symbol of their culture that its harvest traditionally marks the start of the year. Resembling kiwi fruit in taste, the saguaro fruit is boiled into syrup, candy, and jam. The Keeper of the Smoke, who is the village headman and ceremonial leader, oversees the fermentation of the juice produced from the fruit, a three-day process that occurs in the "rain house."

At night, while the juice ferments, men and women dance to the singing of rain songs, and on the third day, the headman partakes of the wine as he recites ritual orations describing the desolation of the desert without rain. The ceremony is intended to bring rains up from the Gulf of California, thus ensuring the well-being of crops planted after the festival. By saturating his body with saguaro wine, the headman simulates the rain's saturation of cracked and dusty land.

Today, the Tohono O'odham are a contemporary people who are using income generated by their casinos to build and staff their own college, as well as a nursing home where tribal elders can be cared for by those who speak their own language. Casino profits have also provided scholarships for higher education. The work of tribal members helps to revitalize and preserve the Tohono O'odham language, one of only a handful of Native American languages still in vigorous use that is predicted to survive into future generations.

TOHONO O'ODHAM Tohono O'odham youth hit the midway at the tribe's annual fair in Arizona.

Plains

A high plateau stretches throughout the central region of North America, roughly from the Rocky Mountains on the west to the Mississippi River on the east, and from the Rio Grande in the south to the Arctic Ocean's Mackenzie River delta in the north. The region includes part or all of ten U.S. states—North Dakota, South Dakota, Montana, Wyoming, Nebraska, Kansas, Colorado, Oklahoma, Texas, and New Mexico—as well as two Canadian provinces—Saskatchewan and Manitoba, with parts of Alberta and the Northwest Territories. Major rivers running through this area include the two longest rivers in the United States, the Missouri and the Mississippi. The climate in this region consists of cold, possibly snowy winters and warm, mostly dry summers.

Blackfeet

Population: 105,300
Location: United States
Language family: Algonquian

Among the westernmost Algonquian-speaking peoples are the Blackfeet tribes. The Southern Piegan have a reservation in northern Montana, and the Northern Piegan, Blood, and Blackfeet proper live on reserves in southern Alberta. Once dependent on the bison for food, shelter, and tools, the Blackfeet saw their way of life transformed after acquiring horses in the mid-1700s. Families following bison had used dogs to carry belongings, but the dogs could carry only about 75 pounds, which meant women had to carry large loads and the old and the infirm who could not walk had to be left behind. After they obtained horses and guns, the Blackfeet became fierce warriors who were known for raids far from home. Extending their hunting and raiding territory, they eventually forced the Shoshoni, who lacked guns, southward and westward into the Rocky Mountains, and they drove the Kutenai westward across the Canadian Rockies.

A people without history is like wind on the buffalo grass.

PLAINS INDIAN PROVERB

Today, like many other tribes, the Blackfeet are sharing their land through "edu-tourism." In the form of tribally run tours, edu-tourism allows visitors to see some of the most beautiful wilderness in the West, and it may also help bring down the extremely high unemployment rates found on reservations. Such programs enable people to work as guides near their homes, telling tourists what they want them to know about their cultures. In Montana, Blackfeet Historical Site tours take visitors to centuries-old places, teaching them about the significance of sacred sites and how to travel with a spirit of respect. The Blackfeet themselves still smudge their bodies with sweet grass and pine boughs for purification before going into the

BLACKFEET A young Blackfeet girl participates in North American Indian Days in Browning, Montana.

mountains, and they leave tobacco offerings to the trees and animals that they find along their way. They also continue to hold an annual Sun Dance, take ritual sweat baths, and maintain sacred medicine bundles.

Cheyenne

Population: 19,000
Location: United States
Language family: Algonquian

Around A.D. 1700, the ancestors of the Algonquian-speaking Cheyenne moved onto the northern prairies from the eastern woodlands, where they had long been gatherers of wild rice. By 1800, they had moved farther west onto the Great Plains, becoming nomadic bison hunters who depended on animals they killed during the summer to sustain them through the long winter.

Cheyenne warrior societies, such as the Dog Soldiers, enforced strict control over hunters and, together with the Council of Forty-Four, provided order for the physical and spiritual survival of their people. The council head was the Sweet Medicine chief, who protected the sacred sweet-grass bundle; just below him were four subchiefs representing supernatural beings. The next level of power was vested in 39 chiefs known for both generosity and wisdom in settling disputes. If, for example, the council decided that a case of murder threatened tribal welfare, they banished the accused person, along with those who chose to accompany him or her, for at least five years. While their decisions today may often conflict with those of the officially recognized tribal government, the Southern and Northern Cheyenne chiefs, the religious leaders, and the military society members continue to operate as the traditional tribal government authority.

The Northern Cheyenne Reservation is in southeastern Montana. The Oklahoma reservation of the Southern Cheyenne was eradicated by the U.S. federal government, but this people formed the Cheyenne-Arapaho Tribes of Oklahoma, an organization with its own tribally elected business council and president. Both the Northern and Southern Cheyenne teach their language and culture in their schools, and many continue to perform traditional ceremonies, some of which have been revived—the Sacred Arrows and Sun Dances, for example. Women participate in the War Mothers Association, honoring Cheyenne veterans from all wars. For those skilled at working with porcupine quills, there is the Quillwork Society, whose members have sacred rights to use particular quill and beadwork designs.

Today, the Northern Cheyenne and 41 other tribes participate in the InterTribal Bison Cooperative. By purchasing bison, which are indigenous to the plains, and learning to manage their herds, the tribes are helping to restore the ecological balance that was disturbed when cattle ranching led to overgrazing. They are gaining needed income from bison meat, a leaner source of protein than beef. Restoring bison to the plains helps accomplish two other important goals: reestablishing connections to the Creator and reviving ancient traditions.

Crow

Population: 15,200
Location: United States
Language family: Siouan

In the struggle for control of the northern plains, the Crow were such fierce enemies of the Lakota Sioux that they served as scouts for Lt. Col. George Custer in his 1876 campaign against Sitting Bull on Crow lands near the Little Bighorn River in Montana. Even so, the Sioux defeated Custer, whose name was subsequently used in the battlefield designation. In 1991, following lengthy congressional hearings, the site was renamed Little Bighorn Battlefield National Monument, and a memorial for Indian warriors was commissioned to correct a one-sided view of history and commemorate the efforts of those who had fallen there. After two centuries of animosity, people from both native nations met in 1995 for reconciliation ceremonies at the battlefield, along with other tribes and representatives of the Seventh Cavalry.

CHEYENNE Wild mustangs respond to the calming techniques of a man on the Cheyenne River Indian Reservation in South Dakota.

CROW Shawl swirling, a teenage girl joins in the women's Fancy Shawl Dance on the Crow Indian Reservation in Montana.

Historic accounts of the Crow record their tall, striking appearance, noble bearing, and horsemanship. The men placed elaborately ornamented blankets and saddles on the backs of their horses, and women used beadwork on trade cloth to create collars that adorned the animals' necks and chests. Women also painted designs on the hide coverings of tipis, cone-shaped dwellings that they made even larger after they acquired horses to carry heavier loads.

Today's Crow, whose reservation is in south-central Montana, retain their unique vision of themselves. They continue to perform the traditional Sun Dance in summer, medicine men conduct healing ceremonies throughout the year, and a majority of children and adults still speak the ancestral language. At the Crow Fair each August, people renew their kinship and social ties while celebrating with parades, feasts, rodeos, giveaways, and powwow drumming, singing, and dancing. For competitions, Crow artists use vivid colors and both geometric and flower designs to decorate horse-riding gear and dance clothing. Also called "the Tipi Capital of the World," the fair boasts as many as 1,500 tipis pitched near the banks of the Little Bighorn.

Dakota

Population: 170,000
Location: United States
Language family: Siouan

For many, the image of a Dakota warrior galloping across the Great Plains still captivates, as it appears to represent a time-honored Indian way of life. In fact, the nomadic horse culture was a comparatively short-lived phase in the Dakota tradition. Not until the 1750s or so did the Dakota acquire, through raiding or trading, the horses that allowed them to hunt bison more efficiently and gave them greater freedom in their other endeavors.

Dakota is an umbrella term, recognized by the tribe members themselves, encompassing the groups that are more popularly known as Sioux: the Dakota, the Nakota, and the Lakota. The word *Sioux* itself has multiple linguistic origins; the term is a shortened French corruption of a Chippewa (Ojibwe) word meaning "lesser adder," denoting an enemy. Dakota, Nakota, and Lakota dialects belong to the language family known as Siouan; the blanket term *Dakota* does not apply in this case.

The Dakota, the Nakota, and the Lakota, also known as the Santee, Yankton, and Teton, respectively, represent eastern, central, and western distributions of an Indian population with some cultural variations, in addition to distinct dialects. These Indians perhaps started out in the area of the lower Mississippi River centuries ago and then moved to the Ohio Valley, before ending up in the upper Mississippi region.

As eastern Indian tribes began a general westward movement away from white settlers, they displaced other native groups in a domino effect. By the late 1700s, Lakota tribes had moved from the prairies onto the western High Plains. The Nakota also were soon on the move; the Dakota were the last to leave the prairies. Out on the plains, all the Dakota Sioux used strategy and horsemanship to develop warfare into an art.

Horses also transformed the bison hunt, because men no longer had to stampede entire herds over cliffs; instead, riders could bring down several bison with rifles or bows and arrows. The Dakota considered the bison a sacred gift from the Creator, and through the Sun Dance, their annual ritual of sacrifice and thanksgiving, they honored the bison for the food, shelter, tools, and fuel it provided. By 1875, the bison was nearing extinction because of overhunting by white settlers, who often took the animals' hides and left their carcasses to rot. A popular pastime for those on frontier hunting trips was to shoot the animals en masse from moving trains as they passed.

Extermination was encouraged by the military, which believed the end of the bison would help to bring the Indians under their control. They continued pursuing the Dakota until 1876, when these Indians, led by Lakota Chiefs Crazy Horse and Sitting Bull, and their Cheyenne allies, defeated Custer at the Battle of the Little Bighorn, the greatest Indian victory but a sad step toward the finish and a losing war.

Now numbering about 150,000 and living on reservations in Minnesota, North and South Dakota, Nebraska, and Montana, the Dakota tribes today are grouped into the three divisions of Dakota, Lakota, and Nakota. For all of these tribes, pipe smoking remains an essential marker of their Dakota identity and culture. In Pipestone, Minnesota, both Dakota and Ojibwa artists continue to fashion the dusty-red catlinite stone into the traditional T-shaped calumets. Highly prized, these pipes are then smoked to bring a spiritual dimension to human interactions, for as the smoke rises from the pipe, it carries the smoker's prayers to the spirit world.

DAKOTA The Pine Ridge Indian Reservation in South Dakota hosts the Oglala Nation Pow Wow, where this Dakota man dances.

Today, mounted Dakota warriors embody pan-Indian identity for all Native Americans, many of whom travel great distances to participate in annual powwows featuring Plains-style dancers. In reality, the nomadic hunters have been replaced by workers in a multitude of occupations, although unemployment plagues many Dakota tribes. On many reservations, the Dakota own and operate bingo halls and casinos, and some are involved in the development of natural resources, including oil, on tribal lands.

Mandan

Population: 1,200
Location: United States
Language family: Siouan

Not all Plains people became nomadic after they had acquired horses. The Mandan of the upper Missouri River are an example of a path that turned in another direction: They continued to live in fixed settlements and did not change their way of life to follow bison by horseback. They retained the agricultural lifestyle of their forebears and simply went on seasonal bison hunts, every summer and every autumn, in the neighboring countryside.

Mandan villages therefore developed more intricate topographies. Each Mandan village featured a formal town plan with a central plaza, where ceremonies and competitive games were held. Surrounding the plaza were dome-shaped earth lodges for families. Partly subterranean, lodges were snug in winter and cool in summer; their slightly flattened roofs also served as porches and proved an ideal place for relaxing at the end of the day. In addition, Mandan women were able to use the roofs as a place to dry corn and to store goods.

In the rich soil near the Missouri River, women raised corn, tobacco, sunflowers, pumpkins, beans, and squashes. But rich soil was not all that was needed for growing crops; the Mandan also regularly invoked divine assistance through elaborate agricultural rituals. Playing important roles in ceremonies of renewal were sacred ceremonial bundles of crops, each considered vital to the people's continued existence. The owner of

MANDAN A Mandan man watches smoke rise in an earth lodge on the Fort Berthold Indian Reservation in North Dakota.

the sacred corn bundle, for example, was responsible for calling the spirits of the corn back from the south before seeds could be distributed to women for planting.

Traditional Mandan culture was illustrated in paintings by 19th-century artists George Catlin and Karl Bodmer. They were also described in the journals of explorers Meriwether Lewis and William Clark, who spent the winter of 1804-05 with the Mandan. As portrayed in these works, the Mandan people lived in settled villages, and this tradition left them especially susceptible to European-introduced diseases such as smallpox, which killed more than 800 of their people in 1837.

Today, the Mandan share North Dakota's Fort Berthold Indian Reservation with the Arikara and Hidatsa. Known as the Three Affiliated Tribes, all look to a single tribal government, although each tribe maintains its distinct ethnic identity. The Mandan strive to keep their culture alive with language classes and with traditional religious ceremonies, beadwork and quillwork, and powwows.

Plateau & Great Basin

Rare rainfall and snowmelt in the mountains of Nevada and western Utah drain into valleys but go no farther, earning this area the name of Great Basin. Treacherous north-south mountain ranges thwarted many westward settlers. Even if they managed to make it over the Wasatch Range on the eastern side, they had to contend with peaks of up to 9,000 feet and long desert valleys before encountering the Sierra Nevada at the western end. The area is bordered by the Mojave Desert and the Grand Canyon's Colorado Plateau to the north.

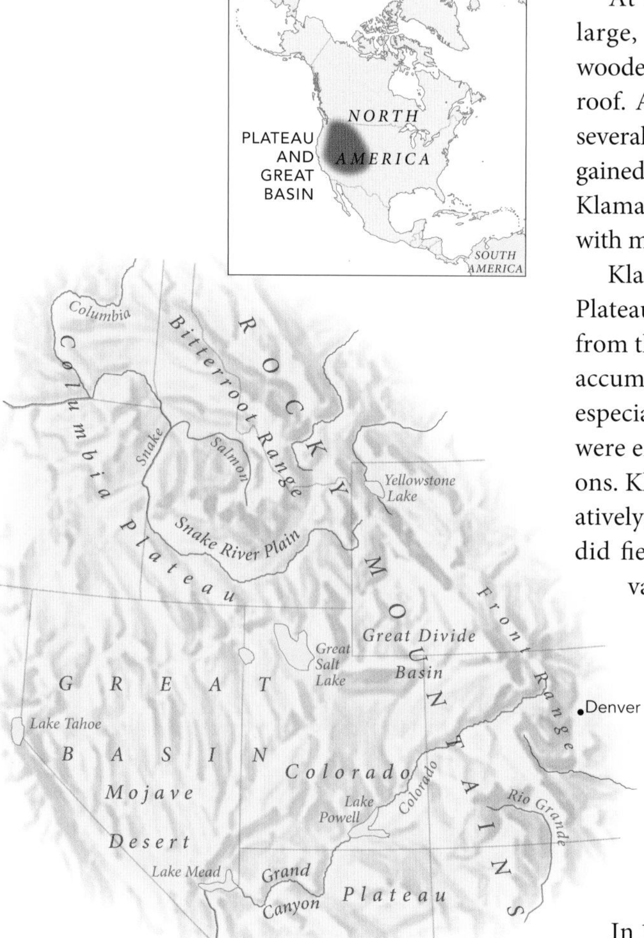

Klamath

Population: 4,400
Location: United States
Language family: Klamath-Modoc

The Klamath traditionally lived at the southern edge of the plateau region in southern Oregon and northern California. They were not a large, unified tribe but a series of tribelets that set up winter camps along the upland lakes, marshes, and streams that form the headwaters of the Klamath River. In the summer, they caught freshwater fish, hunted waterfowl and game, and gathered wild foods such as roots, berries, and *wokas*—water lily seeds that they ground into flour. All individual Klamath were encouraged to be industrious and to amass enough food to sustain them through the long winters.

At their winter camps, the Klamath constructed large, circular, semisubterranean lodges that had wooden posts for support and earth-covered logs for a roof. A single family typically lived in a small lodge; several families could fit into a larger one. Entry was gained through a hole in the roof. In the summer the Klamath erected smaller wooden structures covered with mats.

Klamath culture bore elements of the aboriginal Plateau culture combined with later elements added from the Northwest coast tribes, such as conspicuous accumulations of wealth, and also the Plains peoples, especially after the introduction of horses. Individuals were encouraged to stockpile shells, skins, and weapons. Klamath tribelets did not come together cooperatively, but they did join forces to wage war, which they did fiercely and frequently, taking slaves from the vanquished tribes.

Klamath shamans acquired powers through prayer, fasting, and visions, and they used their powers to stage dramatic healings among their people. Individual Klamath also went off by themselves to acquire spiritual insight. Adolescent boys were sent on five-day coming-of-age vision quests. Currently, they practice a religion that combines traditional and Christian beliefs.

In 1864, following years of hostilities, the Klamath ceded more than 13 million square miles of land to the

KLAMATH Preparing for the annual celebration of the Restoration of the Klamath Tribes in Oregon, a Klamath man adjusts his costume.

U.S. government and began reservation life, joined by the Modoc and Yahooskin to become the Klamath Tribes. They channeled their industriousness into new economic ventures such as cattle raising, freight hauling, and timber production. Tribe members took advantage of training opportunities and were able to find skilled local jobs.

In 1954, the government terminated recognition of the Klamath, sending into rapid decline one of the most successful and self-sufficient Indian tribes in the country. Almost immediately Klamath individual and collective economic fortunes plummeted, and within a few decades, more than 70 percent of the Klamath lived below the poverty level. But the tribe's traditional work ethic did not disappear with its recognition, and the Klamath organized legal efforts to regain it. They prevailed in the 1970s, when hunting and fishing rights were recognized, although no land was returned. In 1986, their tribal status was restored, but again without a land base. Today, the Klamath, with only a few thousand enrolled members, continue to press their claims and in the meantime have created many programs to benefit their people and the population of Klamath County, Oregon.

Nez Percé

Population: 6,700
Location: United States
Language family: Plateau Penutian

Now residing on a reservation in north-central Idaho, the Nez Percé are best known for their leader named Hinmuuttu-yalatlat ("thunder rolling down the mountain"), popularly known as Chief Joseph, whose valiant attempt to lead 200 to 300 warriors and their families to sanctuary in Canada is one of the most heroic—and tragic—feats of American history. Forced to relocate in 1855, when Oregon settlers wanted their land, the Nez Percé again were pressured to move eight years later, after gold was discovered on their reservation. Tempers flared dramatically in 1877, but when the Army refused to negotiate with

Chief Joseph, he and his people pushed north in a 15-week, 1,700-mile trek over the steep Bitterroot Range. Seeing his people cold and starving, Joseph stopped and delivered his legendary surrender pronouncement: "Hear me, my Chiefs! I am tired; my heart is sick and sad. From where the sun now stands I will fight no more forever."

The Nez Percé homeland—where the borders of Washington, Oregon, and Idaho meet and the Clearwater, Salmon, and Snake Rivers flow—provided the people with grassy valleys and forested mountains. In the spring, summer, and autumn, men hunted and fished while women gathered camas bulbs, roots, and berries. After the autumn bison hunt on Montana's eastern plains, the Nez Percé would return to winter camps in the sheltered valleys of their homeland.

The most renowned horsemen of their area, the Nez Percé carefully worked to improve the bloodlines of their herds. Men prized Appaloosas for their endurance and distinctive speckled markings, and women fashioned elaborately beaded horse trappings such as collars, bridles, and saddles. Although the people were forced to give up their 1,100 horses when they surrendered in 1877, today's tribe now has a few Appaloosas; they were a gift from a New Mexico Appaloosa breeder who donated ten mares to the tribally sponsored Chief Joseph Foundation in 1991.

The Nez Percé are very proud of their heritage. In 1988, they began the Cultural Resources Program as a means of preserving their language, history, and arts, and every July descendants of Chief Joseph's band travel to Oregon to celebrate Chief Joseph Days.

NEZ PERCÉ An Idaho Nez Percé mother leads her children on an Appaloosa, the breed favored by the tribe.

Northern Paiute

Population: 13,800
Location: United States
Language family: Uto-Aztecan

The homeland of the Northern Paiute once encompassed more than 70,000 square miles of Nevada, southern Oregon, western Idaho, and eastern California. It was part of the Great Basin, the only area in North America where nearly all groups spoke languages from the same language family; such uniformity came about probably because this resource-poor region never attracted a large, diverse population.

Today, Northern Paiute tribes have reservations in Nevada, California, and Oregon. Descendants of a cultural tradition that stretches back more than 10,000 years, the Northern Paiute thrived by developing an intimate knowledge of their environment. With limited precipitation and no rivers that drain to the ocean, the Great Basin forced its early inhabitants to establish and maintain an even more delicate balance with their surroundings than did people in more hospitable areas. Resources and terrain still vary widely here, from the freshwater marshland of western Nevada to the deserts and salt flats that make up the majority of the region.

The rugged environment supported small groups who could survive on intermittent harvests of roots, seeds, and berries. Seminomadic family groups endured by gathering a wide variety of plants that ripened at different times of year and by supplementing this activity with hunting and fishing. Groups with relatively richer, more specialized environments were able to remain for a season or part of a season in one location, while other groups had to follow a more nomadic or generalized seasonal routine, relying on an immense range of plants and animals. Each family belonged to a flexible camp group, which expanded and contracted according to the changing seasons and availability of resources. In autumn, groups came together to harvest pine nuts and undertake communal activities such as rabbit and antelope hunts.

Although each tribe has its own government today, not all groups have been able to develop viable tribal businesses. Some individuals raise cattle and grow hay and grain, while others make a living through wage work in a variety of jobs. In the 1980s, to preserve an important part of their heritage, some Northern Paiute tribes began conducting classes in traditional basket making, a highly developed art form that had been necessary for survival.

NORTHERN PAIUTE A Northern Paiute elder holds an eagle-wing fan and other tribal implements.

Most of the people now participate in some branch of Christianity; even so, some individuals retain aspects of their heritage through Paiute songs, stories, and prayers, and a few traditional healers continue to practice. The Ghost Dance Movement, which originated with two late 19th-century prophets who envisioned a revival of native culture, has come and gone. But since the 1930s, people have adopted a number of other tribes' religious movements and activities: the Native American Church, the Sweat Lodge, and the Sun Dance, for example. Because so few of their own traditions have survived, many young Northern Paiute join in these pan-Indian rituals as a way to strengthen their ethnic identity.

Northeast

The northeastern region of North America runs roughly from eastern Illinois in the United States up to Labrador and Newfoundland on the northeastern coast of Canada. States included in this region are Illinois, Indiana, Ohio, Pennsylvania, New Jersey, New York, Vermont, New Hampshire, Massachusetts, Connecticut, Rhode Island, and Maine; Canadian provinces are Ontario, Quebec, Newfoundland and Labrador, New Brunswick, Nova Scotia, and Prince Edward Island. Running from southwest to northeast like a crooked backbone through this entire region is the Appalachian Mountain range, which takes in smaller ranges, including the Alleghenies and the Catskills. Much of this area enjoys four distinct seasons, with cold, snowy winters, verdant springs, warm summers, and colorful autumns. Annual rainfall varies from more than 80 inches in Newfoundland to as little as 20 annually in Ohio and Indiana.

Americans

Population: 321 million
Location: United States
Language family: Germanic

Spanish settlers arriving in Florida in the mid-16th century represented the vanguard of the European presence that would create the nonindigenous American people. The establishment of permanent settlements of European colonists seeking economic opportunities and political and religious freedom in Virginia, Massachusetts, and New York in the early 17th century began the wave of colonization and immigration that continues in the United States to this day.

That colonization wrested the vast area that would become the United States and its bountiful resources from indigenous peoples, perhaps 12 million strong in 1500, who had settled its length and breadth. They produced cultures as diverse as the Anasazi of the Southwest and the Iroquois of the Northeast. By the late 19th century, most Native American groups had been decimated by conflict or disease or relocated to far smaller and inferior nontraditional lands. Today, Native Americans, also commonly called American Indians, make up only 1.2 percent of the population.

Although the arrival of each subsequent ethnic group contributed to the cultural fabric of the United States—creating the widely held characterization of the "melting pot"—the early influx of northern and western Europeans established many of the social, legal, linguistic, and religious foundations of American life. Subsequent immigrants streamed in from southern and eastern Europe, Latin America, Africa, and Asia. Today, Latinos make up about 17 percent of the U.S. population.

The importation of Africans to work as indentured servants and slaves is as old as permanent European settlement in North America. Slavery officially endured in the United States until 1863. In the more than century and a half since then, the circumstances of the 13 percent of the population that is African American remain a concern.

Even in the context of a diverse population, there are widely shared core beliefs tied to the notion of American nationality. They include freedom, equality,

AMERICANS Entrepreneurship starts at an early age for these young Maryland boys.

the rights of the individual, and the ability of anyone willing to put in the effort to achieve a self-defined good life—the so-called American dream. Many core beliefs reflect the ideals that prompted the late 18th-century revolution against British control and became the cornerstone of the U.S. Constitution. These ideals, coupled with opportunities, fuel both legal and illegal immigration to the United States. And while it is hard to ascribe shared personality traits to 321 million people, Americans in general are often characterized as open and direct; informal in manner, dress, and speech; and quick to connect with other people. These traits are tempered by a strong sense of personal space and privacy, which is being eroded by social media.

Although the United States has no federally mandated national language, English serves as the primary language, spoken by 98 percent of the population. Other languages commonly spoken include Spanish, Chinese, Tagalog, and Vietnamese. Religious freedom and the separation of church and state are hallmarks of American culture. Some 70 percent of Americans identify as Christian—46 percent of those as Protestant and 21 percent as Catholic. Other major religious affiliations include Judaism, Islam, and Buddhism. Nearly a quarter of all Americans say they have no religious affiliation, a proportion that has been rising steadily in recent years.

Americans celebrate a range of national holidays that are largely patriotic or commemorative, such as Memorial Day, Independence Day, and Thanksgiving. Christmas and Easter are prominent religious holidays, and many school systems officially recognize the Jewish High Holidays and Muslim holy days as well.

The American economy, still the world's largest by most measures, prospers from the country's extensive natural resources and strength in the areas of agriculture, manufacturing, finance, and the military. Business is rooted in the private sector, with more than a quarter of the world's largest companies headquartered in the United States. However, the disparity between rich and poor Americans continues to grow, with some 30 million people living below the poverty level. As consumers, Americans rank first in the world in many categories, from energy to food.

Sports, especially football, baseball, and basketball, engage huge segments of the American population. American cultural influence is pervasive worldwide, embracing sports, as well as the realms of food, fashion, and media and mass entertainment—particularly music. American-grown musical genres include jazz, blues, gospel, country and western, bluegrass, rock and roll, rap, and hip-hop.

Amish

Population: 81,000
Location: United States
Language family: Germanic

The Amish get their name from Jakob Ammann, a 16th-century Alsatian reformer who led a breakaway group of Mennonites, or Anabaptists, into a stricter interpretation of that radical European Reformation tradition. Anabaptists, found in Switzerland, Germany, and France, were persecuted for rebaptizing adults, believing that a mature confession of faith should precede baptism. In search of religious freedom, the Amish immigrated to North America in the 18th and 19th centuries, and the European communities eventually died out. Today there are 200 Amish settlements in 22 states and Ontario, Canada. The largest clusters are in

Pennsylvania, Ohio, New York, Indiana, and Illinois.

The Amish way of life is tied to *Gelassenheit,* or submission to a higher authority. Their code of conduct requires plain living, humility, obedience, community spirit, and suppression of individual ambitions. The Amish live apart in farm communities of some 20 to 35 families who make up congregations. They usually meet for services in homes or barns, led by a bishop, two preachers, and deacons. Those who disobey the *Ordnung,* or law, may end up being shunned by the entire community.

The Amish are perhaps better known as the Pennsylvania Dutch, which is a misnomer from the word for German—*Deutsch*—or the name they call themselves, Deitsch. All Amish speak a German dialect, and they also speak English and often standard German.

Though the Amish dress plainly, they wear colors—green, blue, and lavender shirts and dresses are common—but not patterns. Men wear dark pants, vests, and hats and grow full beards, but they keep the upper lip clean-shaven. Women wear dresses with aprons and caps or bonnets. No makeup or jewelry is worn, not even wedding rings.

Old Order Amish ban the use of electricity, telephones, cars, and many other modern technologies and amenities. Travel and transport occur by horse and buggy, although Amish may ride as passengers in cars. New Order Amish may use phones installed in rural areas and draw electricity off 12-volt batteries; some use tractors. Variations in what is allowed can occur among different communities in the same region.

AMISH An Ohio Amish woman pours raw milk into a can on the family dairy farm.

The Amish operate their own one-room schools that mostly advance only to the eighth grade, a practice that was upheld by the U.S. Supreme Court. They hire their own teachers from within the community. Secondary and higher education are deemed unnecessary for Amish life and unwise for the temptations they may pose to Amish adolescents.

Nevertheless, Amish young people are given a culturally sanctioned opportunity called *Rumspringen,* or running around, to explore forbidden activities such as driving, dating, drinking, and unsupervised travel, beginning at age 16. Some of their exploits now end up on TV and the Internet. Despite the taste of freedom, six out of every seven return home and opt for full incorporation into the Amish community.

The Amish tend to marry young and to select their own spouses. Most marriages occur on Tuesdays or Thursdays during the postharvest wedding season, from October through December. Weddings often draw in the whole community. The couple's parents may give furniture, farm equipment, and other substantial gifts.

Large families are common among the Amish—on average they have seven children—and houses are correspondingly large. The family usually sits down to eat meals together. During the school year, the main meal is taken in the evening; it shifts to midday in summer. The Amish eat plainly but not blandly, and pickled vegetables are a favorite food.

The Amish celebrate Christian holidays but without the secular trimmings. In addition to religious observances, Christmas involves festive meals and some gift giving, but no Christmas tree or Santa Claus.

The Amish work hard, but they also like to have fun—a capella singalongs are a popular pastime. They also play softball and volleyball and go hiking, fishing, swimming, skating, and sledding. Amish travel to visit relatives and attend weddings, and some escape to Florida for a few weeks in winter to stay in the Amish village in Sarasota.

Land pressures have made it harder for the Amish to stay in farming, and so they have turned to cottage

CANADIANS Canadians support ice hockey at all levels. Here, they cheer on the junior national team in a game against Finland.

industries such as furniture making, shed building, and shopkeeping. New ventures mean new situations that often take the Amish a bit further from their traditions.

Canadians

Population: 35 million
Location: Canada
Language family: Germanic (English), Italic (French)

The second largest nation after Russia, Canada encompasses diverse terrain and climate, including temperate coasts, windblown plains, the northern reaches of the Rockies, a vast expanse of boreal forest, and challenging Arctic ice and tundra.

Canada, with a population of only 35 million, is enormously underpopulated for its size. It has slightly more territory than the United States, but only one-ninth the population. Most of the population clusters in Ontario and Québec, and 90 percent of all Canadians live within 100 or so miles of the U.S. border. A push to encourage settlement began in the 1890s, not long after Canada achieved dominion status from Great Britain. Canada continues to attract immigrants and has experienced recent influxes from Asia and Africa.

Canadian culture has been formed from a mix of British, French, and native traditions, with a significant dose of American influence, although Canadians view themselves as more tolerant, more polite, and less aggressive than Americans. Ethnically, Canadians identify as 28 percent British Isles origin, 23 percent French, 15 percent other European, 6 percent Asian, 6 percent African, and 2 percent First Nations, the name used for indigenous peoples. Some 28 percent of Canadians identify themselves as "mixed."

Canada recognizes two official languages, English and French, spoken at home by 64 percent and 21 percent of the population, respectively. The French speakers derive mostly from Canada's French colonial past. Other common languages are Chinese, Italian, and Punjabi. Regarding religious affiliation, 40 percent of Canadians are Catholic, 20 percent are Protestant, and there are sizable numbers of Muslims, Hindus, and Sikhs.

In recent decades, the Canadian government has addressed past wrongs done to native populations. From the first European contact in the 17th century, native groups have suffered land theft, decimation, relocation, and discrimination, and were dragged into conflicts between the British and the French.

Canada's ten provinces and three territories sort themselves into regions that exhibit distinctive general personalities and evoke strong allegiances. Broadly speaking, the Atlantic provinces have a reserved and somewhat provincial character, while western Canada, especially British Columbia, is more open and progressive. Ontario skews conservative as the center of government and business. The North displays a pioneer spirit, with an economy rooted in natural resources. And Québec forms the epicenter of French culture and nationalism. The Québec independence movement is a hot-button issue that flares periodically. In a fairly short time, Canada has transitioned from a rural to an urban economy, which now centers 70 percent on services, 28 percent on industry, and only 2 percent on agriculture.

It is good to tell one's heart.

CHIPPEWA PROVERB

Canadians rally around Canada's iconic institutions, including its national emblem, the maple leaf; its distinctive branch of law enforcement, the Royal Canadian Mounted Police; its national animal, the industrious beaver; and its national sport, ice hockey. The Mounties, known for their bright red tunics and flat-brimmed hats, police at the federal as well as the national, provincial, territorial, and municipal levels. Not all jurisdictions contract for their services, but many do.

National celebrations include Canada Day on July 1, a date commemorating the union of British colonies in 1867. Typical events include picnics and fireworks. Canadians also recognize their connection with other nations associated with Great Britain on Commonwealth Day in March. Thanksgiving occurs on October 1. Like its U.S. counterpart, the holiday focuses on a gathering of family and friends and a festive meal, often showcasing a turkey.

Canadian cuisine incorporates contributions from numerous ethnicities, including traditional English (Sunday roast and two veg), French Canadian (poutine, a heap of French fries with gravy and cheese curds), and First Nations (bannock, a flatbread adapted from the Scots by the Inuit). Multicultural Canada strives to be tolerant and inclusive of all its constituent peoples.

Chippewa

Population: 175,000
Location: United States, Canada
Language family: Algonquian

One of North America's largest Indian groups, also called the Anishinaabe or the Ojibwe, the Chippewa live on reservations or reserves in Michigan, Wisconsin, Minnesota, North Dakota, Montana, and the Canadian provinces of Ontario, Manitoba, and Saskatchewan. They all once spoke an Algonquian language and dwelled in dome-shaped wigwams covered with birch bark. Because wild plants and game in their region did not support large villages, they formed widely scattered, autonomous bands of approximately one hundred or so people. In the fall, hunters traveled as far as a hundred miles to track moose, woodland caribou, and bears—animals with which people also sought, through prayer and ceremonies, to establish spiritual connections central to their religion.

Joining other families in maple groves near rivers and lakes, the Chippewa marked the end of winter with sugar-making activities. From there, they proceeded to summer villages, where they spent the warm months tanning hides and building the birch-bark canoes for which they became famous. At summer's end, families harvested corn, squash, and perhaps wild rice, a seed-bearing grass that grows along the muddy shores

CHIPPEWA A child reads to her mother and siblings on the Fort Bois Chippewa Indian Reservation in Minnesota.

DEEP ANCESTRY

THE WANDERER

Ricardo Hernandez-Bonifacio heard about the Genographic Project from a co-worker, and he was intrigued by the idea that his DNA could reveal details about his family history. His wife also tested herself, and he was interested in seeing how his migratory patterns differed from hers. He swabbed his cheek and sent his sample in, hoping to get more information.

Ricardo hails from El Salvador, and he moved to the United States when he was a child. He knew that the Spanish had conquered Central America in the 16th century, so he expected a mix of European and Native American markers. What he found surprised him. His highest regional percentage was Native American, at 48 percent. "I would have expected the European components to be larger than my Native American," but Ricardo retains a strong genetic link to the first settlers of the Americas.

DNA Results

The story of his family's journey to Central America is told in his Y-chromosome result. Ricardo belongs to haplogroup Q-M3, which is the major Y-chromosome haplogroup in the Americas. It was among the first male lineages to enter the Americas, around 15,000 to 20,000 years ago, via a land bridge connecting Alaska and northeast Asia. This was during the time of the last glacial maximum, the coldest part of the last ice age. Sea levels were as much as 300 feet lower than today, exposing a desolate finger of land that connected the two continents. Following caribou, the first Native Americans walked across the land bridge from one hemisphere to the other.

It is likely that the initial groups were relatively small, and consistent with this, there are a limited number of genetic lineages found in the Americas. There were likely four founding maternal lineages (mitochondrial haplogroups A, B, C, and D) and one major male lineage (Q, the ancestor to Q-M3). The M3 marker defining Ricardo's lineage likely appeared early on as the first migration was taking place, and men carrying the M3 marker spread rapidly south and east over the next thousand years. Today, it is the dominant male lineage throughout the Americas.

Ricardo Hernandez-Bonifacio, Y-chromosome haplogroup Q-M3

PATERNAL MIGRATION
HAPLOGROUP Q-M3

Paternal Q-M3: During the last ice age, present-day Alaska and Asia were connected, creating an entryway for early Americans.

The Next Generation

Ricardo said that the result "has given [him] a point of reference so I can research my history further, and it has changed how I will tell my story to my children." While the tale of the arduous journey made through the coldest weather the world has ever experienced resonates with him, it is the tangible genetic evidence of his connection to Native Americans that is the most important to Ricardo—that, and how his own children will carry that connection into the next generation.

—*Spencer Wells*

of some streams and marshes. Wild rice often made survival possible in regions where a short growing season made agriculture unreliable.

The Chippewa performed the ceremonies of the *Midewiwin* (Great Medicine Society), a secret group of men and women healers whose primary purpose during the rituals was healing the sick and initiating new members who had served long, demanding apprenticeships. Today, the Chippewa continue to carry out the two- to five-day ceremonies, as well as traditional forms of hunting, fishing, sugar mapling, wild rice gathering, and beadworking—all important markers of their ethnicity. They use drum circles, dances, canoe building, social media, and Anishnaabemowin language classes to sustain their worldview and traditions.

Cree

Population: 90,000
Location: Canada
Language family: Algonquian

The Cree were a widely dispersed people who at one time ranged from Labrador in eastern Canada all the way to Alberta in the west and as far north as the Northern Territories and south into North Dakota and Montana. Their territory encompassed the northern boreal forest, the tundra, and part of the northern plains. This harsh environment had short summers and long, cold winters followed by strong melts and flooding.

The name *Cree* probably comes from *Kristeneaux,* a French corruption of the name of a particular Cree tribe. The Cree language belongs to the central Algonquian group and today is spoken by tens of thousands of people.

Cree men were skilled at hunting and trapping. The hunters roamed together in small bands, which formed larger communities in the summer. In addition to fishing, these groups of men hunted moose, elk, caribou, deer, and bear. Cree women contributed to the food supply by gathering roots, berries, tubers, and eggs.

The Cree also kept domesticated dogs that they used as pack animals and for hauling. The dogs could carry parts of dismantled housing, with long poles strapped across their chests. When the travois, a kind of runnerless sled, was introduced in the area, dogs were harnessed up to pull loads.

The Cree way of life began to change irrevocably when French and English trappers and traders arrived in their territory. The Cree were often pressed into service as guides for the new arrivals. They also worked as trappers and middlemen. They traded pelts for desirable goods, including household items, tools, and especially firearms, which aided in expanding their territory. Cree communities sprang up around the trading posts, and tribe members became dependent not only on the goods but also on alcohol, which, along with disease, took a heavy toll on the population.

The Cree succumbed to the major outbreaks of smallpox in the 18th century, and to subsequent epidemics of influenza and tuberculosis in the 20th century. Each time disease struck, their population plummeted. Today, all the different groups of Cree join together to represent one of Canada's largest groups of indigenous people, living in Ontario, Alberta, and Quebec.

As some Cree began to acquire horses in the 18th century, they took on elements of Plains Indian culture, including hunting bison, the use of horses, as well

CREE William White Buffalo of the Cree Nation poses at the Talking Stick Aboriginal Arts Festival in Vancouver, British Columbia.

as large-scale intertribal warfare with scalp taking.

Traditionally, the Cree people were animists, and they believed that spirits inhabit all living beings as well as some nonliving objects. Their belief in the *manitou,* or a singular Great Spirit, may have developed after the influence of European contact, which introduced Christianity and monotheism. Cree individuals by tradition sought to enhance their own power through visions, sometimes induced by a sweat lodge ceremony, in which they tapped into the spirit world. In times of misfortune or illness, they consulted medicine men who used their own enhanced powers for divination and healing. The Cree were heavily missionized through their trading relationships with the French and the British, and most became at least nominally Roman Catholic or Anglican.

In the 19th and 20th centuries, the Cree in both Canada and the United States were collected into reserves and reservations. Some halfhearted attempts were made to adapt their lifestyle to farming, and if a number of Cree did succeed, it was through their own efforts. As they lost their traditional way of life, the Cree shared the problems that plagued other native groups, including rampant unemployment and alcoholism. These issues continue to confront the Cree to the present day. Their way of life also faces problems from negative effects of oil drilling and hydroelectric projects, both of which threaten to further erode their quality of life.

Iroquois

Population: 81,000
Location: United States
Language family: Iroquoian

Now living on reservations and reserves in upstate New York and Canada, the Iroquois people are best known for the powerful league formed by five Iroquoian-speaking nations to peacefully resolve conflicts. The league, which gave fair representation to each group, was established sometime between A.D. 1400 and 1600 after a long history of bloody intertribal feuding. Within the member tribes—Mohawk, Oneida, Onondaga, Cayuga, and Seneca (and later, the Tuscarora)—clan mothers chose sachems to represent their family lineages on the basis of their integrity, wisdom, vision, and oratorical ability. The 49 sachems composed the governing council of the league, also served on their separate tribal councils, and settled disputes through diplomacy and ceremonies.

IROQUOIS Dancing forms part of an annual Mohawk festival in Fonda, New York. Mohawks are one of the six Iroquois nations.

Inspired by the Iroquois model that put values of unity, democracy, and liberty into action, Benjamin Franklin and other leaders of the American colonies sought the help of the Iroquois in establishing a fair and democratic colonial union. Their meeting in 1754 produced the Albany Plan of Union, a precursor of the Articles of Confederation. Today, ideals that the Iroquois League espoused continue to guide such organizations as the United Nations.

The Iroquois call themselves the Haudenosaunee—People of the Longhouse—in reference to their traditional homes. Accommodating as many as a dozen families, each bark-covered dwelling was as long as 200 feet, with a width of 25 feet; families had separate apartments but shared a fire with others. Iroquois settlements held between 30 to 150 longhouses and were protected by palisades.

Large numbers of Iroquois men and women now pursue urban lifestyles, holding jobs in industry, education, and other fields, but they return to the reservations for ceremonies. They also still play lacrosse. The Iroquois version of this Native American game was the basis for today's sport, and it is a mainstay of ethnic identity.

To control land claims, counter environmental pollution, and keep their traditions alive, the people have revitalized the Six Nations Iroquois Confederacy, as it is known today. The confederacy participates in international forums related to the rights of indigenous peoples and even issues its own passports.

Southeast

Known for its summertime heat and humidity, the Southeast states are Maryland, Delaware, Virginia, West Virginia, North Carolina, South Carolina, Kentucky, Tennessee, Georgia, Florida, Mississippi, and part of Louisiana. The region also contains Washington, D.C. Florida, jutting southeast from the mainland between the Atlantic Ocean and the Gulf of Mexico, has a subtropic climate and is highly vulnerable to hurricanes and tropical storms, as are the coastal regions of the other Gulf Coast states. Coastline along the Atlantic in this region features wide, sandy beaches, mild waters, and tame surf. The Appalachian Mountains, which run from Maine to Georgia, form the largest mountain range in this region and one of the oldest in the United States.

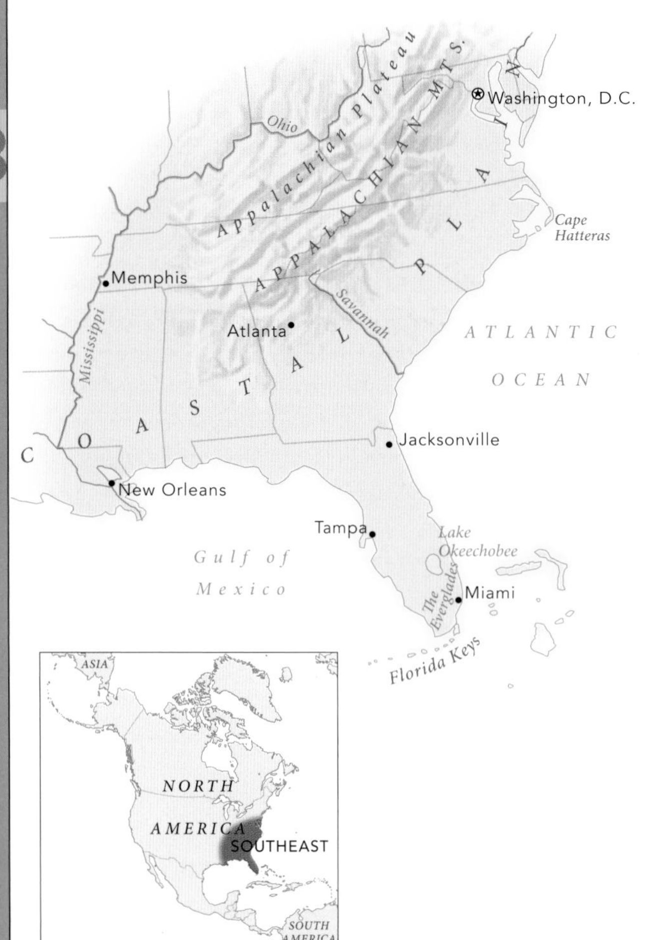

Cajuns

Population: 600,000
Location: United States
Language family: Romance

In order to understand the Cajuns, it is necessary to know their backstory—the long journey that brought them to southern Louisiana. The French had settled in Acadie (present-day Nova Scotia) in 1604 and became a colony of farmers, trappers, and fishers whose settlements resembled those of their northern French homes.

But in 1755, after the British had won France's Canadian possessions and were themselves becoming established there, they abruptly exiled the Acadians, an event known as the *Grand dérangement,* or Great Displacement. Some 8,000 were rounded up and loaded onto ships, with no attempt made to keep families together, and sent on their way to destinations on the Atlantic coast. Those who made it all the way to Louisiana, sometimes after a lapse of 20 years, received a warm welcome from the French government there. The Acadians (whose name eventually was corrupted to Cajuns) were given land and guidance on settling it.

Some Acadians became farmers, concentrated on the banks of the Mississippi River, the Bayou Teche and Bayou Lafourche, and on the prairies to the west, where they also raised cattle. Those who settled in the swamps and marshes and along the coast became hunters, trappers, and fishers. They plied the waters in small boats called pirogues, which they often dug out from cypress logs. The French, Spanish, and Creoles the Acadians encountered in Louisiana were of a higher class than the newcomers, but intermarriages occurred, as they did also with local Indians. In general, though, Cajun communities tended to be isolated and also insular, because people were encouraged to marry within them.

Cajun French is a somewhat archaic form of French with a simplified grammar, combined with contributions from English, Spanish, and Indian and African languages. The speaking of Cajun French was outlawed for a time in Louisiana and English made compulsory. Cajun French is spoken less often now, especially by the younger generation. Most Cajuns use English regularly.

The Cajuns are largely Roman Catholic, a fact that differentiated them from many other Louisiana settlers. Religious observance is tempered with a sanctioned

CAJUNS What's a little mud to a dancing Cajun couple? They dip at the Festivals Acadiens et Creoles in Lafayette, Louisiana.

joie de vivre, expressed in the saying "*Laissez les bons temps rouler!*—Let the good times roll!" Most Cajun life events call for elaborate celebration, usually with music and dancing. The highlight of the calendar is still Mardi Gras (Fat Tuesday), a days-long festival prior to the beginning of Lent.

The Acadians brought a musical tradition that included *complaintes,* a capella songs describing the trials and tribulations of life. They took up the fiddle, creating spirited accompaniment. The accordion joined the mix in the early 19th century, and eventually other percussion instruments, such as the washboard and triangle, were added. Dancing was and is synonymous with Cajun music, especially the two-step and waltz. Most Cajun communities held a weekly late-night dance called a *fais do-do*—literally, "go to sleep"—because children were put to bed there. These dances offered opportunities for supervised courtship, with the young men sometimes held in a pen called *une cage aux chiens,* a dogs' cage, unless they were actually dancing.

Cajun food fuses Acadian French with local cultures and ingredients, such as bountiful shrimp and crayfish. Many Cajun dishes start with a roux—butter, lard, or other fat mixed with flour and cooked over low heat. Typical dishes are gumbo (from an African word for okra), a spicy mixture of meats and vegetables that often includes ground sassafras filé for seasoning, and jambalaya, a rice-based dish that may include ham, sausage, poultry, and seafood.

Today, Louisiana's Cajuns operate in all social and economic spheres. Many are employed in commercial fisheries and in offshore oil and gas drilling and exploration. Cajun landholdings have declined over time; much of their former land is now owned by large corporations. Mostly spared by Hurricane Katrina in August 2005, Cajun Country was hit by flooding from Hurricane Rita less than a month later.

Cherokee

Population: 819,000
Location: United States
Language family: Iroquoian

For centuries, the Appalachian highlands in North Carolina and Tennessee were home to the Cherokee, the largest tribe in the southeastern United States. Then white settlers pushed across the Appalachians, taking over the fertile farmland of the Cherokee and other Indian groups in the region and forcing their removal to Indian Territory (now Oklahoma).

The Cherokee, who with the Chickasaw, Choctaw, Creek, and Seminole were known as the Five Civilized Tribes, had adapted to white culture and owned houses, farms, livestock, and slaves. By the 1820s, thousands of the people had learned to read and write their own language, using a syllabary devised by a Cherokee named Sequoyah, the only man in history known to have created a writing system single-handedly. His writing system had 85 symbols, most of which signified a consonant followed by a vowel sound in the Cherokee language. Tribal doctors recorded their medical beliefs and practices with Sequoyah's invention, and as a result much is known today about traditional Cherokee medicine.

In 1838, President Andrew Jackson enforced the Indian Removal Act of 1830, evicting 18,000 Cherokee from their homes in the Southeast, even though the law had said the people must give their consent and the Supreme Court had affirmed their rights to remain. Under conditions so brutal that 4,000 died, the Cherokee were marched westward; over the next decade, 60,000 Native Americans would walk the Trail of Tears to Oklahoma.

CHEROKEE A Cherokee man appears in full regalia for a powwow in Cherokee, North Carolina.

Some Cherokee managed to elude capture by hiding in North Carolina's mountains, and eventually their right to remain in the East was recognized by both the federal and state governments. In 1889, they became known as the Eastern Band of Cherokee Indians. Members of this band still tell the story today of Cherokee removal in their revered outdoor drama, *Unto These Hills.*

When the majority of Cherokee arrived in Oklahoma in 1839, they adopted a new constitution and began to establish schools, churches, and businesses. They also started to print their own newspaper and periodical. Soon the Cherokee had a higher literacy rate than their white neighbors did.

In 1984, the Eastern Band of North Carolina and the Cherokee Nation of Oklahoma met in council for the first time in nearly 150 years; since then, they have met every two years. Both groups have elected women as principal chiefs, evidence that women play important roles in Cherokee society. Both have also worked to keep their culture strong through the continued use of their own language and by participation in such age-old activities as the stickball game. Traditional basket making, long a vital part of the identity of Cherokee women, is still carried out in North Carolina, where, in 1996, the Eastern Band used their casino profits to buy farmland encompassing Kituhwa, the Cherokee origin place and sacred burial ground.

Creek & Seminole

Population: 88,300 (Creek), 32,000 (Seminole)
Location: United States
Language family: Muskogean

The Creek of Alabama and eastern Georgia probably were descendants of prehistoric Mississippian temple mound builders who had migrated eastward. They maintained a complex and loose confederacy of both Creek and non-Creek tribes, anchored by the core Creek group, the Muskogee.

GENOGRAPHIC INSIGHT

EARLY IN THE AMERICAS

Life-forms are made up of organic matter, all their molecules containing carbon. Two stable, nonradioactive states, or isotopes, of carbon occur throughout the natural world: carbon-12 and carbon-13, so named for the combined number of protons and neutrons in each atom.

Fortunately for archaeologists, a third isotope, carbon-14, differs from the others in that it is unstable and over time decays into a different atom entirely. Carbon-14 results when cosmic rays bombard Earth at 30,000 to 50,000 feet above sea level and force nitrogen in the atmosphere to undergo a nuclear reaction.

Like other carbon atoms, the resulting carbon-14 can combine with oxygen to form carbon dioxide and then enter plants through the process of photosynthesis. Plants, and the animals that eat plants, ingest minute amounts of this harmless unstable isotope—and thereby incorporate into their bodies a timekeeper that will keep going tens of thousands of years after their bodies expire.

Carbon-14 atoms decay at a known constant rate into stable, nonradioactive nitrogen-14. Because their life cycle follows a predictable pattern, the proportional amounts of carbon-14 and nitrogen-14 that can be found in decayed life-forms can be measured and used to infer how much time has passed since the organism was living.

Telltale Atoms

Although the stone tools humans used are inorganic, they tend to accumulate in places where humans congregated—places where they eat, cook, hunt, and are buried. These are the areas where archaeologists can find abundant organic deposits, such as fire hearths scattered with charred seeds or hunting grounds replete with animal bones and spear points. If the sites are not too old, the organic material left behind, analyzed with radiocarbon dating techniques, can help anthropologists date human activity at a given location.

Native Americans often carve alabaster into implements and decorative sculptures.

In 1908, a deadly storm known as the Folsom Flood swept across southwestern North America, killing 17 people and causing property damage that took months to repair. That year a cowboy mending a fence on his New Mexico ranch noticed a few bones protruding from the exposed soil below. Having seen his fair share of bones from years on the range, George McJunkin quickly deduced that these were much larger and older than what he typically came across and conjectured (correctly) that they belonged to an extinct species of bison. He also discovered an arrowhead, about 13 feet down. Owing to the depth of the find, he believed both bones and point must have been quite old.

For decades, that arrowhead remained on McJunkin's mantel. Then a number of similar findings from New Mexico increased interest in the region. Dating the sites sent shockwaves through the anthropological community: Results showed that humans had been in New Mexico for at least 11,000 years. The spearheads from these sites came to characterize a prehistoric Native American culture known as Clovis, named for the town in New Mexico where the first site was systematically excavated.

Doubts

Clovis sites have since been identified throughout the contiguous United States, Mexico, and Central America, leading to the presumption that all indigenous cultures of North and South America descended from the Clovis people. New doubts have since arisen. Sites such as Monte Verde in southern Chile have been dated to 12,500 years old, indicating that humans arrived there far earlier than the first Clovis people.

Though the Clovis debate lives on, the Southwest's importance in Native American history is undisputed. The region saw several later cultures flourish, including that of the ancient Pueblo or Anasazi people, who lived along on the Colorado Plateau before climate change in the 12th and 13th centuries turned the landscape dry and forced a people who had thrived for more than 2,000 years to quickly abandon their settlements.

Creek towns tended to be large—from 100 to 1,000 residents—headed by a *miko,* or chief. The miko presided over tribal council meetings (which began with participants ingesting a heavily caffeinated "black drink") where decision-making was based on consensus. Within the towns were family compounds of one to four buildings, depending on the owner's wealth. The Creek led a very active ritual life, including a number of traditional "stomp" dances performed on the community's stomp ground.

In colonial times, the Creek became allies of the English, although they were not above playing Europeans off each other. In the 1700s, after the English and Creek had all but destroyed the Apalachee and Timucua tribes of Spanish-held northern Florida, some Creek began moving into that region. They then became known as the Seminole, from the Muskogee word *simanóli,* meaning "runaway" or, more appropriately, "pioneer," which refers to the tribe's move to a new homeland, away from encroaching planters.

Border friction between white slaveholders and the Seminole intensified, though, because the latter treated African Americans as equals, giving haven to runaway slaves, intermarrying with them, and heeding their advice in council. By the 1830s, when President Andrew Jackson signed an act to relocate the Seminole into Oklahoma, small bands were waging fierce guerrilla warfare in what came to be considered the longest and least successful Indian war in U.S. history.

Determined to preserve their freedom, the Seminole managed to carry on even after Osceola, their most famous leader, was captured under a flag of truce in 1837 and died in captivity the following year. From 3,000 to 4,000 Seminole, along with the remaining Creek in their home territory and other tribes, were forcibly marched to Oklahoma along the Trail of Tears. A few hundred Seminole managed to escape by fleeing south into the Everglades, eluding pursuers who in time gave up looking for them.

By the early 1900s, Florida's Seminole were trading alligator hides and egret feathers to non-Indians for cotton cloth, pots, and metal tools. In 1928, they started selling their crafts along the recently built Tamiami Trail, which linked Tampa to Miami via the Everglades.

SEMINOLE Choosing a favorite necklace is a difficult task for this Seminole girl in Florida.

Over the next three decades, as increasing numbers of tourists began to flock to Miami, non-Indian entrepreneurs invited the tribe to create and work in Seminole "villages." Alligator wrestling, popularized at this time, and the machine-sewn patchwork clothing women began to make around 1900 have continued to promote a strong sense of tribal identity.

In 1957, the Seminole Tribe of Florida received federal recognition. Four years later, the Miccosukee, a tribe long incorporated into the Seminole, broke away from them. More conservative in social and religious values, the Miccosukee continue to follow the political leadership of their medicine men and council. Today, both tribes attract European and American visitors to their re-created villages, where museums tell the history of their people. They also operate successful ecotourism enterprises that take visitors through the Everglades in hovercrafts and swamp buggies. In contrast, most Oklahoma Seminole rely on jobs in the oil industry or in construction, manufacturing, and agriculture. Oklahoma Creek reside mostly on reservations and in tribal towns, and hold tribal jobs or work in bingo halls or on farms. Oklahoma Seminole and Creek maintain some of their traditions, including dances, on tribal stomp grounds there.

Lumbee

Population: 73,700
Location: United States
Language family: Algonquian (extinct)

When European settlers in North Carolina first came face to face with the Lumbee, sometime around 1730, they found a people who looked like members of the other Indian tribes they knew, yet they spoke only English. Moreover, the Lumbee were settled farmers without any vestige of an Indian lifestyle whose appearance also showed both Anglo and African features.

The detailed facts of Lumbee origin remain a mystery to the Lumbee themselves and to those who have studied them. One popular explanation tied the Lumbee to the original English settlers of Sir Walter Raleigh's Lost Colony of Roanoke Island who vanished some time between 1587 and 1590; it has been pointed out that the Lumbee share about 20 surnames with those settlers. More recent scholarship suggests that Lumbee ancestry incorporates remnants of Cherewan and other Siouan tribes who once lived in the area; some Cherokee, Tuscarora, and Croatoan Indians; and some Algonquian and Iroquoian speakers as well. Clearly there were also white and African contributions to Lumbee population genetics.

LUMBEE Lumbee law student Anthony Locklear worked as an intern at the U.S. Department of the Interior.

For nearly three centuries, the Lumbee have had an ongoing presence in the former marshland of the southeastern corner where North and South Carolina meet. The area is watered by the Pee Dee River and the Lumber (formerly Lumbee) River, from which the Lumbee get their name. Most Lumbee land is now in Robeson and surrounding counties. There are also a few scattered Lumbee populations elsewhere in the Southeast and in Baltimore, Philadelphia, and Detroit.

The Lumbee lived an isolated yet parallel existence as Protestant farmers with the English, Highland Scots, and Scots Irish settlers, until land competition began to draw attention to their status. In 1835, the state of North Carolina took away the voting rights of all "free persons of color," including the Lumbee.

The now disenfranchised Lumbee endured further oppression during the Civil War. After conscription into forced labor, they were starved and mistreated while building fortifications near Wilmington. This abuse created a fired-up Lumbee avenger: Henry Berry Lowry, who had also watched the North Carolina Home Guard shoot his father and brother. He and his triracial followers took 18 lives during an eight-year spree, which made Lowry a man wanted by every conceivable branch of law enforcement. He was caught and escaped a number of times until finally, in 1872, he disappeared and was never seen again. Lumbee today acknowledge Lowry as a tribal hero.

What the Lumbee lack in specific details about their distant Indian origins they make up for with detailed knowledge of their personal genealogies. Family is a cornerstone of contemporary Lumbee life, and extended family members keep in constant contact with each other. Lumbee marry other Lumbee at a rate of some 70 percent, and most tribe members in Robeson County attend one of 130 Lumbee churches there, mainly Methodist and Baptist, which also serve as community resource centers. Children attend many Lumbee-majority schools, and higher education needs are met in part at the University of North Carolina at Pembroke, once a Lumbee teacher-training college. The Lumbee annual homecoming each summer brings thousands of Lumbee to Robeson County for parades, cookouts, competitions, and informal get-togethers.

Seek wisdom, not knowledge.
Knowledge is of the past.
Wisdom is of the future.

LUMBEE PROVERB

Contemporary Lumbee have entered almost all areas of the local economy, although some still farm. As a tribe with more than 70,000 members, they zealously pursue both the Lumbee cause and a pan-Indian identity. For more than a century, they have tried to gain full federal recognition as an Indian tribe. In 1987, the Lumbee submitted to the Bureau of Indian Affairs an exhaustive, three-volume petition believed to cover all the criteria required for complete recognition by the federal government. They were then surprised to learn that a loophole in the 1956 act giving them partial recognition prevented the petition from even being heard. The Lumbee continue with their cause and have lobbied for the legislative change necessary for the petition to be considered.

Gullah & Geechee

Population: 250,000
Location: United States
Language family: English and African-based Creole

The barrier islands and tidewater area from the Carolinas through Georgia into northern Florida are the traditional home of the Gullah and Geechee. Also known as the Sea Islanders, they descend from freed slaves who formerly worked the rice fields and cotton plantations in this low-lying region.

The Gullah and Geechee arrived in the Southeast as slaves in the 1700s. Linguistic and other research has determined that they most likely came from Sierra Leone and nearby countries in West Africa. Some spent time in Barbados before reaching the Carolinas, where many entered through St. Helena Island near the port of Charleston, South Carolina.

Sierra Leone lies on the "Rice Coast" of Africa, and experience with that difficult and labor-intensive crop meant that slaves from the region were highly prized by low-country plantation owners. These West Africans also possessed a high tolerance to diseases rife in the semitropical coastal areas, such as malaria. When the owners fled inland to escape the heat, humidity, and bugs, as they often did from April to October and sometimes permanently, the Gullah and Geechee remained in charge and alone, giving them a degree of freedom unknown to slaves elsewhere.

At the end of the Civil War, the Gullah and Geechee were freed and offered the chance to buy island land in parcels rarely larger than ten acres. The ability to farm their own plots allowed them to escape the fate of other freed slaves, many of whom became sharecroppers. Still isolated by the rivers, swamps, and marshes separating them from the mainland, the Gullah and Geechee have been able to retain more aspects of their traditional culture than any other former American slave population.

The Gullah-Geechee language, which has drawn scholars to the islands since the mid-20th century, shares vocabulary and syntax with Krio—an English-based Creole—spoken in Sierra Leone. It also contains elements of English as it was spoken in the colonies in the 17th and 18th centuries, and some Barbadian influences. In addition, Gullah and Geechee oral

GULLAH & GEECHEE Gullah and Geechee ride the ferry to Georgia's Sapelo Island to attend a church celebration.

traditions include songs and fragments of tales passed down from non-Creole West African tribal languages, notably Mende and Vai. (Probably the names *Gullah* and *Geechee* are adapted pronunciations of West African tribal names.) A 1979 survey identified 100,000 speakers of Gullah-Geechee, with fully 10 percent speaking only Gullah. Although Gullah-Geechee remained an unwritten language until recently, the colorful animal characters of its folktales, such as Br'er Rabbit and Br'er Fox, were immortalized in the 19th-century Uncle Remus stories written by Joel Chandler Harris.

While continuing to work as farmers, the Gullah and Geechee have practiced traditional crafts and sold their ware to outsiders. Women weave fine, circular sea-grass baskets that are identical to those found in Sierra Leone. Men traditionally carve, making grave monuments, human figures, and walking sticks.

The Gullah and Geechee are Christians who nevertheless tend to regard the individual as possessing a body, a soul, and a separate spirit that may linger in the earthly realm after death. They traditionally placed eating utensils and other practical items on graves to meet the spirits' needs. Both voudou and hoodoo are part of the Gullah and Geechee tradition; good and evil spirits are summoned for divination, retribution, or curing. They practice folk medicine that incorporates the use of herbs and other natural remedies.

Gullah and Geechee build their meals around the rice they have traditionally grown. The grain often is consumed for breakfast, lunch, and dinner, although grits also make an appearance at the first meal of the day. Rice combined with chicken and okra is a favorite dish, as is Frogmore stew (named for the Frogmore plantation), which contains a combination of small beef sausages, crabs, shrimp, and corn.

Gullah and Geechee communities typically are kin-based clusters of homes built on family lands. They tend to hold their property communally, without division for inheritance, and over the generations this has created many stakeholders in the land, not all of them local—a situation exploited recently by resort developers. They often convince a few of the heirs to force a division sale and then grab the property for development.

The plight of the Gullah and Geechee and their culture in the face of aggressive development and natural attrition has engendered programs to preserve and protect Gullah and Geechee culture and interests. The former mission school on St. Helena, stronghold of the culture, serves as the center of these efforts.

ACROSS CULTURES

WRITTEN WORD, ORAL WORLD

Since long before the printed word, stories, wisdom, history, and knowledge were collected, remembered, and shared as part of a worldwide oral tradition, still alive today.

We live in a world inundated with written words. On paper and on the screen, in trashy novels and in great works of literature, in texting, field guides, instruction manuals, and e-mails—we transmit and receive much of our knowledge through writing.

This was not always the case. For thousands of years, millions of people existed in a purely oral world, where information was conveyed without the aid of scripts. In our literate world, it's difficult to conceive that most of human knowledge, for most of human history, existed entirely in memory. Oral traditions range from one-word names to epic adventures that take weeks to tell. They have been shaped by, and they have shaped, human cognition. Oral traditions have been a vital tool for human education, expression, and survival. Though oral cultures are speedily being transformed by literacy today, the verbal arts persist.

Teller of Tales

In a felt-covered yurt in Mongolia, a *toolju* prepares to begin his tale. He arrived in the afternoon, and now as the sun sets and the last chores are finished, the herders settle in to listen. The storyteller has donned a special robe and made a tea offering to the spirits, and he begins to sing.

The toolju has performed this tale before, as have the bard he learned it from and countless others across the steppes. There are as many variations of the story as there are tellings, but the core remains the same. This time, the audience reacts vocally as they hear of a sister's quest to bring her brother back to life, contests of strength, disguises, an evil king, a talking horse, a beautiful princess—and a happy ending.

Although the enthralled listeners here are Tuvan nomads, a similar scene could be taking place in many other places in the world: an Irish pub, an Australian desert, the rain forest of America's Pacific Northwest.

Oral Traditions and Knowledge

The study of oral traditions in the 20th century started with epic tales like these—and particularly in scholars' attempts to understand the oral roots of classic epics like Homer's *Iliad* and *Odyssey*. We remember those stories, echoes of a tradition long past, only because they were passed down to us in books. But epics elsewhere survive, never written down and still passed from teller to teller.

In the mountains of Kyrgyzstan, the epic tale of the hero Manas has been passed down orally for a thousand years. When transcribed, the tales of how Manas and his companions united the warring tribes can exceed 500,000 lines, more than *The Iliad* and the *Odyssey* combined. Still a living epic tradition, it is recited today by Manas-tellers, all of whom learned the stories by ear.

But epics do not represent the entirety of oral traditions. In a completely oral society, everything is preserved by memory—not just the imaginative folktales, epic histories, riddles, creation myths, jokes, and bawdy stories, but also factual information about the world, medicine, plants, astronomy, anatomy, and law.

An oral tradition can convey information simply—by naming a poisonous plant "baneberry," for example—or with dizzyingly complex symbols. The Hai||om people of Namibia (|| represents a click sound), long famed for their pathfinding skills, employ an almost constant back and forth of "topographical gossip" on their travels, an intricate system of landscape feature names they use to orient themselves.

Students learn to read and recite the ancient Sanskrit language at an ashram in northern India.

Elements of the mythic and the immediately practical are often one and the same. A creation myth from the Chehalis people of Washington State describes the travels of Honné, the primordial spirit, and his magical transformations. As the tale progresses, Honné carefully assigns a dozen varieties of salmon their names, color, season, place and manner of spawning, and life span. To recall and retell that story is to pass on significant zoological information.

The Wayampi people of Brazil tell creation stories featuring the golden-tailed japu bird whose snot, spread over the jungle as it sobs, became vines and creepers. This may not seem at first glance to include practical information, but when a Wayampi man sings at the japu festival, he transmits as much detailed information about this culturally important bird's habitat, nest, mating rituals, foraging behavior, preferred foods, and flight patterns as any field guide. The narrative matrix used to deliver information is itself a mnemonic device that enables the audience to achieve greater retention than if the information was presented as a bare list of facts.

Persistence of Memory

You may remember the game in which a whispered phrase is passed along a line of children and emerges at the end completely garbled. It gives a simple analogy for the way great bodies of knowledge have remained intact for thousands of years in a purely oral milieu.

Some scholars draw a parallel with evolutionary selection. As information is passed down, only what is memorable can survive. What remains after dozens of generations is a body of knowledge that has adapted to fit the niche of human memory. The irrelevant falls away and the core elements remain. Certain similarities emerge in oral traditions formed by this process. Successful strategies for being memorable include rhyme, alliteration, and rhythmic stress patterns, as well as repeated stock phrases, characters, themes, and episodes that a teller may draw on.

Structural constraints can function as mnemonic mechanisms that greatly aid memory. The most familiar poetic device in English is rhyme, but rhyme is fairly rare worldwide. For the toolju, rhyme would be pointless. Tuvan, like many other languages, makes heavy use of suffixes, making rhymes both trivially easy and useless as an error-checking or mnemonic device. More common worldwide are rhythmic stress patterns and the regular repetition of certain sounds by alliteration or assonance.

A repertoire of formulaic characters—the clever young woman, the magical companion, the evil king—and events—setting out on a quest, reviving the dead, feats of strength—recurs in many narrative traditions and aids the bard's task of memorization and composition. These general motifs combine with specific characters (the wise Odysseus, perhaps, or the mighty Manas) and episodes in those characters' lives that the teller can stitch together, depending on the particular occasion.

Storytellers draw on voice, gesture, and performance to link the present to the past, as shown by this Koryak woman of Siberia.

Change and Tradition

Despite all of these strategies for remembering, oral traditions are organic systems, and forgetting also has its place. When the tales of Ndwera Jakpa, the founder of the Gonja state in Ghana, were first transcribed by the British, they featured the seven sons of the founder, each credited with beginning one of the seven divisions of Gonja. Sixty years later, after boundaries had shifted and only five divisions remained, bards sang of the five sons of Ndwera Jakpa.

The adaptivity of oral transmission is also evident in nonnarrative traditions. The Natchez tribe, who lived along the lower Mississippi, named their lunar months after important foods gathered. Their eighth month was "Turkey Month," their ninth "Corn Month." By the 1750s, their calendar included "Watermelon Month" and "Peach Month"—two foods introduced by European settlers. Contrast this with our modern English calendar, which has existed in a literate context for centuries and preserves a jumble of meaningless month names derived from Latin.

Even when oral traditions are not responsively changing their core elements, there is constant low-level change. The Tuvan toolju's tale is unique: It's different every time he tells it. Although bards may claim to be repeating a tale exactly as it was told to them and although their memories are exceptional by literate standards, all of these strategies produce an oral performance that falls somewhere between a recitation and an improvisation.

Certain other elements are shared by oral traditions. As with the toolju's tea offering, ritual often surrounds tales, providing a necessary cultural context. Ritual can extend from a special hat worn only while storytelling, to the trancelike state of a Manas-like storyteller channeling the epic. As live performance, the words are accompanied by gestures—subtle or full body, idiosyncratic or stylized. Driven by its purely sequential nature (you can't glance back at the preceding line), oral expression is often highly repetitive for the sake of both the bard's memory and the audience's comprehension.

Oral to Literate

India, one of the most populous and linguistically diverse countries in the world, is also one with an ancient history of writing. In the perhaps 4,000 years that India has had some form of writing, more than a dozen scripts have been used. But over that time, India has remained home to primarily oral cultures. Among other great oral traditions, it is home to Vedic oratory, remarkable for its antiquity, size, and attention to accuracy (to such a degree that practitioners train in singing complex variations, forward and backward, in an effort to retain the exact order of the words).

India's long history of oral traditions is rapidly changing, however. In the 50 years between 1951 and 2001, its literacy increased from less than 20 percent to 65 percent. Literacy increases worldwide during the last century are laudable, but something is clearly lost when oral traditions fall silent. Devoid of ritual, participation, gesture, tone, and organic fecundity and disembodied from their human source, even those oral traditions that are transcribed and preserved are no longer living but petrified. The *Odyssey,* frozen in typeface, is not the Manas, still told dynamically from memory.

This is not a new observation. More than a thousand years ago, ambivalence toward a wholesale shift to literacy can be seen in early written English. Images of fire and bookworms haunt the books of Anglo-Saxon riddles, reflecting their authors' anxiety at relegating their oral culture to paper. In some ways, memory is the more durable medium. Stories told by Australian Aborigines may date back 20,000 years, while no written sources we have today even approach such antiquity.

What Might Be Lost

At a fundamental level, all of language is an oral tradition, and it all shares the same fragility, more petrified than preserved by written records in the absence of person-to-person transmission. We are currently in the throes of a massive extinction of the world's languages as speakers abandon indigenous tongues for global megalanguages. Global language extinction threatens to disrupt oral traditions, risking the loss of an enormous body of ancient knowledge preserved in the world's oral traditions.

—Robert E. Hart and K. David Harrison

A Bedouin man entertains his companions after the evening meal in the Sinai Desert.

Chapter 7

CENTRAL & SOUTH AMERICA

THE CARIBBEAN
African Caribbeans • Amerindians • East Indians
Euro-Caribbeans • Haitians • Jamaicans • Kogi

MESOAMERICA
Garifuna • Highland Maya
Lowland Maya • Huichol • Mixtec • Nahua
Otomí • Tarascans • Zapotec

AMAZON & ORINOCO BASINS
Amahuaca • Bororo • Brazilians
Caribs • Desana • Kuikuru • Mundurucú
Nukak • Saramaka • Shuar • Tapirapé
Warao • Xikrín • Yanomamí

ANDES
Aymara • Kallawaya • Mapuche
Otavaleños • Q'eros • Quechua

CENTRAL SOUTH AMERICA
Guaraní • Kadiwéu • Mataco • Tupinambá

PATAGONIA
Argentines • Chileans

esoamerica, the Caribbean, and South America are geographical neighbors, with each region sharing at least part of its history and culture with the other. Each area, however, has also experienced a long, complex past quite separate from the other, and this has resulted in fundamental differences between them.

Anthropologists recognize Mesoamerica as a distinct region whose peoples, although ethnically diverse, have shared significant cultural characteristics across varied terrains and a broad expanse of time. Beginning in the semidesert landscapes of north-central Mexico, the region of Mesoamerica reaches south and east toward the tropical terrain of upper Central America, where it then stretches through Guatemala, Belize, and portions of Honduras, El Salvador, and Nicaragua.

In the late 15th century, when Europeans began arriving in the Western Hemisphere, Mesoamerica was dominated by the vast empire of the Aztec. But within decades of the European arrival, Spanish conquistadores had overwhelmed that powerful people and all other major groups. Native Aztec, Zapotec, Mixtec, and others were congregated into European-style villages and towns and forcibly converted to Roman Catholicism. They were introduced to new farming methods and were taught how to create new products of fiber and metal. In short, they were forced to conform to a new economic system.

To meet their growing labor needs, the Spanish and later colonial powers also turned to Africans, who were forcibly brought to the New World to work on fruit plantations along the Caribbean coastal plain. Today, many people in both Mesoamerica and along the Caribbean are of mixed background—mostly European Amerindian, European African, or Amerindian African. Spanish is the dominant language in much of the region. Many indigenous peoples speak only their own language.

Cuisine, religion, festivals, and arts also reflect the European influence on indigenous cultures. Beans, maize, chili peppers, and other local staples are augmented by wheat, fruits, pork, and chicken introduced from Europe. In religion, the phenomenon of folk Catholicism fuses traditional native cosmology with Christian practices.

To the east, extending through the Caribbean Sea from the northern coast of South America to the southern tip of Florida, are the islands called the West Indies, including the Lesser Antilles, the Greater Antilles, and the Bahamas. The Spanish controlled all these islands from 1492, when Columbus arrived, until the 1620s, when northern Europeans established permanent settlements. Here, the plantation economy took hold, and millions of Africans were brought across the Atlantic in bondage. For much of the 17th and 18th centuries, these islands witnessed wars and piracy carried out by the dominant colonial powers. The 19th century brought emancipation to the slave population, and soon immigrant labor arrived from India and elsewhere.

The feather earplugs worn by a young Yanomamí child are part of the group's personal adornments.

The ethnic composition of the West Indies is difficult to describe because culture, ethnicity, class, and race have combined through the generations. Whites and blacks are about equal in number; a slightly smaller percentage is of mixed ancestry. The linguistically diverse West Indian region includes colonial languages, tongues brought by waves of immigration, and Creole forms such as Haitian Kreyol and Aruban Papiamento. Throughout the Caribbean, Roman Catholicism, Protestantism, Hinduism, Islam, and syncretic religions like Santeria, Voudou, Orisha, and Spiritual Baptism all thrive. So does music, such as calypso and reggae, which blends European, African, and native traditions and instruments.

SOUTH AMERICA The South American continent is a land of dramatic environmental contrasts, with a remarkable diversity of indigenous human populations to match. In the north, a wide coastal plain borders the Caribbean Sea and is backed by a branch of the Andes mountain range, the backbone of the continent for almost 5,000 miles. The celebrated Altiplano, or high plain, surrounds Lake Titicaca, on the border of Bolivia and Peru. East of the Andes, humid tropical forest stretches across the enormous Amazon Basin. Along the Pacific coast, west of the mountains, lie some of Earth's driest deserts.

A vendor totes her wares to market in Panajachel in Guatemala's southwestern highlands.

In the central part of South America, the Amazon Basin gives way to the Brazilian Highlands. Here is the huge sandstone plateau of the Mato Grosso region. Southwest of the Mato Grosso is the Gran Chaco, a series of arid plateaus descending from the Andes. To the south lie the vast grasslands of the pampas, which stretch to the arid plateaus of Patagonia. A mountainous landscape of islands, fjords, glaciers, and densely forested valleys lies to the west and extends into Tierra del Fuego. Here, at the remote southern end of the continent, human groups arrived more than 8,000 years ago.

South America's environmental diversity led to a wide range of human societies, which adapted brilliantly to their surroundings by developing an intimate knowledge of them. For example, at the time of the Spanish conquest in the 16th century, Andean peoples were cultivating 70 indigenous plant species—including the staple, potato—a number nearly equal to all the cultivars of Europe and Asia combined at the time. Shamans fulfilled essential spiritual needs in many communities.

The Andes hosted the rise of a great ancient civilization, culminating in the Inca Empire of the A.D. 15th century, centered in Cuzco, Peru. The empire eventually fell apart in the face of civil war and the arrival of rapacious conquistadores, beginning with Francisco Pizarro in 1532. Still, many people continued their traditional lifeways in remote valleys and desert communities.

At the time of the Spanish conquest, many South Americans made their living by hunting and gathering or by simple subsistence farming. Much of the diversity of indigenous life was destroyed by Spanish colonization, which brought not only European farming methods and technology but also horses and the pervasive influence of Catholicism. Thousands died from infectious diseases, maltreatment, and forced labor. Today, many once powerful Indian peoples eke out an existence in city slums or peasant communities. A surprising number maintain their traditional ways, despite the arrivals of land-hungry prospectors, multinational corporations, and modern technology.

The Caribbean

A narrow string of hundreds of islands, the Caribbean arches more than 2,500 miles around the Caribbean Sea. The arch starts off the southeast coast of Florida and runs to roughly the northeast coast of Venezuela in South America. The climate is tropical and subtropical, cooled by tradewinds, making it a highly desirable vacation, sailing, and scuba-diving destination. Major islands of the Caribbean include Cuba, Jamaica, the Cayman Islands, Turks and Caicos, Hispaniola (containing Haiti and the Dominican Republic), Puerto Rico, Virgin Islands, Leeward Islands, Windward Islands, Barbados, Grenada, Trinidad, and Tobago.

African Caribbeans

Population: 16.5 million
Location: Caribbean island nations
Language family: Many languages spoken

Africans have had a significant impact on shaping the Caribbean ever since the 16th century. From across vast stretches of the African continent, slavers brought millions of people with diverse backgrounds to the New World. That experience, as well as contact with other African peoples and Europeans, has created a unique culture in the West Indies. To be sure, there are numerous variations among islands.

Jamaicans, for example, differ in many ways from Guadeloupeans. Nevertheless, a unique and coherent African American Caribbean culture transcends national boundaries in important ways.

Approximately 42 million people live in the Caribbean region, and about half of them have some African ancestry. They descend from people who either brought with them or later developed unique family structures, religions, cuisines, musical forms, dances, and festivals. They even had credit associations known as *susus*—revolving credit arrangements whereby participants pooled their money and then withdrew individual lump sums when needed. African peoples also adopted and transformed traditions that they encountered in the New World, thereby producing new forms of culture. Among their contributions are the syncretic religions of Voudou, Santeria, and Rastafarianism; musical forms including calypso, reggae, rumba, and zouk; and festivals such as Carnival, Jonkanoo, and Crop-Over.

In almost every part of the Caribbean, Africans rebelled against their enslavement. In some cases, communities of escaped slaves, called Maroons, established themselves in difficult landscapes and held off repeated attempts by the colonial forces to root them out. In other cases, organized rebellions were attempted; the most notable success was the 1804 overthrow of the French in Haiti.

In Maroon societies and in other parts of the Caribbean region, aspects of several African languages were preserved to greater or lesser degrees. Many tongues spoken in the Caribbean today, whether Kreyol of Haiti, Patwa of Jamaica, or Papiamento of the Dutch Antilles, owe some portion of their vocabulary and syntax to African languages such as Yoruba and KiKongo.

AFRICAN CARIBBEANS With a smile as wide as her headdress, a Haitian woman marches in Port-au-Prince's Festival of Flowers.

Amerindians

Population: 78,000
Location: Caribbean island nations
Language family: Germanic

When Columbus arrived in the New World, he encountered three distinct native peoples: the Ciboney, or Siboney, had settlements on Cuba and Hispaniola; the Arawak controlled the Bahamas, Trinidad, and much of the Greater Antilles; and the Caribs were settled mostly in the Virgin Islands and along the Lesser Antilles. Some Carib groups lived on the South American mainland, from the Amazon River northward to the Guianas and the sea that would be named for them.

The Caribs occupied small autonomous settlements, grew staple foods such as cassava and other root crops, and hunted in nearby areas with their blowguns or bows and arrows. Although much of their history has been lost, the people were reputedly fierce warriors whose numbers were decimated by war, disease, and the brutal conditions of slavery in the early decades of European colonization. For the most part, divisions between the Carib groups ran along linguistic lines, but the languages they spoke seem to have been mutually understandable.

The Arawak, often described by the Spanish as gentle and hospitable, lived a life that was technologically simple yet ingenious. They and the Caribs practiced *conuco,* a kind of ecologically sound farming method that involved burning a small patch of land to enrich the soil. Farmers then formed mounds of earth where they planted manioc and other crops that had high yields but a low environmental impact. Because they lived in a tropical climate, the Arawak usually felt the need to construct only simple shelters. The people also were known for creating elaborate stone sculptures, painting their bodies, and making jewelry.

According to accounts of contemporaries, both the Arawak and the Caribs were more centrally organized than the Ciboney. The Arawak had complex, intertwining forms of religion and government: Shamans were responsible for controlling *zemis,* good and evil spirits that they sometimes captured and housed in icons made of gold; caciques, or local chiefs, ruled semiautonomous provinces within the larger territory

AMERINDIANS The Emberá people of Panama are working to revitalize their Amerindian culture and language.

of an island. Carib organization was less centralized, with villages remaining largely independent of one another. Although European settlers considered Caribs to be savages who ate human flesh, most scholars today say that this image was exaggerated to justify the people's conquest. This is not to say that cannibalism played no part in the culture; it was very likely practiced in ritual situations following battles. The bravest enemy warriors were tortured and killed, and parts of their bodies were ceremonially consumed.

While no "pure" native populations are left, several Caribbean communities preserve their age-old traditions and celebrate native ancestry in such venues as the Santa Rosa Carib Festival in northern Trinidad. A concerted effort is also under way to remember contributions to Caribbean life made by people such as the Taino, a subgroup of the Arawak in Puerto Rico.

East Indians

Population: 466,000
Location: Caribbean island nations
Language family: Germanic

In 1838, slavery ended in the British West Indies. But the plantation economy still required workers, and planters were often unwilling or unable to pay fair wages to former slaves, the islands' only source of manpower. Besides, many Africans were understandably loath to work for their former masters. In an attempt to undercut African labor power, planters asked the colonial government to help find cheap labor in other parts of the empire. Various solutions were tried—including the importation of European workers, free Africans, and Chinese laborers—before the English found their greatest success in indentured workers from India.

Between 1838 and 1917, more than half a million people were brought from the Indian subcontinent to colonies in South America and the Caribbean. The vast majority came to settle in Guyana (formerly British Guiana) and Trinidad, with many others going to Guadeloupe, Jamaica, Suriname, Martinique, French Guiana, and Grenada. Smaller numbers landed in Belize, St. Vincent, St. Kitts, and St. Croix. Today, descendants of those people from India and its former lands still make up a significant portion of the Caribbean population.

The Indian presence has had a profound impact on the local Caribbean culture, with cuisine, dance, music, festivals, politics, enterprise, art, and religion in many Caribbean regions all possessing elements of Indian culture. These forms, though, are not "purely" Indian. Through the years a "creolization" has occurred, resulting in something unique to the region. A fine example of Caribbean Indian culture can be found in Trinidad's annual Hosay Festival, which commemorates the deaths of the Prophet Muhammad's grandsons Hassan and Husain. Called the Muharram Festival by Shiite Muslims, this event in Trinidad is celebrated with a colorful parade of large floats, or *taziya,* in the shape of elaborate mosques. The floats are accompanied by *tassa* drummers—groups of men with sticks, beating drums slung over their shoulders.

Indo-Caribbean culture has also produced "chutney" music—a fast-tempo, modernized version of folk tunes—and other musical forms; it has itself become mixed with African-derived calypso and soca (soul meeting calypso) to produce a highly popular dance music called soca-chutney.

Finally, Indian cooking has greatly influenced the cuisines of Trinidad and Guyana with such mainstays as roti: flat bread filled with a stew of curried potatoes, chickpeas, and shrimp, goat, or chicken; "doubles": curried chickpeas served between two pieces of flat bread; and *pholouri:* deep-fried balls of yellow split-pea batter served with a mango chutney—to name just a few. Hindi words pepper the local languages, especially when it comes to food: Spinach is *bhaji,* potato is *aloo,* long string beans are *bodi,* and chickpeas are *channa.* Sometimes overlooked when considering Caribbean culture, Indian culture has evolved into a major element of the region's ethnic mix.

EAST INDIANS A cross-wearing man drums at the Hosay festival, a primarily Muslim observance in Trinidad, which draws East Indians of all faiths.

Euro-Caribbeans

Population: 12 million
Location: Caribbean island nations
Language family: Germanic

The permanent European presence in the Caribbean began with the arrival of Columbus in 1492 and continued as the Spanish Empire sent colonists to extract mineral wealth from the region. Because islands without precious metals were of little interest, early colonists moved fairly rapidly from one place to another in search of riches, all the while enslaving local populations and unknowingly exposing them to diseases they brought from Europe.

With the arrival of other European powers through the 16th and 17th centuries, and the subsequent growth of plantation economies, additional labor was needed.

One hand washes the other; both get clean.

CURAÇAOAN PROVERB

But landowners did not turn immediately to African slavery as the solution; instead, they looked to European sources for workers. For a while, they relied on indentured servants: Some had been kidnapped and brought to the New World, while many others left their homelands to escape religious persecution, war, or even prison sentences. These early white laborers often suffered harsh treatment, succumbed to disease, and were eventually replaced by Africans.

As African slavery began to dominate the islands, white servitude declined. Even so, poor whites were not viewed kindly, and for a time local governments considered many of them to be persona non grata: idlers, thieves, and ne'er-do-wells. Owners and overseers, on the other hand, continued to do well.

Through the years, both the elite and far-from-elite classes of Europeans have had a significant effect on the Caribbean region. They brought holidays and festivals—New Year's and Carnival, for example—as well as masked balls and various cuisines, musical forms, and dances. Like the people who brought them, many of these customs, activities, and traditions have been radically transformed over time to emerge as uniquely Caribbean.

EURO-CARIBBEANS Béké, a Creole term for descendants of European colonizers, pause along the road near Fort-de-France, Martinique.

HAITIANS Many Haitians still are affected by the destruction from the country's 2010 earthquake.

Haitians

Population: 10.1 million
Location: Haiti
Language family: French-based Creole

Two hundred years ago, escaped black slaves known as Maroons staged a forceful and persistent rebellion against French colonial planters and imperial powers in Haiti, the western one-third of the island of Hispaniola in the Caribbean Sea. Their success in 1804 resulted in the establishment of the first independent black republic in the modern world.

Since then, Haiti's fortunes mostly have followed a downward spiral. Today, more than ten million Haitians find themselves the poorest people in the Western Hemisphere, with high infant mortality and a life expectancy of only 52 years. Their mostly mountainous island has been stripped of its forests for building, farming, and to make charcoal for fuel. The island also presents double natural perils: an earthquake-producing tectonic fault running through the center and a prime location in the Caribbean hurricane belt.

Nearly all Haitians descend from Africans, although 5 percent of the population is mulatto or white. French was Haiti's official language until the constitution of 1987, when it was pushed into second place by Kreyol, a local language with French pronunciation and vocabulary but syntax similar to other Caribbean Creoles. The country's elite still speak French, and higher status accrues to those who are fluent in it.

Haitians may be poor, but nearly all of them own a little land, much of it not very suitable for agriculture. Nevertheless, most Haitian homesteads have gardens planted by the male head of the family. They produce food for the household and, more important, a surplus that the women sell on the streets and in local markets, an important source of family income.

Visitors to a family compound announce their arrival in the yard by calling out "*Onè*—Honor," which is answered by the host with, "*Respè*—Respect." It would be unthinkable not to offer something to drink and perhaps eat—or, if nothing is available, a profound apology. Rice and beans appear at the main meal of the day, along with yams, sweet potatoes, manioc, plantain, corn, and sometimes chicken. Mangoes and other fruits are eaten as snacks.

Haiti at one time was an overwhelmingly Catholic country—some 90 percent—another legacy of French colonialism. In the last few decades, Catholicism has lost ground to widespread Protestant missionizing, which has captured the religious affiliation of perhaps 20 percent of the population. These Haitian Protestants refer to their faith as *levanjil,* or evangelism. At the same time, the practice of Voudou has declined somewhat. Voudou is a syncretic tradition fusing West African and Catholic beliefs and rituals. In Haiti, it can concern itself as much or more with healing as it does with causing harm to one's perceived enemies, and it involves the use of a number of different specialists, men or women.

As if to make up for the relentless hardships of their lives, Haitians spend a good deal of time and many resources planning their funerals, which are elaborate events involving days of socializing, feasting, and processions. When death occurs, relatives gather at the home of the deceased and settle in for several days. The trend currently favors above-ground burial in a *kav,* a kind of multioccupant tomb that costs a great deal to build—often more than the average home. Many Haitians think of the afterlife as going to *lafrik gine,* or Guinea Africa.

Haitians in general believe in *lwa,* a constellation of spirits who must be served in order to obtain their favor. Baron Samedi, for example, guards the crossroads where souls pass to the other side; believers offer him his symbols of a top hat, sunglasses, and cigar. Christians still usually acknowledge the lwa, although more mainstream Catholic and Protestant adherents may drop the custom.

The political instability of the past decades has shattered an already fragile Haitian economy, and the country's crumbling infrastructure likely makes Haiti's overall situation worse than it was a century ago. Several million Haitians have left the country, many as boat people headed to the United States. In January 2010, a magnitude 7 earthquake hit the area around Haiti's capital, Port-au-Prince, killing upward of 100,000 people and affecting millions of others. Internationally supported relief and restoration efforts continue there.

Jamaicans

Population: 2.6 million
Location: Jamaica
Language family: English-based Creole

Some 2.6 million people inhabit Jamaica, the third largest island in the Caribbean. It has a mountainous interior and flat coastal area that supports large urban settlements such as Kingston, the capital, on the south, and trendy, tropical playground resorts on the north. More than 90 percent of the people in Jamaica are of African ancestry; the rest include small numbers of East Indians, Chinese, and whites. Wealth is unevenly distributed in Jamaica, with a wide gap between rich and poor.

Originally the homeland of Arawak Indians, who were decimated on contact with Europeans, here as elsewhere in the Caribbean, Jamaica endured more than 450 years of Spanish and British domination before obtaining independence in 1962. English is Jamaica's official language and has retained some Elizabethan influences. Upon leaving a friend, a person might say, "Peradventure I will see you tomorrow." Most Jamaicans also regularly speak Patwa, a Creole language combining English with elements of French and Spanish that, when paired with a laid-back demeanor, serves to differentiate Jamaican identity from European authority.

Family life in Jamaica takes different forms depending on class. The middle and upper classes usually aspire to formalized, monogamous marriages and nuclear families. The poor tend to enter informal relationships that may eventually be legally sanctioned, often after the birth of several children and the attainment of financial stability. Often, though, a woman will reside in a household with her children from different relationships, her mother, and perhaps other kin on her mother's side. The bonds of kinship reckoned through women are frequently the relationships that count most among the poor. In hard times, children may be passed from relative to relative to be cared for until circumstances improve. Regardless of class, child

JAMAICANS A gleeful extended Jamaican family gathers at the window of their home.

rearing stresses politeness and courtesy in children. Elders are to be respected at all times. Children, and especially girls, acquire responsibility at an early age, often taking charge of younger siblings.

Religion permeates Jamaican life, taking myriad fluid forms. Mainstream Protestant denominations, notably Anglican, Baptist, and Methodist, made early inroads in Jamaica. In recent decades, Pentacostalism has flourished; its emphasis on the power of the Holy Spirit against devils and demons resonates with the African-based belief in *obeah,* or sorcery, which many Jamaicans regard as the root cause of most misfortunes. "Obeah men" are thought to commandeer the spirits of the dead, turning them into "duppies" who perform malevolent acts. As a result, funeral rites take on great significance, as they are geared toward appeasing spirits of the dead.

Jamaicans also participate in a number of syncretic African Caribbean movements such as Zion Revivalism and Rastafarianism. The latter venerates the late Ethiopian ruler Haile Selassie, incorporates ritual marijuana use, and worships Jah, a god within each individual. Rastafarians often wear the red, green, and gold of the Ethiopian flag; the men frequently grow dreadlocks. Reggae musician Bob Marley embraced Rastafarianism, bringing it global recognition.

Jamaican celebrations include Festival, a multiweek event that leads up to Independence Day on August 6 and features performances and competitions. Jonkanoo (John Canoe) occurs at Christmastime, when people don elaborate costumes and masks to parade and dance in the streets. Most ceremonial occasions include rice in some form, often served with goat curry. Jamaicans also eat tropical fruits and tubers—bananas, plantains, breadfruit, sweet potatoes, yams, and cassava—along with salted fish and jerk chicken or jerk pork, highly spiced and roasted over a wood fire.

About a quarter of Jamaicans are engaged in agriculture, growing coffee, sugarcane, tropical fruits, cacao, spices, and other products. Some labor in the country's valuable bauxite mines, extracting the mineral that goes into aluminum manufacture. Thousands of women called "higglers" traditionally buy produce and inexpensive imported goods for resale at local markets. And tourism has for years brought foreign exchange and provided many jobs for Jamaicans.

Kogi

Population: 9,900
Location: Colombia
Language family: Arhuacan

Colombia's Kogi people live in an environment of striking contrasts: Within a distance of only 22 miles, the land rises from sea level to peaks more than 18,000 feet high in the Sierra Nevada de Santa Marta. Here, land suitable for cultivation occurs in small pockets up to 11,500 feet, so the Kogi live vertically, traveling up and down among the hot, temperate, and colder zones of their environment. They cultivate bananas and manioc in the tropical lowlands, bananas and beans up to 8,200 feet, potatoes and onions in the temperate zone. No one zone can meet all their dietary needs. In earlier times, they staved off malnutrition by also eating fish from the Caribbean, but this resource became scarcer as the European population rose.

When the character of a man is not clear to you, look at his friends.

HAITIAN PROVERB

Ironically, the Kogi would fare much better if they could cultivate maize in the fertile, terraced fields left by their prehistoric predecessors, the Tairona. But their ritual leaders, or *mámas,* forbid it because they believe evil spirits occupy these lands. The Kogi belief system thus limits their agricultural productivity and diet—despite their reputation as people who make full use of their environment, an image they cherish in their dealings with the outside world. This image is in jeopardy, for their homeland's plains and slopes are eroded and degraded after centuries of slash-and-burn agriculture, and the people now fear that their environment will not feed them adequately.

To sustain themselves during hard work and arduous climbing, Kogi men rely heavily on coca leaves, chewed with lime to release narcotic alkaloids. They eat only small amounts of food. Women and children seem to be well fed and do not chew coca. Their diet is always uncertain, though, because insufficient rains and the degraded soils often lead to poor harvests. In good years, people move frequently from one zone to another, with a preference for lower, more productive elevations. At these times, the Kogi trade with mestizos and others on the coast. During hungry periods, the people move up the mountains and shift little from one elevation to another.

The largest settlements lie at different elevations, most of them between about 3,000 and 6,000 feet above sea level, but they are occupied only a few days a week when families come in from the fields. Village dwellings are circular, with thatched, conical roofs, and are scattered around a central building where the men usually reside. Women and children reside in family dwellings. Every family also has three or four field houses at different elevations that they use seasonally. Typically two dwellings face each other across an open area, the husband and adolescent boys living in one, the wife and younger children in the other. In hard times, the Kogi tend to settle close to ritual centers, near the leaders who can provide them with spiritual guidance.

Kogi society is divided into two main groups, with one comprising all the men and the other all the women, a pattern reflected in their housing arrangements. There are smaller kin groups too, resulting in marriage arrangements that are so complex only *mámas* can decipher them. After marriage, a son-in-law works for his in-laws until he is deemed capable of starting fields on his own.

About 2,000 Kogi lived in an area of about 386 square miles in the 1940s, when their society was closely studied. Now that European settlement has blocked their access to the coast, many are becoming peasant farmers or laborers as their traditional lifestyle becomes less productive.

KOGI A Kogi family relaxes in the shade of a typical conical thatched house.

Mesoamerica

Also called Central America, this region forms the narrow connection between North and South America. Mesoamerica coasts the Gulf of Mexico and the Caribbean Sea to the east and the Pacific Ocean to the west. Countries in this region include Mexico, Guatemala, Belize, El Salvador, Honduras, Nicaragua, Costa Rica, and Panama. Temperatures in this region are fairly consistent throughout the year. For example, the average temperature in Managua, Nicaragua, is 79°F in January and 89°F in July. Annual rainfall throughout much of the region is significant, especially in some of the low-lying areas, which receive more than 80 inches a year. In elevation, Mesoamerica reaches from long stretches of swampland along the Mosquito Coast of Nicaragua and Honduras to the Guatemalan highlands, peaking at 13,845 feet above sea level in its highest mountain, Tajumulco, an inactive volcano.

Garifuna

Population: 195,800
Location: Honduras, Belize, Guatemala
Language family: Maipurean

The Garifuna people, once known as Black Caribs, are descended from African slaves and the Carib Indians, who once gave escaped slaves refuge on St. Vincent Island in the Lesser Antilles. This group is now spread throughout parts of Mesoamerica (primarily Belize, Honduras, Guatemala, and Nicaragua), and a significant population resides in the United States.

The genesis of the Black Carib population can be traced to French and English policy in the 17th and 18th centuries. Initially, the island of St. Vincent was set aside as a refuge for Carib Indians, who later sold land to French settlers in exchange for arms. The Caribs were joined by Maroons, African slaves who had escaped from neighboring islands or been shipwrecked, and these two groups intermarried.

In 1763, the Treaty of Paris transferred control of St. Vincent from the French to the English without recognizing Carib land rights, and more settlers moved in. The island switched hands again, and then again. As the English sought more land for sugar plantations, local resistance grew until both sides were engaged in

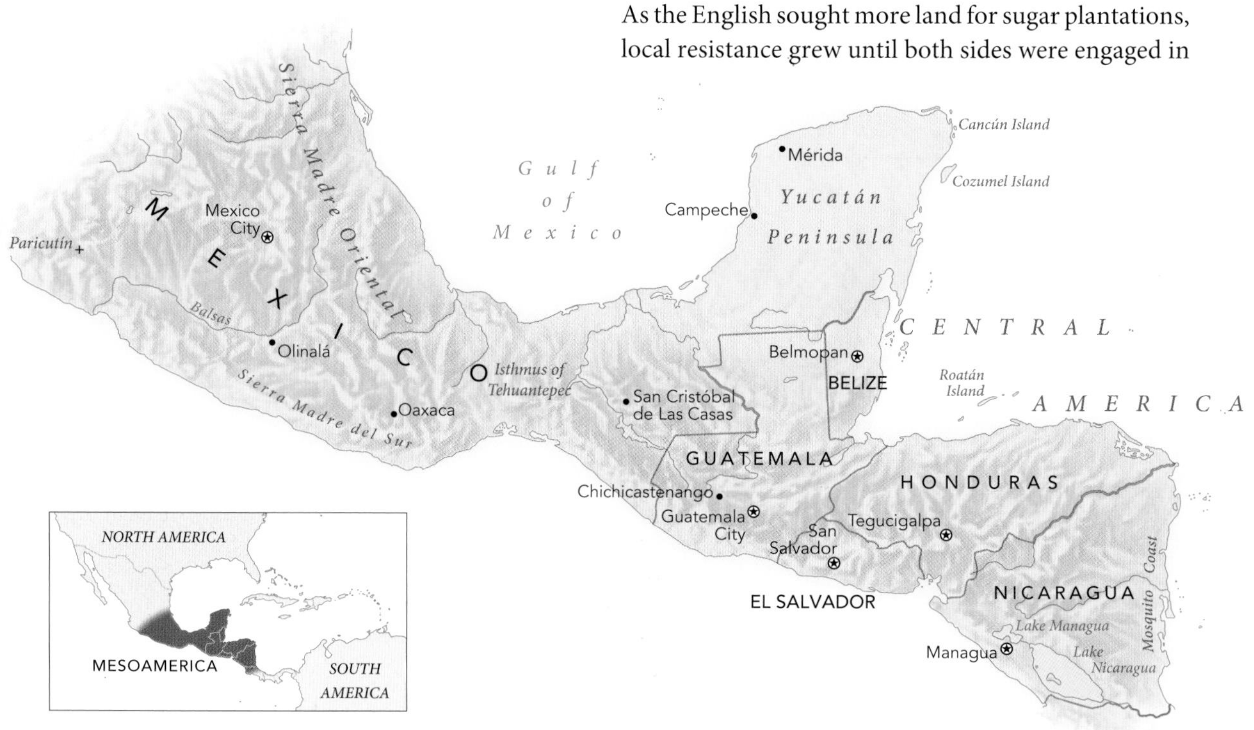

GARIFUNA Father and son Garifuna fishers pull up on the beach with the day's catch.

war. The Black Caribs won the war in the sense that their lands were preserved by the terms of a 1773 treaty with the English. But that treaty also acknowledged the King of England as the people's ruler and was later used to justify their deportation. In 1795, another outbreak of hostilities led to the demise of Black Carib sovereignty and to the group's removal to Roatán Island off the Honduran coast. From there, they settled in mainland areas of Central America, where they came to be known as the Garifuna.

Today, the Garifuna thrive mainly in small towns and villages, but many have migrated to large U.S. cities. The Garifuna tongue shows influences of many languages, including Arawak, Yoruba, Swahili, Bantu, Spanish, English, and French. It is distinct from others in Central America in that many words are gender specific, with men and women using different words for the same things.

Over the years, the Garifuna have developed a unique musical style that owes much to the group's various roots. The music builds from a foundation of two or three *garaon* drums. The *primera* (first) or heart drum improvises against the counterrhythmic *segunda* (second) or shadow drum; a steady bass drum called the *tercera* (third) holds the ensemble together. Sometimes several guitar strings or wires (snares) are stretched across a drumhead to create the buzzing sound traditional in some West African music. Turtle shells, claves (hardwood sticks), bottles, and a variety of shakers and scrapers are also used for percussion.

Highland Maya

Population: 3 million
Location: Mexico, Guatemala
Language family: Mayan

The Tzotzil, Quiché, and Cakchiquel, along with 15 or so other highland Maya groups, occupy their ancestral lands in southeastern Mexico and adjacent Guatemala. Each of these groups may be defined by its own distinctive version of the greater Mayan language family, made up of some 30 related tongues used in both the highlands and the lowlands. From the lofty, volcano-studded mountain ranges along the Pacific coast to the shores of the Yucatán Peninsula, five million Maya speak at least one of these languages.

The highland Maya known as the Tzotzil (literally, "people of the bat") live with their neighbors in Chamula, Zinacantan, Tenejapa, Oxchuc, and other towns lying in a region centered by the colonial city of San Cristóbal de Las Casas—the commercial focus of Ladino, or non-Indian, culture in Mexico's Chiapas state. Primarily maize farmers, these Maya maintain a relatively isolated existence that warily mixes the demands of national bureaucracy with an adherence to traditional customs and beliefs dating back 2,000 years or more. Their cosmology is dominated by sacred geography, with features such as mountains and water bodies receiving special devotion, and their rituals may be a blend of Maya and Christian religious practices.

Highland Maya women of Chiapas and Guatemala are particularly noted for their traditional weavings, skillfully made with back-strap looms supported by a strap across the weaver's lower back; distinct from town to town, weaving patterns often incorporate ancient Maya concepts of time and the cosmos. A written account of creation and the gods and heroes of sacred myth is found in the famed *Popol Vuh,* or *Book of Council.* Produced by ancestors of today's Quiché people, near the Guatemalan town of Chichicastenango, it is perhaps the most notable surviving work of ancient American literature.

Generations of highland Maya have been scorned by the local Ladinos as barriers to development and commerce, and they have suffered unusual degrees of oppression and negligence by their national governments. In recent decades, military repression of the

HIGHLAND MAYA In the highlands of Chiapas, Mexico, a Maya woman weaves *huipil,* a traditional blouse.

Guatemalan Maya, punctuated by torture and executions, fostered mass movements of refugees to camps in Mexico. But many people have since returned under a truce—a situation resulting in part from the leadership of Rigoberta Menchú, a Quiché woman who in 1992 received the Nobel Peace Prize.

In Mexico, the Zapatista revolutionary movement, which began in San Cristóbal de Las Casas, has spread to many of the Maya towns. It demands immediate action by the local and national governments to alleviate poverty and end the long-standing indifference toward the Maya, whose traditional culture has endured to a remarkable degree.

Lowland Maya

Population: 768,000
Location: Mexico, Guatemala
Language family: Yucatecan; Cholan-Tzeltalan

The lands occupied by the Yucatec, Lacandon, and other speakers of lowland Maya languages reach from Mexico's state of Tabasco east to the Caribbean Sea and from the highlands and piedmont of Guatemala to the northern shores of the Yucatán Peninsula. Except for the Yucatán's Puuc, or hill area, and Belize's Maya Mountains, the northern Maya lowlands consist of a vast slab of limestone that is still emerging from the sea.

Quarrelsome people harm themselves.

MIXTEC PROVERB

Lowland Maya mainly speak Yucatec or closely related languages such as Lacandon or Itzá, often mixed to varying degrees with Spanish. For subsistence, the people depend largely on the raising of maize, beans, squash, and chili peppers in *milpas,* fields selectively cleared, burned, and planted when the rainy season begins each year in early June.

During the Classic period, from A.D. 250 to 900, Maya architects designed great stone palaces, temples, and public buildings. They were part of a cosmopolitan society also known for painting, sculpture, astronomy, and a hieroglyphic writing system. Many traditions of lowland Maya life have survived, from house types to ancestral beliefs in the significance of the cardinal directions and in supernatural figures connected to landscapes and rain. The lowland groups associate numbers with gender: four for male, representing the corners of the milpa, or cornfield; three for female, the number of stones in the traditional hearth. Towns usually feature a mix of dwellings, including rounded rectangular houses with walls of stone or bound poles sealed with mud and topped with palm thatch. In addition, they may have *mamposteria,* flat-roofed cement structures topped with tanks for conserving rainwater.

LOWLAND MAYA Swinging a censer to make smoke, Lowland Maya make an offering near Lake Atitlán in Guatemala.

In 1847, the northern Maya began a general uprising against mestizos and whites. Known as the Caste War of Yucatán, this sporadic conflict was later spurred by a local religious movement. Hostilities finally ended in 1902, when the Mexican army occupied Chan-Santa Cruz, the Maya capital of the rebellion.

Modern times have seen important changes in the lives of the northern Maya. In the mid-20th century, for example, collapse of the international demand for Yucatán's sisal fiber eliminated an important source of local employment and income. This prompted even more movement of rural people to the urban centers of Mérida, Campeche, and the coastal resorts, dominated by Cancún, where age-old cultural traditions are fast disappearing.

Huichol

Population: 20,000
Location: Mexico
Language family: Uto-Aztecan

The traditional home of the Huichol lies in the Sierra Madre Occidental, a rugged landscape where the Mexican states of Nayarit, Jalisco, Zacatecas, and Durango come together. Partly because of their relative isolation from the colonial Spanish capital and other main European settlements of the central plateau, the Huichol and their immediate neighbors—among them the Tepehuans and Cora—have succeeded more than most others in resisting changes that came in the wake of the Spanish conquest.

HUICHOL A Huichol man wears the group's traditional embroidered shirt and hat with dangles.

In fundamental ways, the modern Huichol culture differs very little from its pre-Hispanic form; this is particularly true in terms of worldview and religion, despite the introduction of Catholicism by the Europeans. The leaders of family lineages customarily lead rites held in family shrines, while native priests, officiating from sacred seats, customarily oversee community rituals on special occasions. These rites and rituals involve the nearly one hundred supernaturals in the complex Huichol pantheon, among them the messenger god who takes the form of a winged deer and acts as an intermediary between people in the physical world and beings in the supernatural realm.

The Huichol may be best known for religious objects. Some items include featherwork, beads, and woven string; others are made from painted stone disks and gourds. "Prayer arrows," dangling feathers affixed to shafts, help shamans send messages to the gods. *Ojos de Dios,* or eyes of God, are made of colored yarn stretched between cruciform sticks and used to keep away the souls of the dead and ward off malevolent spirits. Perhaps the most remarkable form of Huichol art are the *tablas,* paintings made by pressing colored yarn into beeswax coating on boards. Tablas range from geometrical to pictorial and depict supernatural animals and other beings.

The creativity exhibited in religious art extends to Huichol handicraft and clothing, including men's trousers, shirts, and capes, as well as shoulder bags, earrings, and other adornments. Each item displays an intricate geometry of subtle forms and colors.

Mixtec

Population: 399,000
Location: Mexico
Language family: Mixtecan

In the Nahuatl language of the Aztec, Mixtec means "cloud people," an apt description, for most of the Mixtec live in the mountains of Mexico's western state of Oaxaca; others occupy lower areas such as the hot, humid coastal region to the south. To the east, centered in the

GENOGRAPHIC INSIGHT

PEOPLING OF THE AMERICAS

Once the first villages became established in the Arctic Northwest of North America, their inhabitants quickly adapted to life in a foreign land, spreading to all corners of the Americas in a few thousand years. How they did so along the corridor from present-day Alaska to the southern tip of Tierra del Fuego—a passageway spanning 12,000 miles and covering a vast longitudinal and ecological spectrum—remains one of the great mysteries of human migration.

East-west corridors like the Eurasian Steppe, a homogeneous grassland connecting the vast plains of Central Asia and eastern Europe, are thought to have given people a jump start in the rapid development of advanced technologies. Climate tends to be similar along similar latitudes, making it easier to find new resources. Cotton and silk produced in the east can be traded for iron and bronze smelt in the west, both areas able to develop more quickly than populations living in isolation.

Introducing Maize

But such geography is not compulsory fare for advancing societies. While major ancient civilizations developing in the Indus Valley, the Tigris-Euphrates basin, and along the Yellow River in China all benefited from this geographic advantage, surprisingly the fifth, and arguably largest, grew directly along the American corridor.

It all could have started with a gust of wind. Nine thousand years ago, there existed in and around Mexico a number of species of large grasses, teosinte, five of which survive today. These plants' numerous branches produce small bunches of flowers, which then mature to form tiny clusters of seeds encased by a hard shell. Each species exhibits a unique combination of genetic and physical characteristics that could be mixed into a cocktail of functional types ideal for plant domestication. While the exact evolutionary process that produced maize is still debated—in fact, the role of teosinte itself is in dispute—genetic studies show that some degree of hybridization between wild species, mixing perhaps in the winds of the Mesoamerican highlands, produced offspring more attractive to humans.

The Aztec created this brazier, with a painted representation of their corn goddess.

By 6,000 years ago, groups of migrants along the north-south passage had taken the biggest and best of these maize hybrids and begun growing them in lowland Mesoamerica, choosing new ear and kernel forms and plants showing the best yield potential. The diversity of these crossbreeds made them adaptable to new environments and enabled humans to bring seeds from southern Mexico to surrounding areas for the first time. With maize as a staple, pre-Columbian societies flourished.

Settlers quickly established small communities, and over the next several thousand years, independent, complex societies began popping up throughout the region. But the agricultural intensity was initially minimal, and a flyover of southern Mexico 4,000 years ago would have shown a landscape largely unchanged by humans.

Pyramids Rise Up

In 1200 B.C. the isolated cities between Jalisco and Oaxaca coalesced into the first Mesoamerican civilization, the Olmec, whose pyramids and walled cities soon dominated the landscape. The culture flourished for 800 years, producing jade and clay figurines and the earliest form of writing in the Americas, until it suddenly broke apart, the ruling class unable to maintain its grip on the empire.

People added squash, beans, and peppers to their diet. Soon larger, more powerful civilizations replaced the fragmented Olmec. The Maya, and later the Aztec to the north, burst onto the scene with sophisticated mathematical systems and written languages, which enabled the construction of monumental architecture, dwarfing the pyramids that already dotted the Mesoamerican landscape. The cultures expanded and at their zenith could boast the largest urban populations anywhere on Earth. Despite Spanish colonization, the Aztec and Maya never fully disappeared. Many of today's Mesoamerican populations directly descend from these earliest settlers.

MIXTEC A Mixtec woman prays. Both Catholics and Protestants have made converts among the group.

valley of Oaxaca, live the Zapotec, the other great people of the region. The intertwined histories of the Zapotec and Mixtec peoples reach deep into the past, particularly at the famed archaeological site of Monte Albán, the ancient mountaintop capital and ceremonial center.

Archaeological evidence indicates that the Zapotec flourished at Monte Albán from about A.D. 250 to 750, when the neighboring Mixtec invaded the valley and took over the city. In the centuries before the Spanish conquest, the Mixtec achieved greatness through art and craftsmanship—mainly gold jewelry, turquoise-inlay work, and deerskin codices (books). These books held pictographic histories and royal genealogies, particularly relative to the life and exploits of the king and culture hero Eight Deer Jaguar Claw.

Modern Mixtec still occupy the lands where their ancestors once dwelled—highland and coastal regions in western Oaxaca and adjacent areas of Guerrero and Puebla. Their lives, like those of other traditional Mesoamerican peoples, revolve around the cultivation of maize, chili peppers, beans, and squash, as well as the raising of livestock.

The Mixtec and Zapotec vary somewhat in terms of their language, culture, and crafts. Stylistic differences are evident primarily in their pottery, basketry, and woven or embroidered *huipils,* the traditional blouses or tunics worn by women in the region.

Nahua

Population: 1.3 million
Location: Mexico
Language family: Uto-Aztecan

The Nahua of Mexico's central highlands are defined mainly by languages that are closely related dialects descended from Nahuatl, which was spoken by the Mexica, or Aztec, when Spanish conquistador Hernán Cortés entered their imperial capital, Tenochtitlán (now Mexico City), in 1519. Nahuatl speakers may also converse in Spanish, but use of this language varies greatly according to location in rural and urban areas. Today's 1.5 million Nahua live in and around Mexico City and throughout most of the central plateau, an area more or less equal to the main part of the Aztec Empire under Moctezuma II.

The various groups of Nahua share the fundamental Mesoamerican subsistence pattern of cultivating maize, beans, chili peppers, tomatoes, and squash in the rich volcanic soils of their heartland, but to do this, they rely on the European-style plow drawn by draft animals. In steep mountain areas where the soil can be poor, farmers burn the fields before planting. They also rotate maize and bean crops to help increase fertility. Other major crops are maguey (the source of the fermented drink known as pulque), sugarcane, rice, and coffee.

A church and a plaza sit at the center of a typical Nahua town. The plaza often serves as a site for the local market. The church also functions as the center point for dividing the settlement into four parts, or barrios, each of which bears a Nahuatl name—a settlement pattern that dates to before the Spanish conquest. A good example of this ancient pattern can be seen in the town plan of Milpa Alta, just south of Mexico City.

A grain at a time,
the hen fills her craw.

MEXICAN PROVERB

Perhaps the best known of all Nahua population centers, the Milpa Alta area is famed for its geometrical weavings of *quexquemitls* (triangular capes) and other garments made on Aztec back-strap looms. Additional household industries and crafts include adobe bricks, maguey-fiber ropes, and numerous items fashioned for the tourist trade. Among the most popular Nahua objects are pottery, copies of ancient figurines, and the remarkable lacquered boxes, chests, and gourds from the town of Olinalá in Guerrero.

NAHUA A Nahua man plows his hilly bean field in the state of Morelos, Mexico.

Otomí

Population: 280,000
Location: Mexico
Language family: Otomian

A group known as the Otomí occupy an area of central Mexico lying just to the northwest of Nahua lands. Like most of their close neighbors, the Otomí are highland people who seldom dwell below an elevation of 3,000 feet above mean sea level. The language they speak is distinct from Nahuatl and other Uto-Aztecan tongues used in adjacent areas and is more closely related to those of the Mixtec and Zapotec peoples living to the south.

The ways of Otomí farmers vary among populations. The more acculturated groups grow wheat, barley, and coffee, while more traditional farmers mix crops of maize, beans, and squash in the same field, often using the ancient Mesoamerican digging stick—or a modern version tipped with iron—for planting. Otomí settlements range from compact towns to dispersed settlements. House types include plank constructions thatched with grass or maguey leaves and adobe houses with flat or slanted roofs.

OTOMÍ Embroidery can be men's work among the Otomí. This man works on a *tenango*, a cloth drawn with flowers or animals and then embroidered.

Well known for creating a variety of items sold to both locals and tourists, Otomí craftspeople produce pottery, stone metates (maize grinders), baskets, and woven clothing and bags. Some Otomí, particularly women in the town of San Pablito in the state of Puebla, are noted for their manufacture of *amate,* paper made from the bark of the fig tree *(amatl);* the tradition dates back more than a thousand years. In pre-Hispanic times, amate was used in rituals or sometimes as the pages of screenfold books called codices, which were filled with hieroglyphic texts and illustrations to record important events and divination practices. Some modern Otomí cut supernatural figures from this special paper and use them in rituals connected with water and agricultural fertility. Dark paper cutouts of Otomí deities, mounted on lighter background sheets, have become a popular form of folk art in central Mexico.

Tarascans

Population: 180,000
Location: Mexico
Language family: Tarascan

A highland region in western Mexico is home to the Tarascan people, most of whom live in the state of Michoacán. Often called a "land of fire and water" because of its volcanoes and lakes, Michoacán is the location of the Paricutín volcano. From deep crevices in a farmer's field, Paricutín quickly grew a 400-foot-high cone and destroyed two villages with its lava flows in 1943.

The Tarascan language, relatively homogeneous over the people's territory, is apparently unrelated to other linguistic groups in Mesoamerica. But it may have tenuous ties to language families in both North and South America, which suggests a very early presence on the Mesoamerican landscape. While modern Tarascans generally share in the traditional Meso-American ways of maize farming, pottery making, and weaving, other circumstances attending their possible origins and cultural history make them unique.

The story of Tarascan culture in pre-Hispanic times derives mainly from the archaeology of Tzintzuntzan, the ancient regional capital on the shores of Lake Pátzcuaro, and from various chronicles set down during the time of the Spanish conquest. After studying illustrations of the unusual clothing used in the Tarascan area around that time, art historian Patricia Anawalt suggests that the people originated in the south, near present-day Ecuador, and migrated north along a well-known ancient trade route along the Pacific coast.

Today, each Tarascan village specializes in one or two kinds of handicraft, a practice that may have begun in the mid-16th century with Vasco de Quiroga, first bishop of Michoacán. The bishop not only organized social and religious life in the villages but also strengthened the occupational specialization so common in pre-Hispanic times by introducing new products and techniques. Among the distinctive output of Tarascan craftspeople today, one finds featherwork "paintings," guitars, and lacquerwork gourd and wood products, including elaborate masks worn

TARASCANS Perhaps waiting for a bus, an elderly Tarascan couple occupies a bench in Patzcuaro, western Mexico.

ZAPOTEC An elderly Zapotec woman honors her ancestors during a Day of the Dead observance in Oaxaca, Mexico.

during the traditional dances that form an essential part of native Mesoamerican religion. Tarascans are also notable for their work with copper, bronze, and other alloys—a tradition that appears to date back to at least A.D. 1000, when knowledge of metallurgy first appeared in the region.

Zapotec

Population: 441,000
Location: Mexico
Language family: Zapotecan

Today's Zapotec population, the largest indigenous group in Mexico's Oaxaca, occupies most of the eastern portion of that state, as well as parts of neighboring ones. The varied lands of this region range from the drier, flat landscapes of the Isthmus of Tehuantepec to the high ridges and deep canyons of the Sierra Madre del Sur. The major valleys of Oaxaca—Zaachila, Tlacolula, and Etla—lie within the Sierra Madre.

During the Classic period, between about A.D. 250 and 900, ancestors of the modern-day Zapotec built some of the largest and most important regional centers of ancient Mesoamerica. The greatest of these, Monte Albán, occupies an artificially flattened mountaintop overlooking today's city of Oaxaca. There, a grand series of palaces, plazas, pyramid-temples, and other structures, including a great pentagonal observatory, were constructed over the ruins left by settlers who had occupied the site as far back as 1000 B.C.

Modern Zapotec share the belief systems of many of their Mesoamerican neighbors. They practice Catholicism while continuing to acknowledge a pantheon of supernatural beings associated with sacred geography and natural phenomena. They have also maintained their ancient belief in animal guardians, companion spirits that individuals acquire at birth.

Through the years, economic activities have included traditional Zapotec weavings and ceramics, with different settlements' wares having distinctive shapes and colors. They are also known for music, as well as for cultivating the language arts, which has given rise to locally famous speechmakers and a native literary tradition stretching back to the 16th century.

Amazon & Orinoco Basins

The Amazon may not be the world's longest river, but more water flows through its system than the combined total of the next ten most voluminous rivers. The world's highest waterfall, Angel Falls, plunges 3,212 feet in the Orinoco region. The mouth of the Amazon lies in north-central Brazil, and its tributaries spider throughout the country, plus Colombia, Peru, and Bolivia. The Orinoco River area encompasses the top part of South America, north of the Amazon. Starting in the Guiana Highlands, the Orinoco runs 1,700 miles, mostly through Venezuela, to the Atlantic Ocean.

Amahuaca

Population: 500
Location: Peru, Brazil
Language family: Panoan

Hunters of peccaries and rodents and farmers of maize and manioc, the Amahuaca are a remote group who live in tiny settlements of thatched dwellings in the border region of southeastern Peru and western Brazil. Their rugged homeland in the tropical forest is difficult to access, so even today they have only occasional contact with the outside world. The Amahuaca would prefer to continue their traditional ways, but international logging interests and prospectors threaten their homeland and their very simple material culture. Before steel axes and adzes arrived in the late 19th century, the people relied on stone axes, digging sticks, and tortoise shell hatchets.

As in other forest groups, the boundaries between the physical and spiritual worlds are blurred, with shamans acting as intermediaries between them. To induce trances during group sessions, the Amahuaca use hallucinogens, especially ayahuasca—a narcotic beverage made from the yagé vine—which produces vertigo and visions. The people believe that the partaker's soul leaves the body to question spirits who appear.

This group is also remarkable for having practiced endocannibalism—consuming the flesh of their own kind. In 1960, anthropologist Gertrude Dole witnessed the death of an Amahuaca infant during the night. The body was buried under the house, then disinterred, and burned in a pot. The baby's mother carefully removed the bones from the ashes and, after several days of grieving, ground maize and made gruel, which she mixed with bone powder and drank. Her period

He who plays the role of redeemer ends up getting crucified.

MEXICAN PROVERB

AMAHUACA The future will likely bring increased contact with the outside world for this young Amahuaca boy.

of mourning was intended to appease the deceased's spirit, just in case it had lingered. Once the mother had consumed all the bones, she reportedly became voluble and happy.

Bororo

Population: 1,300
Location: Brazil
Language family: Bororoan

In the mid-20th century, French anthropologist Claude Lévi-Strauss studied the Bororo of the southern Brazilian Highlands and found them to be a group with a highly complex social life. During the more productive times of year, the people came together into a large settlement, but during the drier months, they dispersed in small groups, searching for food.

In his work, Lévi-Strauss focused on 140 Bororo living in the riverside village of Kejara. A line running parallel to the river divided the village into moieties known as Tugaré and Cera; another line divided the community into upstream and downstream halves, with further subdivisions into clans and groups of rich and poor, each with its own religious symbols. The Tugaré moiety honored mythical heroes responsible for water, rivers, vegetation, fish, and human artifacts. The Cera heroes dated from the time of the creation, when they brought order to the living world. According to the natural hierarchy of things, at least in Bororo terms, the Tugaré were closer to the physical universe and therefore "strong," whereas the Cera were closer to the human world and "weak."

To Lévi-Strauss, the Bororo seemed to delight in making the affairs of a small-scale society extremely complex. Today, both moieties of society are still intricately engaged: They bury each other's dead, exchange women in marriage, and live, as it were, one for the other. Paraphrasing Lévi-Strauss, one could say the

BORORO A Bororo man shows his game face as he prepares for Brazil's Indigenous Games.

Bororo delight in social complexity, for it preserves the social order and ensures that they stay in tune with their vision of the cosmos.

Brazilians

Population: 204 million
Location: Brazil
Language family: Italic

The lone Portuguese-speaking country of South America, Brazil has for the better part of 500 years represented a vast frontier of opportunities and potential. Soon after the Portuguese claimed Brazil in 1500, they began exploiting its resources by exporting hardwood. Within a few years, the plantation economy was established with sugarcane, using indigenous peoples as labor until the first African slaves were imported. One by one Brazil's natural resources were tapped: gold, gems, rubber; growing terrain ideal for coffee, rice, and cocoa; and vast, grassy expanses suitable for raising cattle. The Portuguese continued to arrive, along with waves of other Europeans, Middle Easterners, and Asians.

The European settlers encountered indigenous peoples representing at least four different language families. Those living near the coast have had the most influence on Brazilian culture. Words from the Tupí-Guaraní languages for flora, fauna, and place-names contribute to Brazilian Portuguese, which is spoken by nearly the entire population, although the literacy rate lags behind at some 86 percent. Today, more than 250 individual Indian groups live mainly on reserves established by the government.

Racial mixing occurred early in Brazil's colonial history. Single Portuguese men looked for women among the Indians and slaves—the more than two million Africans contributed to a Brazilian population now of at least 45 percent African descent, the largest outside Africa. Recent DNA studies demonstrate even more of a mixture in the general population than expected.

While most Brazilians are nominally Roman Catholic, the church vies for influence with African traditions, particularly Candomblé, Macumba, and Umbanda. Other diverse faiths contribute to the religious landscape, including the growing popularity of Pentacostalism and Charismatic Catholicism, both of which appeal to the Brazilians' emotive nature.

While many Brazilians point with pride to their "racial democracy," the situation on the ground shows the struggles of people of color and the widening gap between rich and poor, a situation out of sync with the high ranking overall of the Brazilian economy. Although each locality has its own economic divide, the country as a whole is differentiated between the economically disadvantaged north and the more prosperous south, with similar distinctions separating the interior from the coast. Former slaves, emancipated in 1888, became the foundation of the favelas, or shantytowns, that sprang up at the edge of Rio and other cities. Municipal governments have not succeeded in eliminating these makeshift communities and in some cases have provided them with amenities such as water, sewers, and electricity.

I shall sing the sun up, once the stars have been washed from the sky.

SURINAMESE PROVERB

Some 80 percent of Brazilians live within 200 miles of the ocean, with about 18 million clustered in the coastal cities of São Paolo, Rio de Janeiro, and Salvador. Brazilians flock to their beautiful beaches where minimalist swimwear is the rule even if the wearer is deep into middle age or carries more than a few extra pounds. New Year's Eve, which occurs during Brazilian summer, also draws people to the beaches. At midnight they walk to the sea and toss in white flowers as offerings to the African goddess of the sea, Yemanjá, with hopeful prayers for the new year.

Brazilians come together for Carnival, the days-long pre-Lenten national holiday marked by parades, extravagant costumes, dancing, and musical competitions between samba schools—favela organizations. Social and economic distinctions become blurred and even reversed during Carnival, taking people, especially the poor, out of the routines of their daily lives. *Futebol* (soccer) also unites the Brazilians; World Cup play typically overtakes the entire country.

Parentela—the extended group of all maternal and paternal relatives, including in-laws—represents the core of many Brazilians' social world. Brazilians prefer to live near their kin and to socialize with them on a regular basis. Young adults tend to stay at home until they marry and not to stray too far from the family base when they do. Combined with compadrio, or godparenthood, the parentela provides nearly all the social support an individual needs.

Caribs

Population: 11,200
Location: Venezuela
Language family: Cariban

Expert canoe navigators and warriors, the Carib Indians of the Caribbean's Lesser Antilles drove out the indigenous Arawak about a century before Columbus and settled on the northern and northwestern shores of South America. They were originally known as Galibi, a word corrupted by the Spaniards to *cannibal;* hence, they were thought to be eaters of human flesh. If the Caribs were cannibals, their consumption of enemies may have had purely ritual

BRAZILIANS Cariocas, as residents of Rio de Janeiro are called, crowd the trolley that services Rio's historic area.

CARIBS A Carib mother and her children live on the island of Dominica in the Caribbean's Lesser Antilles.

significance. Decimated by Europeans fighting for control of the Caribbean, these people survive only in small enclaves today, notably in Venezuela and Guyana.

For all their reputed ferocity, Caribs were, and still are, expert fisherfolk. They built dugout canoes, equipped them with sails, and ventured far offshore. Blowguns as well as bows and arrows provided them with forest game, and they farmed maize, cassava, yams, and squash. Today, coastal Caribs live in spacious thatched houses close to riverbanks. The largest Carib settlement, Galibi, is a fishing town in Suriname; a cooperative there sells fish locally to help the community remain self-supporting.

The Barama Caribs of Guyana live in small farming communities of thatched, timber houses near the Barama River and along forest trails. They are hunters, fisherfolk, and subsistence farmers whose lands are now threatened by mineral prospectors and international logging companies. For centuries, they have relied on simple, traditional methods for catching fish, but they have not hesitated to use European fishhooks and nets when they could obtain them through trade. Much of the time, though, they resort to their ancient strategies: Men stand on large rocks and try to spear fish or shoot them with bows and arrows. Barama Caribs also set fishing lines, using hooks made from thorn trees or even bent domestic pins, but their most effective method uses poison made from haiari roots. Although the Guyana government forbids the poisoning of the country's streams, this method is still commonplace. Almost every creek has a wickerwork fence lying across its mouth, placed there to catch the dead or stupefied fish that float downstream. Waiting women and children kill the sluggish fish as they float to the surface.

The Caribs are now Catholics. Until recently, however, shamans played an important role in Carib rituals after serving long apprenticeships. They used tight ropes to master the movements of their souls. Hands bound, they swung through the air and induced vertigo as a way of transporting themselves to the supernatural realm.

DEEP ANCESTRY

THE MIGRANT

Asiye Elevli took the ancestry test through a college class where they were using the Genographic participation kit as part of the curriculum. She thought she knew all about her family history. Her father was from Turkey, where his family had lived forever and where people traditionally married within their clan, and her mother, who was born and raised in Mexico, was of Spanish descent. She expected to see predominantly European and Southwest Asian components in her DNA. But what she learned surprised her, revealing a story of recent migration as well as ancient wandering.

DNA Results

Asiye's results showed that she was predominantly Native American, with additional components from European and even Northeast Asian populations. While her mother had grown up in Mexico and lived there prior to moving to Spain, she had always been told that even her Mexican side of the family was of predominantly European ancestry. Although Mexico is home to more than ten million self-identified indigenous people, the vast majority of Mexicans speak Spanish and consider themselves to be of predominantly European ancestry. Only recently have Mexicans, as well as many other Latin American groups, began to identify with their strong indigenous roots. And, as it was with Asiye, her results revealed a much stronger connection to the New World that she anticipated.

Founding Lineage

Her mtDNA results were the clincher—she was haplogroup A2h1. This rare maternal lineage is found in indigenous Mexicans, as opposed to Europeans, consistent with her grandmother's indigenous maternal heritage. A2h1 is a Mesoamerican-specific sublineage of haplogroup A2, one of the founding mitochondrial DNA lineages in the Americas. Along with Y-chromosome haplogroup Q and maternal lineages B2, C1, D1, and X2a, A2 likely arrived in the Americas with some of the earliest trekkers who migrated from Asia across the Bering Strait some 15,000 to 20,000 years ago. The "Ice Maiden," an Inca mummy discovered in 1995 by anthropologist Johan Reinhard at an altitude of 20,000 feet on a mountain in southern Peru, also belongs to the A2 lineage. The A2 branch is so important among indigenous American groups that it occurs throughout North, Central, and South America, sometimes at frequencies of 75 percent or higher. But the height of diversity for haplogroup A2 is Mexico and Mesoamerica, the home of the Aztec Empire and ancient Maya and the birthplace of maize (corn) domestication. And that is where Asiye's ancestral matriarch, the originator of haplogroup A2h1, lived several thousand years ago.

Asiye Miraj Elevli, mtDNA haplogroup A2h1

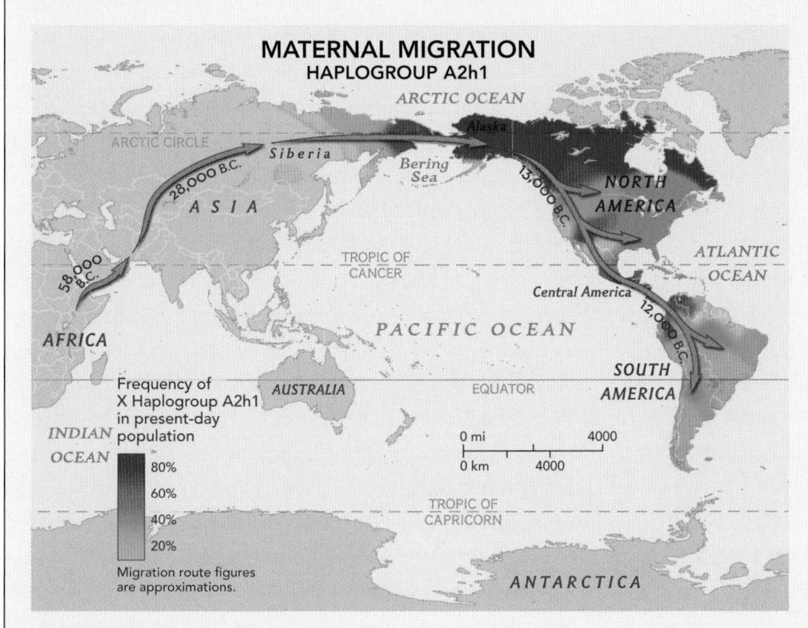

Maternal A2: Once they reached the Americas, humans found abundant resources and a comfortable climate.

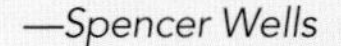

—Spencer Wells

DESANA Crushed berries make the red paint that decorates the face of this Desana woman.

Desana

Population: 1,500
Location: Brazil, Colombia
Language family: Tucanoan

The rain forest of Colombia's southeastern lowlands is home to the Desana, hunters and fisherfolk who also obtain much of their diet from manioc. Each household has to fend for itself and get food from a carefully delineated area surrounded by the territories of other groups. Because fish and game animals are such important foods, people try not to take too many fish from the rivers or to overhunt their prey. They also try to manage the size of their families by limiting the frequency of sexual relations.

The Desana live in local kin groups. Their dwellings are long, rectangular houses composed of wood and thatch and occupied by four to eight related families. Each family practices strict birth control, relying on oral contraceptives made from herbs and on sexual abstinence, especially before and during a hunt. For the Desana, the consequences of sex can present grave threats to their survival. People and animals share homelands with finite resources, they say, and thus both are limited in their reproductive potential. If the human population were to increase too much, then the animals and supernatural Master of Animals would grow envious of the humans' misuse of the limited sexual energy available to all. In this culture, men who sire more children than the norm are regarded with scorn, and their families are considered little more than those of dogs. The strong social controls, reinforced by the Desana rebuke of those who ignore them, are a way to keep in equilibrium with their food supplies.

Kuikuru

Population: 500
Location: Brazil
Language family: Cariban

The Kuikuru survive in a single group near the Kuluene River, a tributary of the Xingu in central Brazil's rain forest. They are famous for slash-and-burn, shifting agriculture, practiced amid a large forest tract near their one village. Before 1900 and the coming of iron tools, Kuikuru farmers cleared their gardens with stone axes and the sharp-toothed jaws of piranhas. Their material culture was marked by skill in making feather ornaments and baskets. Like other Brazilian groups, they clustered their thatched houses in a circle, with the men's house a major center of ceremonial activity.

Now, as then, Kuikuru farmers begin clearing land at the end of the rainy season. They let felled trees and other vegetation dry for a while; just before the rains come, they fire the debris in piles atop the fields. As rain falls, wood ash washes into the soil and fertilizes it. The Kuikuru plant at least 11 varieties of manioc, a tropical root crop that is poisonous unless pounded and boiled before being eaten. Although they harvest about four or five tons of manioc per acre, they lose more than half the crop to peccaries and ants.

Most other slash-and-burn farmers must move their villages regularly, but the Kuikuru have so much land within a comfortable four-mile walking radius that they never have to leave their long-established traditional settlement. Under ideal conditions, each man spends only about three and a half hours a day working to feed his family; about two hours go to farming and an hour or so to fishing. Kuikuru men spend the rest of the day dancing, wrestling, and gossiping. Today, with modern agricultural methods and fertilizers slowly coming into use, many people are choosing to leave the traditional village and settle on small family holdings nearby. Nevertheless, the Kuikuru have shown that the simplest methods of manioc cultivation can provide enough food for people to remain in the same village permanently.

Over time, social tensions and factions develop in the community, but the Kuikuru find strength in traditional religious beliefs. They seek the help of their shamans, who play important roles in a society where secular and spiritual worlds are deeply intertwined. The shamans defuse quarrels that could lead to violence. They also investigate accusations of sorcery, attempting to identify the perpetrator of evil spells so they can neutralize the magic. The Kuikuru believe in a creation god, Kanassa, and he is said to have trapped a king vulture, the master of fire, and thereby brought fire and life to the people.

Mundurucú

Population: 10,000
Location: Brazil
Language family: Mundurukú

In 1770, a Mundurucú war party traveled more than 400 miles from the group's homeland on Brazil's upper Tapajós River to attack Portuguese settlements along the Amazon. Afterward, the Portuguese promptly engaged their services as mercenaries to help them fight other hostile Indians. At home, the Mundurucú lived on gently rolling grasslands away from the river. But by 1920, many of them had moved closer to the riverside forest, where they tapped rubber trees during the slack months of the agricultural year to satisfy a burgeoning demand from the distant industrial world.

Today, the Mundurucú number about 10,000, their earlier population reduced by disease, maltreatment,

KUIKURU Kuikuru dance at an exhibition during the opening ceremony of the XII Games of the Indigenous People in Brazil.

and the abuses of gold and rubber hunters. They are mainly subsistence farmers and rubber traders. Between September and late May, the people cultivate manioc and other crops, but both hunting and fishing are important throughout the year in a climate where dried fish lasts just two weeks.

> *You can hide your grandmother, but you can't prevent her from coughing.*
>
> SURINAMESE PROVERB

Traditionally, Mundurucú warriors lived in the village men's house and were responsible for guarding their community. Their society was a violent one that made raids on enemies during the dry summer months, primarily for the purpose of obtaining trophy heads. Although hostile in their behavior toward outsiders, the Mundurucú repressed aggressions closer to home; they placed a high premium on cooperation and social harmony between neighboring villages for joint hunting and fishing expeditions.

The people have a simple material culture, making much use of cotton thread, basketry, and poor-quality clay vessels. To hunt, they rely on bows, reed arrows, and wooden spears, and for generations they used war clubs in raids against their enemies. The Mundurucú are famous for their featherwork, which includes aprons, capes, armbands, and other ornaments made by attaching tropical bird feathers to netting. They are also known for body tattoos and for painting lines on their faces that give the appearance of wings spread across their countenances.

MUNDURUCÚ Avon calls at a Mundurucú home, giving the women a chance to try on cosmetics.

Every village has one or two male shamans who are believed to have had supernatural powers since birth. These men cure sick people by blowing tobacco smoke on the skin, then massaging the skin while looking for the magical darts thought to have caused the illness. Shamans are feared for their powers, for they are also believed capable of causing sickness and misfortune. In extreme cases, when they are suspected of causing evil to descend on an individual or a community, they are executed. Although raids and head-hunting have ended, the Mundurucú are still an aggressive and male-dominated society in which violence is sometimes accepted social behavior, especially against women and sorcerers.

Nukak

Population: 700
Location: Colombia
Language family: Puinavean

Not until 1988 did the Nukak of eastern Colombia come into sustained contact with Western civilization. The results of this contact were multiple influenza epidemics and massacres at the hands of growing numbers of colonists. Between 1991 and 1997, intense pressure by Survival, an indigenous people's advocacy group, led to the establishment of a tribal reserve for the remaining Indians. Now, about 700 Nukak occupy a territory about 110 miles long by 38 miles across. They are hunters and foragers who take monkeys, peccaries, and birds, as well as insects, plant foods, and fish, from the rain forest. At the same time, they manage such wild plant species as chontaduro palms, achiote (a shrub with culinary uses), and plantains, growing them in small garden plots established some distance from their settlements. Because of the influence of missionaries and frontier farmers, the Nukak have started farming more systematically in recent years.

After centuries of building campsites in the forest, the Nukak have modified the rain forest environment to their considerable advantage. Every five or six days, the people move to a different area and construct a new

GENOGRAPHIC INSIGHT

ULTIMATE BIODIVERSITY

Most segments of our genome neither code for a gene nor control how active, or expressed, a gene might be. These segments are said to be selectively neutral, meaning they are impervious to the Darwinian effects of natural selection. Though off the hook in the struggle for survival of the fittest, our nonfunctional DNA is at the mercy of a potentially more powerful random process that can cause drastic changes over time. Fortunately, the principles are straightforward.

In math class, we are taught through a basic experiment that though flipping a coin ten times should give equal numbers of heads and tails, it is not uncommon to see the numbers skew to one side, say with eight heads and two tails. Though not conducive to a 50-minute class period, if a student were to take the time to flip a coin one million times—putting probability theory to the test and likely flunking out of school in the process—the tally would almost always come to around 500,000 for each side. There may certainly be long runs in either direction, but even the runs themselves would tend to even out over time. This same effect that can drastically influence different frequencies of heads versus tails also plays a role in determining the frequency of various lineages in neutral segments of our DNA, a process known as genetic drift.

Chancing Migration

Initial migrations into the Americas were made up of perhaps only a few hundred individuals, and of these, only a fraction left the Arctic Northwest to head south. Small populations are more susceptible to the effects of drift, much the way coins are, and perhaps nowhere else on Earth has this played a bigger role in shaping genetic diversity than in the varied Amazon and Orinoco Basins of South America. The region may have been populated by one or multiple waves of early migrants from North America—a question that remains unanswered—but in either case, its first inhabitants quickly spread out into small, isolated pockets, allowing genetic drift to reign supreme. Once there, the jungle kept it that way.

A Cuna woman reflects; her people inhabit the San Blas archipelago of Panama.

Amazonia is roughly defined as the equatorial region east of the Andes between the Guiana and Brazilian highlands, and though people tend to think of Brazil, it actually stretches across parts of Guyana, Venezuela, French Guiana, Suriname, Colombia, Peru, Ecuador, and Bolivia. Despite accounting for less than 2 percent of global landmass, it is home to over half of Earth's rain forests and more than a third of its species. But while the area's high biodiversity makes for fascinating research, it doesn't do a very good job supporting large groups of humans. In fact, the very canopy that sustains this biodiversity robs its soil of practically all useful nutrients, relegating the ground to serve as little more than an anchor for the canopy's root system.

Human Species Drift

While small hunter-gatherer communities exploit the jungle, agriculturalists here are handcuffed by an uncooperative terra firma, since crops do very poorly in the low-light, nutrient-poor environment. Because Amazonian soil is productive for only a short period of time before nutrients are leached from the soil, farmers must constantly work new tracts of land.

This exigency kept human populations small, giving their founders a huge say in the development of a wide range of cultural practices and unique belief systems. Their descendants adopted words to describe a new surrounding, and with no written tradition to keep the linguistic changes in check, the spoken word quickly changed course branching into many different directions.

A great indication of this force of cultural drift as it acts on these populations is the fact that in a bit over 12,000 years of human occupation, more than one thousand unique languages developed in tropical South America—the region's rich biological diversity seemingly matched by the cultural and linguistic diversity of its indigenous populations.

Apparently the human genome wasn't the only thing affected by the Amazon's poor soil.

NUKAK Sacks of donated food provide seating in this Nukak settlement that marks the end of the group's nomadic life.

campsite, and each time they do this, they disturb the ground and leave behind seeds from plants they collect and consume. Those seeds eventually sprout and grow.

The Nukak live in a well-defined world. Like other groups, including Fuegians in the far south, they identify three geographic and social domains of their world: the area habitually used by each band, a carefully delineated wider area around their homeland, and both areas in relation to those of other indigenous people in the region.

Saramaka

Population: 26,000
Location: Suriname
Language family: English-based Creole

Descendants of escaped African slaves, the Saramaka live deep in the forests of Suriname. They are one of several such groups who survived not only the era of slavery but also armed aggression and other efforts to root them out of territory ceded by England to the Netherlands in 1667. Effective raiders of European-owned slave plantations, their numbers grew as other escaped slaves sought sanctuary among them.

Since its very foundation, Saramaka society has been dependent on manufactured goods and food staples brought from Europe, a situation that was addressed by an 18th-century agreement with the Dutch. In 1762, a treaty granted unconditional freedom to the people and required the Dutch crown to pay them tribute every year. Representatives of the crown subsequently brought boats up the winding rivers into Saramaka territory, carrying axes, pots, guns, thread, needles, cloth, and other manufactured items needed by the people. In return, the Saramaka stopped making raids on the Dutch plantations along the coast.

In 1863, after all the slaves on the coast had been emancipated, the Dutch ceased sending tribute upriver and allowed Saramaka men to venture freely into areas controlled by Europeans. Saramaka also traveled to neighboring French Guiana, where they were able to find work and earn money to buy goods that they eventually took back to their villages.

In the 19th century, nearly every able-bodied man spent about half his adulthood living and working outside Saramaka territory, either in Suriname or French Guiana. Men worked for periods that could last as long as several years, and this form of migrant labor helped the Saramaka and their villages to remain semi-autonomous. Today, the tradition continues, but the sought-after goods now include outboard motors, gasoline, chain saws, and a wide variety of foods.

When men eventually return to their home villages, they usually give about half their wealth to their wives, children, aunts, sisters, and various other dependents. The other half is kept in storage to be used during periods of unemployment or in times of special need. If a dependent becomes ill or marries, for example, the accumulated goods pay for the associated expenses.

The Saramaka supplement their economy through swidden horticulture, with the men responsible for cutting and burning trees to clear fields and the women planting and harvesting. Most local food items come from the women's gardens of rice and other staple foods. Men also engage in hunting and fishing, and when a man kills an animal or makes a large catch of fish, his kill or catch is always redistributed along kinship lines within the village.

Because Saramaka society is matrilineal, each person receives his or her identity through the mother's line. Fathers play significant roles in child rearing, but legally the most important connections are on the mother's side of the family: her sisters, parents, and grandparents. At the same time, polygamy is widely practiced, so men may have several wives. Every man and every woman has a separate house. People have houses in their own villages, which means their mothers' villages, and often women have houses in their husbands' villages as well.

SARAMAKA In the matrilineal Saramaka society, a person's identity is reckoned through the mother's line.

In a legal sense, matrilineal groups strongly resemble corporations. If a man commits a crime, for example, his relatives are held responsible too. If that crime is murder, the matrilineal descendants of the murderer are thought to be visited forever by the vengeance-seeking ghost of the victim.

Over the years, the Saramaka have developed a rich polytheistic religious tradition involving sky, earth, water, and forest gods. They also believe in the Akaa, a person's most important spirit. The Saramaka say that when an individual is born, the Akaa comes from the world of the ancestors to accompany the person throughout life; later, it travels with him or her to the land of the dead. There are two kinds of Akaas—good and bad. The "bad" one, as in much of the New World African theology, is not all bad, however. For one thing, it helps protect a person's body from other evil spirits that may try to take up residence there. It also causes the individual to have dreams—which may end in a nightmare if the spirit suddenly leaves the body. The good Akaa is loyal and requires beauty and purity, but it is very timid. At the first sign of a problem, the good Akaa departs, possibly causing sickness. If a person's Akaas are out of balance, he or she is sure to become ill.

Shuar

Population: 35,000
Location: Ecuador
Language family: Jivaroan

The southern lowlands and lower uplands of eastern Ecuador and neighboring Peru are home to the Shuar, or Jívaro. The 20,000 Untsuri Shuar are the most numerous group, and they live across the Macuma River from their traditional enemies, the Achuar.

Killing, warfare, and other violence permeated all aspects of traditional Shuar society, which placed a high premium on acquiring individual power. From youth, every Shuar was prepared to face conflict. Everyone had potential enemies, with the more

SHUAR A Shuar woman demonstrates in Quito, Ecuador's capital.

successful individuals linked to higher potentials for violence. Their objectives were simple: Kill all the members of an enemy household and decapitate them to make *tsantsas,* shrunken heads. The grim trophies not only proved that revenge had been taken but also provided the victor with powerful magic, thereby preventing the dead man's soul from wreaking vengeance in this life or the next.

Today, most Shuar killings involve feuds between members of the same group, and these usually arise from accusations of sorcery, competition over women, or quarrels over crop damage caused by pigs. First, a member of a neighbor's family is assassinated. Then a blood feud begins, resulting in another killing. Feuds can last for generations.

Until recently, these hunters and root-crop cultivators lived in small, autonomous groups that were in a constant state of political disequilibrium from all the feuding, shifting alliances, and warfare. Political power lay in the hands of talented individuals who were expert hunters, good warriors, or shamans. Qualities of physical strength and supernatural knowledge, the courage to protect and avenge family members, and the ability to endure the hardships of war were highly valued in this bellicose society. Lacking a formal political structure, the culture counted personal power and influence above everything else. Kinship ties held society together in loose, fluid alliances that gave people the moral obligation to avenge the death of a relative. Revenge has long been a driving force in Shuar society.

Contact with European colonists, missionaries, and the Ecuadoran government led the Shuar to form a federation that would help them protect their land and their rights against government-sponsored white colonists. Since 1961, the federation has set up permanent centers and associations with communal ownership of the surrounding land. It also gives these entities loans to start cattle cooperatives. The people now govern themselves with a complex mix of traditional and modern political organization.

Head-hunting has long since been outlawed by the government, and although Shuar culture still places a premium on power and revenge, killing is not as widespread as it used to be. Now, nonviolent means such as "strong speaking" are often used to resolve conflicts.

Tapirapé

Population: 500
Location: Brazil
Language family: Tupi-Guarani

Until the early 20th century, the Tapirapé of central Brazil had virtually no contact with Europeans. They were forest farmers living in isolated villages between the Araguaia and Xingu Rivers, and their fate offers an interesting example of the impact of exotic diseases on such communities.

Around 1900, there were five Tapirapé villages, each with a population of 200 to 300 people. Within a decade, one village had disbanded after enemy raids and another had been depopulated by an epidemic, possibly malaria. The first Brazilians arrived in 1910. Although the Tapirapé had only sporadic contact with outsiders in the next few decades, many of them succumbed to new diseases such as influenza, smallpox, and yellow fever. By 1932, another village had been abandoned, and seven years later the last two merged into a single settlement that still survives.

In the 1940s, Dominican friars began meeting families at a Tapirapé River port during the annual dry season, when the people worked the river for turtles and fish. The Dominicans would bestow gifts of salt, tools, and other exotic goods. Eventually Catholic missionaries settled among the Tapirapé; the society was changing profoundly. Today, there is an oversupply of men, which means that many men have no women to perform tasks that women have traditionally done for them. To solve this problem, men make convenient social marriages with very young girls, whose mothers then do all the domestic work. The Tapirapé now sell some crops for cash and are slowly being integrated into Brazilian society, but they live at the margins of a rapidly industrializing state. Families in cities and towns survive under conditions of extreme poverty.

Warao

Population: 36,000
Location: Venezuela
Language family: Language isolate

The Orinoco Delta along the northern coast of Venezuela and Guyana is home to the Warao, a group of traditional fishermen, hunters, and gatherers who occupy the marshes, swamplands, and riverbanks in the vast, watery region. The name *Warao* is one the people use themselves and has been translated as "owners of canoes"; they also are known as Guaraúnos. The different Warao groups speak mutually intelligible varieties of Warao, which many linguists consider a language isolate, but some believe is related to the Chibcha language family of the northern Andes.

Warao have had outside contact from the beginning of the colonial period, when Europeans entered the

TAPIRAPÉ A Tapirapé man shows off his archery skills in a competition.

WARAO Warao build their homes on stilts to escape rising waters in the Orinoco Delta of Venezuela.

delta region to gain access to lands farther upriver, which they believed would turn out to be the fabled El Dorado. While remaining closely tethered to their base in the swamps and marshes, the Warao worked for and traded with the Spanish and Dutch. Then, as now, the Warao plied the delta waters in their low-slung dugout canoes, crafted by men who specialize in the task.

A Warao settlement can consist of a single, large extended family or a multifamily community of some 250 people. After marriage, a Warao man will go to live with his wife's family, where the couple resides in a household of sisters and their families under the ultimate authority of their father.

To build their traditional houses, or *palafitos,* the Warao use parts of the moriche palm, related to the sago, which has played an extensive role in their material culture and economy for centuries. Warao construct their homes on stilts to keep them above the rising tides and build elevated walkways between them. The wall-less dwellings have floors of moriche branches and thatched roofs of the palm's fronds. Warao women also weave moriche fibers into durable hammocks for sleeping that they sell outside the community, along with their baskets.

The versatile moriche provides a source of starch in the Warao diet as well. However, this has been supplanted to a large extent by the Chinese ocumo, similar to taro, which they now cultivate along with a few other crops such as bananas and sugarcane. The Warao continue the long tradition of outside work, laboring in sawmills, palmetto factories, and commercial rice and fishing operations.

Catholic missionaries came to the area almost 200 years ago and converted a number of Warao, as have evangelical Protestants more recently. Most Warao still follow traditional ritual and belief systems, a complex cosmology that includes a cult of ancestor worship. Spirits animate almost all natural phenomena, living and nonliving, some of which can metamorphose into jaguars. The Warao also believe in a world parallel to theirs that exists in the depths of the river. The two worlds intersect at times, as when a river being mates with a Warao woman, producing a monster. Three different kinds of shamans mediate the many connections between the Warao and the supernatural world, which receives blame for all illness and misfortune.

Warao culture is today threatened by ever widening involvement in the regional Venezuelan economy and the popularity of adventure tourism, which brings in groups to the area to sample the Warao way of life for a few days.

Xikrín

Population: 7,100
Location: Brazil
Language family: Northern Jê

Few Brazilian groups still follow their traditional lifeways, for the expanding frontier of industrial civilization has spread deep into the Amazonian rain forest. The Xikrín are among the ones who do. The northernmost group of Kayapó Indians, the Xikrín are farmers, hunters, and foragers living along the Rio Cateté in a homeland rich with brazil nuts. Their material culture is very simple, except for featherwork and basketry, and the people wear few clothes—the men, a penis sheath; the women, short skirts. The people cultivate manioc and maize within a day's journey of settlements made up of thatched houses set in a circle around a central open space. Men and boys hunt and search for land turtles while the women gather wild plants, including the heart of the babassú palm. The most important house in a settlement is the men's dwelling. Inside, young boys receive instruction, and

XIKRÍN These Xikrín live on a tributary of Xingu River in Brazil that may be affected by the completion of the Belo Monte Dam.

preparations take place for major ceremonies that may, for example, celebrate the first maize harvest or honor fish poisons. Only single men reside there; the married men live with their wives' families.

At night, Xikrín communities often gather for orations and discourses, which are important as a way to deliver instructions for the next day, air and resolve disputes, and enhance individual prestige. Little is known of Xikrín traditional religion, but shamans play a major role in mediating between the living and supernatural worlds, as they do in many other native South American societies.

Throughout the first half of the 20th century, contacts with Brazilian rubber interests, nut gatherers, and mahogany loggers were sporadic and hostile. But in the early 1960s, younger members of the tribe elected to live near the mouth of the Rio Cateté in the hope of becoming more closely integrated into Brazilian society. The rest of the community moved farther upriver to keep clear of the intruders. Since then, Xikrín society has been under siege. The downriver village became a rest stop for nut collectors, who used the threat of firearms to obtain women and get help in their search for nut trees; also, infectious diseases decimated the Indian population.

Hide today's anger until tomorrow

SARAMAKA PROVERB

After years of being exploited, the downstream Xikrín realized the true market value of both their resources and their labor, as well as the extent of their exploitation. They reunited with the traditional group back at the original upstream village and welcomed a Catholic mission among them, which has helped control social and commercial relationships with outsiders. In 1998, the Xikrín entered into an agreement with the Brazilian government and the World Bank for a sustainable mahogany-logging project. Nevertheless, their traditional culture remains under threat from the inexorable forces of the global economy.

Yanomamí

Population: 19,700
Location: Venezuela, Brazil
Language family: Yanomaman

Of all the surviving indigenous South American groups, the Yanomamí are among the most numerous. They occupy an area of rain forest in the upper Orinoco highlands of southeastern Venezuela and north-central Brazil. They are also one of the best known groups, thanks to many years of anthropological research.

The studies of anthropologist Napoleon Chagnon, in particular, have made the Yanomamí famous for their alleged fierceness and violent behavior. Recently, however, academic controversy has engulfed the research of Chagnon and other anthropologists who have worked alongside him. Critics have accused them of inciting violence among the people and using inappropriate research methods to form their conclusions. While the extent of violence among the Yanomamí may be exaggerated, there is no question that aggressive, violent behaviors—from wife abuse to raiding—are commonplace in this society. Only carefully controlled, ethical field research will establish whether the people are as violent as several generations of college students have been led to believe.

The Yanomamí are farmers, subsisting on plantains, sweet potatoes, and other carbohydrate-rich crops, as well as game and wild plant foods. Each village is some distance from its neighbors, and each specializes in different products, such as hallucinogenic drugs, arrow points, or cotton hammocks. Communities are not self-sufficient, meaning that they must depend on one another, mingling to trade and to feast. Still, there is widespread mistrust in this volatile environment where the eruption of violence and warfare is always a possibility. It is largely up to the village headmen to not allow matters to get out of control.

Many disputes are settled by aggressive chest-pounding duels or by slapping one's sides. Two or more men may engage in club fights, or there can be a formal war between neighboring villages. The ultimate level of violence involves trickery. On rare occasions, a village may invite its enemies over for a friendly feast, and then try to kill them as soon as they have let their guard

YANOMAMÍ The Yanomamí have long been victims of fortune seekers who have exploited and degraded their lands.

down. Only by resisting ferociously or moving away as far as possible can a weaker community without allies protect itself against enemies.

In villages occupying the lower-lying areas, where populations are larger and fertile land is in shorter supply, behavior is especially aggressive and violence more frequent. Headmen in these communities must be very astute, as well as expert at persuading people to help prepare for feasts and other communal activities. Above all, they must be willing to intervene in the volatile interpersonal disputes and attempt to control the outcomes. Leadership in such an egalitarian society of aggressive individuals comes only from respect and bravery in war. Sometimes headmen are also shamans, expert in the summoning of *hekura* spirits that cure and cause evil. The hekura arrive in the hallucinogenic snuff ingested by nearly all men in many villages, but shamans have a particular ability to use hekura against enemies. As a result, many men go through the long period of fasting and rigorous training to become shamans. In this society, where males dominate and there is a shortage of women, origin myths proclaim that fierce, warlike men arrived on Earth before women. Thus, the men are superior to them.

Andes

The longest mountain chain in the world and second highest only to the Himalaya, the Andes form a distinctive swath running practically the entire length of the continent of South America. More than 45 of the peaks in this mountain range stand more than 20,000 feet above sea level. Between the Andes and the Pacific Ocean to the west lies a narrow strip of coastal plain where can be found the Atacama Desert, the world's driest place. This landscape of extremes is shared by six countries: Colombia, Ecuador, Peru, Bolivia, Chile, and Argentina.

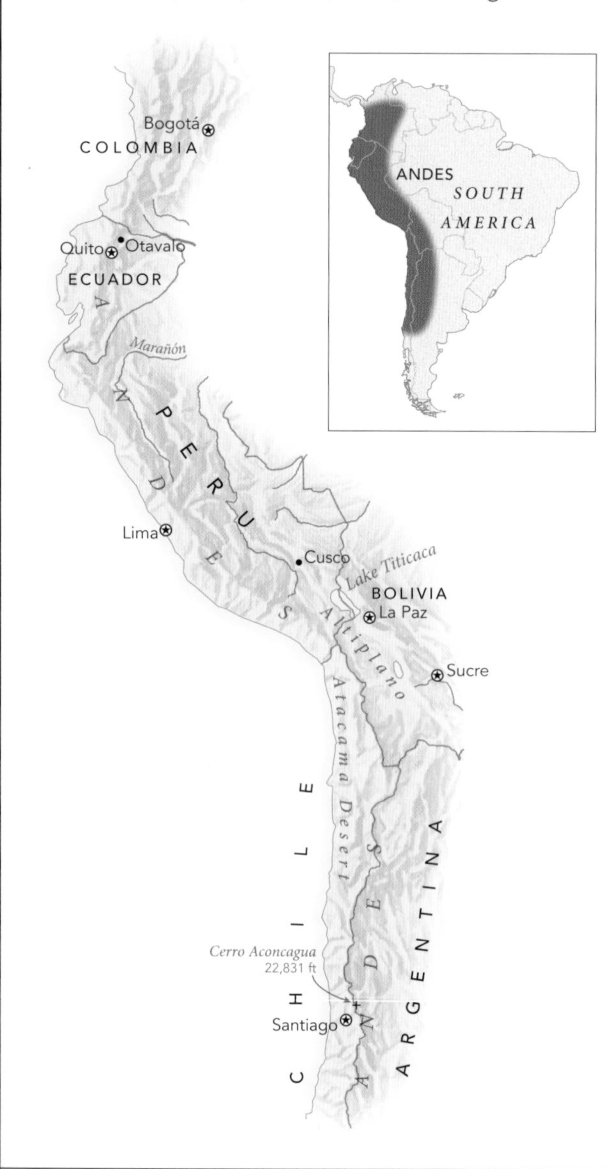

Aymara

Population: 2.8 million
Location: Bolivia, Peru
Language family: Aymaran

Six centuries ago, the Aymara paid tribute to the Inca Empire as they farmed along the shores of Lake Titicaca in highland Bolivia. They grew staple foods such as potatoes, maize, and quinoa. They also herded llamas on the semiarid Altiplano, or high plain, that surrounds the lake. Skilled fisherfolk, the Aymara developed totora-reed canoes and rafts for use on Lake Titicaca. The people are famous for these craft, which are still in use, and for distinctive homespun clothing that echoes colonial Spanish garments. Men wear conical woolen hats with earflaps, the women derby hats with wool wimples for cold days.

After the Spanish conquest, most Aymara were put to work on large estates and held in virtual serfdom. Finally, in 1953, agrarian reform laws abolished labor without pay and provided for land distribution among former estate workers. Most Aymara communities now have difficulty achieving self-sufficiency because of overpopulation, poor soils, and a harsh climate.

Stew without potato is like life without love.

PERUVIAN PROVERB

A thousand years ago, many of the villages around Lake Titicaca grew potatoes in raised, irrigated fields; these fields were highly productive because the water in them provided some measure of protection against frost damage. But the fields fell into disuse some centuries ago, and farmers began to depend more and more on dry-agriculture methods practiced on exposed hillsides. Evidence for the huge raised field systems has been discovered by archaeologists working at the site of the ancient city of Tiwanaku, the capital of a great Andean empire that predated the Inca and may have been founded by ancestors of the Aymara. With the cooperation of local communities, teams of scholars

AYMARA An Aymara woman transports her goods to market in Tarabuco, Bolivia, on a small donkey.

working on both the Bolivian and Peruvian sides of Lake Titicaca have re-created the raised fields and cut canals. While crop yields do fluctuate because of frosts, the surrounding canal water serves as a heat sink, retaining solar radiation.

The experiment has proved that this farming method is quite effective and that such fields can be created and maintained by a small group of people. Much modern agriculture on the Altiplano operates on an industrial scale, but the Aymara and their neighbors do not possess the capital to support large operations. Now that some land has been turned over to them, the Aymara are again turning to ancient, and frequently ignored, indigenous agricultural methods to regain their self-sufficiency.

Kallawaya

Population: Fewer than 1,000
Location: Bolivia
Language family: Mixed language

Living in the Peruvian Andes northwest of Lake Titicaca, the Kallawaya are highland village farmers who cultivate maize, potatoes, quinoa, and other crops in a series of environments "stacked" one above the other on the slopes of Mount Kaata.

The Kallawaya's *ayllu,* an ancient social unit based on kinship and communal ownership of land, is deeply tied to Mount Kaata and treats the peak as a metaphorical human body, whose parts are kept in order by the continued existence of ayllu members. The diviners of

KALLAWAYA In the Bolivian Andes, Kallawaya make offerings to Pachamama, a fertility goddess, on a high pasture.

Ayllu Kaata are responsible for ritually pumping energy into the 13 earth shrines distributed up the mountain. In this way, the mountain "eats" produce from every climatic zone and is thus able to feed its inhabitants in turn. Kallawaya medicine men, who also act as diviners, obtain their powers from ancestors whose mummified remains were once paraded around the fields during fertility rituals.

The Kallawaya are renowned as *curanderos* (healers) and coca leaf diviners. At least a fifth of village men and women are involved in divination, serving visitors who come from hundreds of miles around to seek help. The most powerful diviner is also responsible for the New Earth ceremony. During this important rite, blood and fat are poured into the earth shrines, nourishing Mount Kaata so that it will continue making its resources available.

Like many other Andean groups, the Kallawaya rely on their rituals to sustain them and bring order to their lives. For these people, rituals help maintain and regulate everything that is part of human existence, including the environment, subsistence activities, and interpersonal and intercommunity relations.

Mapuche

Population: 1.5 million
Location: Chile
Language family: Araucanian

"People of the land," the Mapuche are one of the surviving groups of Araucanian Indians who once occupied the fertile valleys and coastal areas of central and southern Chile. Araucanians spoke a common language and fiercely resisted the Spaniards, but the Mapuche resisted the most strongly and were never completely subdued. Many retreated to isolated settlements when newcomers tried to force them to work on large estates.

Today, about 1.5 million Mapuche live in southern Chile, inhabiting rural villages and towns or major urban centers like Santiago, where many are impoverished. For much of the year, rural Mapuche still dwell

GENOGRAPHIC INSIGHT

THE RISE OF THE MESOAMERICAN EMPIRES

Two hundred million years ago, during the middle of the Mesozoic Era, a period of enormous evolutionary change when dinosaurs rose to prominence, flowering plants first appeared, and small mammals emerged in the wake of the dinosaurs' rapid extinction, Earth's landmass was connected as a single supercontinent known as Pangaea. For tens of millions of years this single continent harbored all terrestrial life. Then our planet's crust and upper mantle started breaking apart, causing a series of rifts that divided Pangaea into several smaller plates. As the continents separated, South America headed west while Australia drifted east, a process of plate tectonics that set the Earth's crust on a collision course halfway around the world. Though it moved at about the same speed at which human nails grow, eventually the enormous Pacific plate slammed the smaller Nasca plate into South America. The Nasca plate slid underground, and its subduction, as this mantle downwelling is called, resulted in a north-south corridor of intense volcanic and earthquake activity. These forces created South America's longest mountain range and the Western Hemisphere's tallest point.

Celebrants in Cuzco, Peru, reenact the Inca sun festival, offering thanks for the harvest and prayers for future bounty.

Mountainous Extremes

The Andes run for more than 4,400 miles along South America's western coast and host an area of inland drainage, the Altiplano, that constitutes one of the world's highest plateaus, with an average height of around 11,000 feet. The easterly winds of South America collect abundant moisture over the Amazon rain forest, but they dump almost all of it atop the frozen corridor of peaks, a process so absolute that it creates the driest place on Earth, the Atacama Desert in eastern Chile. With most of the region's precipitation therefore coming from condensation of Pacific fog rather than rainfall, early human migrants encountered a landscape particularly harsh for continual habitation. The llama and the chinchilla were the only animals abundant there.

But those same mountains preserved a vast frozen reservoir that seasonally opened its floodgates to the Norte Chico valleys below. Spring runoff sends enormous quantities of water through this semiarid landscape, and at least 5,000 years ago people began to exploit this resource in an organized fashion. Archaeological evidence from north-central coastal Peru indicates that at around the same time, as many as 25 different communities began terracing the countryside, practicing complex irrigation to transform the barren slopes into productive farming zones.

Abundance

Growing populations moved inland across the Andean plateau to cultivate squash, beans, cotton, and potatoes, reshaping the landscape by terracing and establishing farming communities. Various civilizations rose and fell over the next few millennia, each developing its unique form of worship and government administration. In the 13th century, a central power from the Peruvian highlands rose to become the largest empire in pre-Columbian America. Employing a variety of methods, from brutal conquest to peaceful assimilation, the Inca incorporated almost a million square miles of western South America into a kingdom that was ruled by a single figure believed to descend directly from the sun.

By establishing a system of provincial governments, the Inca used a decentralized power structure to maintain command over great distances, providing security and commerce to communities in its farthest reaches, at times tearing down roads and rebuilding them simply to keep civilians employed. The Inca produced monumental architecture—the city of Machu Picchu found high in the Andes is perhaps the best example—and developed complex mathematical, astronomical, and warfare systems before they ultimately fell to the Spanish conquistadors in the 16th century. Though the empire dissolved, the Inca people lived on, and today, the major languages of the Inca, Quechua and Aymara, are still spoken throughout the Andes.

in comparative isolation in small communities, coming together now and then for formal gatherings in which they honor returning travelers, debate important affairs, or simply enjoy a fiesta. These events are always marked by feasting, drinking, oratory, and, above all, singing.

Mapuche singing has profound social meaning, with every song bearing a message and subtle nuances. Some are prayers and supplications; others air grievances or woo a mate. Chiefs and shamans have their own more formal compositions, while other singers may improvise in their efforts to convey a message or let off steam. Women, in particular, use songs to express their feelings, even to air marital problems, since this method is socially acceptable. This tradition also fills a vital educational role in a society where skills are passed orally from one generation to the next. The songs often have a high poetic quality and help develop the personalities of young and old alike.

MAPUCHE Mapuche chiefs and religious leaders present a petition to the local government in Lumaco, Chile.

Otavaleños

Population: 150,000
Location: Ecuador
Language family: Quechuan

The volcano-surrounded homeland of the Otavaleños lies two hours north of Quito in highland Ecuador. In 1455, Inca armies entered the region, bringing with them the llama and the Quechua language, and before long the Otavaleños, who had woven cotton for generations, were also weaving wool and speaking the new tongue.

Otavaleño weavers traditionally used back-strap looms that required them to lean backward to create the necessary tension. When the Spanish came into contact with Otavaleño regions in the mid-16th century, they introduced treadle looms and sheep's wool, and they encouraged the raising of cash crops, which supplanted the cultivation of maize, potatoes, and quinoa by digging stick and foot plow. Otavaleño weavers prospered until the 19th century, when the industrial revolution introduced cheap mass-produced textiles. Even then, the people persisted with their ancient hand-weaving methods and became famous throughout Ecuador for fine workmanship. The Otavaleños also kept their traditional values and culture, as well as a strong ethnic identity. Although they are now Catholics, they have maintained many traditional religious beliefs that are closely tied to agriculture and the seasons.

Because of their famous textiles, the Otavaleños today are among the most prosperous of all Indian groups in South America. On Poncho Plaza in the town of Otavalo, the handicraft market developed by the Indians in the 1960s and 1970s is a popular and colorful destination for tourists.

Many Otavaleños still dress in traditional costumes. Men put on fedora hats, ponchos, white shirts, and calf-length cotton pants, and they wear their hair in a long braid, or *shimba,* that reaches to the waist. The braid is never cut off and is such a strong cultural tradition that men serving in the Ecuadoran army are permitted to retain it. Women wear white blouses, blue skirts, shawls, and jewelry, including gold and coral necklaces. Proud and traditional, as well as entrepreneurial, Otavaleños value their clothing as an important symbol of their ethnic identity.

OTAVALEÑOS Otavaleño women wash the family laundry and long skeins of wool in a mountain stream.

Q'eros

Population: Fewer than 1,000
Location: Peru
Language family: Quechuan

Among the most isolated of all Andean groups, the Q'eros of Peru live on the eastern, Amazon side of the mountains. There, the terrain is so rugged that roads cannot penetrate the region. The people spend much of each year living in small hamlets of thatched houses, tucked in the upper gorges that extend down from high pastures.

Some 2,600 feet below—about a three-hour walk—lies their ceremonial center. All of its houses, deserted most of the year, have doors facing the rising sun. A Catholic chapel and a schoolhouse also form part of this complex, for the Q'eros' traditional beliefs are now melded with Catholic doctrine.

The homeland of the Q'eros climbs from wooded foothills at 6,000 feet above sea level to a dry, tundra-like landscape at 15,500 feet. Within about 20 square miles, these people exploit the entire range of Andean mountain environments. They offer a classic example of how highland farmers live vertically rather than horizontally. At the highest elevations, people herd alpacas and llamas, whose wool is used in the making of textiles. Operating their traditional back-strap looms, weavers fashion woolen hats, ponchos, and skirts for members of their community.

Below the highest elevations, some locations allow the cultivation of potatoes, ullucu, and other tubers. At 8,700 feet and lower, the Q'eros combine the growing of tubers with more temperate crops such as maize, squash, chili peppers, and sweet potatoes. Until they could obtain European farming implements, people cultivated with the *chaquitaclla,* essentially a digging stick that a farmer drove into the soil with one of his feet. Far below all the fields lies the *sacha-sacha,* a place where the Q'eros never venture. It is the land of endless trees: the Amazonian rain forest.

Life is hard for the Q'eros, who have developed a deep distrust of outsiders over the years. For centuries

Q'EROS High in the Peruvian Andes, the Q'eros manage their herds of llamas and alpacas.

they were exploited by the powerful, far-reaching Inca state. Then the Spanish arrived in their homeland, establishing several haciendas (now abandoned) and treating the people poorly. For many generations, the Q'eros have traded llama wool and potatoes with the Lake Titicaca region to the south, dealing with itinerant merchants known as *runa,* trusted people from the outside world. Today, the Q'eros continue to welcome the merchants, who bring iron tools and additional industrial commodities, but they greet other foreigners with evasion and suspicion.

Quechua

Population: 8.9 million
Location: Peru, Bolivia, Argentina, Ecuador
Language family: Quechuan

Highland Peru's Quechua lived under Spanish rule from the 1500s to the 1800s. In the pre-Spanish times, the Inca Empire imposed a degree of cultural uniformity and the Quechua language on the Andean highlands. But after the Spanish conquest, the Quechuan-speaking peoples broke up into many small groups, which were distinguished as much by their locations as by cultural differences. Today, they still share a common language—it is the most widely spoken native language, with many dialects, in South America—and they are predominantly peasant farmers or workers on large haciendas and industrial farms. Many villagers have moved to larger highland towns, looking for work and living in poverty.

In Peru's highlands, the practice of agriculture has long been limited to mountain valleys and steep hillsides. Maize and quinoa are raised at lower elevations; potatoes are cultivated in terraced gardens higher up. Plows are now common, but farmers in many communities continue to use the traditional foot plow to prepare their small fields. All Quechua groups are also expert weavers who still use back-strap looms and a simple weaving technology to produce the blankets, ponchos, and other garments worn by men and women in the cold mountain environment. Many Quechua dwell on small farmsteads or in village settlements, often grouped around a plaza and a Catholic church. Their houses are often built of mud brick or timber, with thatched roofs.

While the Quechua have adopted Catholicism, they maintain strong traditional spiritual beliefs that govern their view of the world. They envisage the cosmos as a layered entity. An upper world—often called Hananpacha—is a place of abundance where the spirits of people who led exemplary lives dwell, along with the pantheon of Quechua spirits, the Christian God, and Christ. Next is Kaypacha, the world of the living, and it comprises the Earth, humans, plants, animals, and the spirits of good and evil. Then there is Ukhupacha, the Inner World inhabited by diminutive people.

In Quechua belief, gods and the lives of humans are intertwined. The former intervene constantly in daily life, meting out reward and punishment in the guise of fortune and misfortune. The greatest of these is the creator-god Roal, once known as Viracocha; from his mountain home, he keeps an eye on humanity while governing the forces of nature and keeping them in balance. Pachamama is a female deity who resides deep in the Earth and presides over agriculture. Her greatest ceremonies come at the time of planting in August and September, when she receives the seed that will become the next harvest. In a society where agriculture permeates every part of life, it is not surprising that traditional beliefs still flourish despite four centuries of vigorous missionization.

QUECHUA A Quechua woman sells her potatoes at the market in Pisac, a village in the Sacred Valley of the Peruvian Andes.

Central South America

This central region on the east coast of the South American continent includes parts of Brazil, Argentina, and Bolivia, and all of Paraguay and Uruguay. Outside this region's major cities, such as Rio de Janeiro and Buenos Aires, there are fewer than one hundred people per square mile, and sometimes fewer than ten, especially in the arid lowland of Chaco Boreal in eastern Bolivia and northwestern Paraguay, the two landlocked nations of the continent. Topography here ranges from those lowlands to Brazilian highlands, which, in contrast, average a height of 3,300 feet. Mountain ranges in this region include the Serra do Espinhaço, Serra da Mantiqueira, and Serra do Roncador.

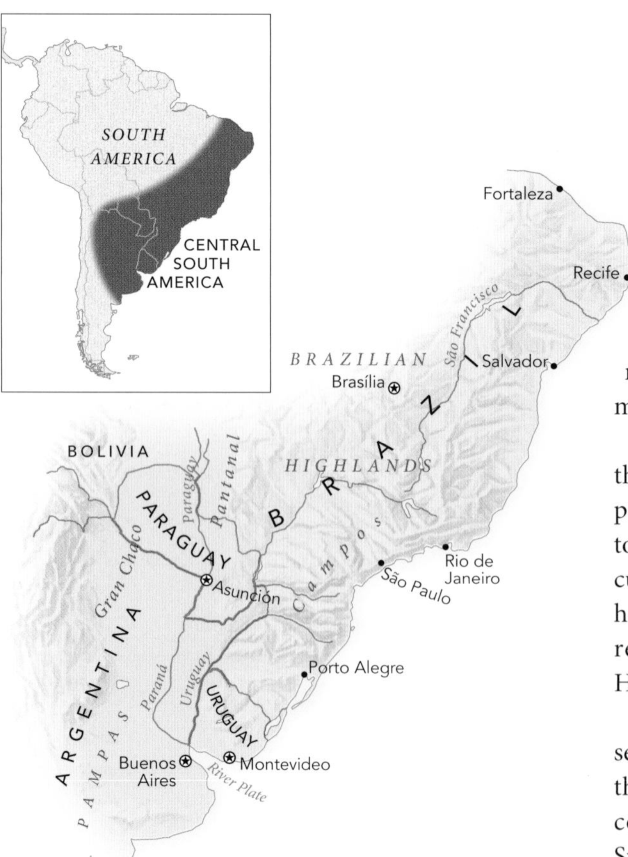

Guaraní

Population: 4.9 million
Location: Paraguay
Language family: Tupí-Guaraní

In pre-Spanish times, the Guaraní peoples migrated into what is now Paraguay, arriving as part of the series of great population movements that took people speaking a common Tupí-Guaraní language southward from the lower Amazon region.

These former tropical farmers migrated in search of the Land of the Grandfather, a mythic sky god. They had sporadic contacts with the Inca civilization of the Andes and were subjugated by the Spanish, who hoped to find silver and gold on their lands. In time, Guaraní shamans spearheaded a strong revivalist movement based on Christian belief in the Redeemer, inciting the people to rise against their oppressors before traveling to a "land without evil."

Children of lazy dogs turn out to be hunters. Children of hunters turn out to be lazy.

ZAPOTEC PROVERB

Revivalism is just a memory now, but the Guaraní legacy is still powerful in Paraguay. Today, more than a million Paraguayans speak Guaraní, most of them living around Asunción, the capital.

Nevertheless, while cultural nationalism glorifies the Guaraní language and heritage, no Paraguayan peasants retain the traditional culture of their ancestors. They do build thatched houses in their ancient cultural manner, and they do grow maté, a Paraguayan herbal stimulant that was a major product of their remote predecessors—but, in essence, they were Hispanicized centuries ago.

In the countryside, people continue to be largely self-sufficient, growing a few cash crops and living at the subsistence level. Some cash is used to finance community fiestas, which here, as elsewhere in Spanish-settled South America, play an important role in rural life and the acquisition of individual prestige.

GUARANÍ A Guaraní girl sits in her family home in Brazil's Mato Grosso.

Kadiwéu

Population: 1,600
Location: Brazil
Language family: Guaicuruan

The Kadiwéu are the only surviving group of the lordly Mbayá peoples, who once dominated an area extending from Paraguay into the Mato Grosso, a western state of Brazil characterized by plateaus, and even farther afield. These people believed that they had a divine right to raid and enslave their neighbors—not only the Indian communities but also Portuguese and Spanish ones. Primarily hunters and foragers, the Kadiwéu demanded tribute in agricultural produce from their victims.

Kadiwéu hunting relied heavily on bolas, two or three balls attached to thongs and used to ensnare guanacos and other game. Always on the move, the people had a simple, portable material culture that had little use for textiles. They clothed themselves in hides, with featherwork and tattoos distinguishing between different kin groups.

The Kadiwéu remained a male-dominated society, which—because sons were preferred—widely practiced infanticide as late as the 1930s. Many newborn daughters were killed by asphyxiation or by breaking their necks and then buried under the beds where their parents slept. They also practiced abortion, for which they used a concoction of bitter roots or pounded on the mother's belly until the fetus died.

Today, the Kadiwéu are peaceful farmers, having reached an accommodation with the authorities, but their social organization still reflects an ethnic division between masters and slaves, the latter being individuals captured from other groups and their descendants. Despite frequent protests, the "slaves" still submit to their "masters" and are expected to provide economic services and food for no compensation. As the

KADIWÉU Once nomadic hunters and gatherers, the Kadiwéu are now settled farmers.

generations pass, however, the gap between slaver and enslaved continues to narrow.

Mataco

Population: 40,000
Location: Argentina, Bolivia
Language family: Matacoan

Since 1628, the Mataco Indians of northwest Argentina's arid Gran Chaco strip of land, on the western edge of South America, have been in contact with Europeans. Because they resisted attempts at colonization and conversion, a large number of them were massacred, and others of their people were forced onto reservations. Today, the Mataco people, much Hispanicized, survive. Many of them can be found working in the lumber business or serving as migrant laborers on large sugar plantations.

The traditional Mataco culture combined the cultivation of maize and beans with the gathering of cactus fruit and other wild plants in the thorny brush of their arid landscape. Men hunted guanacos, an ungulate like a llama, and they also took large numbers of river fish during their rivers' seasonal runs.

Reflecting their mobile lifestyle, most Mataco groups lived in simple brush huts; now, however, many people have become sedentary farmers, living in wattle-and-daub houses on small individual holdings. Today, even though Catholicism is widespread, shamans continue to be important in the more traditional communities. For many people, they are still essential intermediaries for the spirits.

The Mataco try as much as possible to remain isolated from larger society. They place a high premium on social values and reciprocity, as seen in the frequent sharing of food that goes on throughout their community and in their justice system, which punishes offenses committed against the social order.

Theft of field crops, especially tobacco, is treated as an offense against everyone. In small villages, people know each other so well that they can even recognize an individual's footprints, which can offer clues to the identity of any thief. When caught, offenders usually are punished by being expected to pay for what was taken. The punishment suits the crime, and it also serves as a lesson in communal ethics.

Instances of homicide may result from jealousy and marital disputes. If someone commits murder, the relatives of the deceased don war paint and visit the

MATACO A Mataco woman weaves handicrafts using fibers from a bromeliad.

killer in his village, threatening him with their spears and gripping his chest as the women and children take to the forest. Usually the punishment is once again economic, requiring payment in, say, goats, which are then distributed among the relatives of the victim.

In addition to avenging killings by asking for "blood money," the deceased's relatives may begin a formal feud, thereby widening the conflict, or take a scalp from a member of a hostile tribe. In every case, homicide is seen as an offense against the community as a whole.

Even as the Mataco people are slowly being assimilated into wider Argentine society and their language is threatened, they strive to maintain their traditional values. Their distinctive legal practices offer a measure of protection against a rapidly changing world and its modern ethos, in which the individual matters more than the community at large.

Tupinambá

Population: 8,400
Location: Brazil
Language family: Tupí-Guaraní

When the Portuguese arrived in Brazil in the early 16th century, the Tupinambá occupied more than 2,000 miles of the coast, from Ceará in the north to Porto Alegre in the south. This group had settled in their homeland some centuries earlier, after a messianic migration from the north in search of the idyllic Land of the Grandfather, the term they used to identify their mythical sky god.

The Tupinambá people dwelled in large palisaded villages, supporting themselves by cultivating five varieties of manioc, a starchy root vegetable, and by fishing in the waters of both the rivers and the coastal areas nearby.

Each Tupinambá group waged wars against its neighbors in a constant ebb and flow of blood feuds and revenge killings. Among the Portuguese, the Tupinambá soon acquired a reputation as fierce warriors, headhunters, and cannibals, when in fact, the people placed a high premium on social harmony within their own communities.

Following years of wars with the Portuguese, many Tupinambá captives were placed on colonialists' sugar plantations and forced to do work they considered beneath their dignity as warriors. Still, even though they were subdued, the Tupinambá held on to their traditional culture.

TUPINAMBÁ This Tupinambá man represents his people at a summit on sustainability in the Brazilian Amazon.

In the 17th century, under the strain of domination, another powerful messianic movement arose, again causing them to believe that when they reached the Land of the Grandfather, they would achieve immortality, regain their youth, and live at ease.

But this time the Tupinambá did not migrate from their villages. In many communities, they simply stopped working, holding the magical belief that the digging sticks they had traditionally used to cultivate fields would continue to do the agricultural work by themselves. This rich array of spiritual beliefs has now been submerged into the practice of Catholicism, and the surviving Tupinambá of Brazil have become peasant farmers.

Patagonia

Covering most of the southern half of Argentina is a semiarid plateau known as Patagonia. Covering about 260,000 square miles between the Patagonian Andes, the Colorado River, the Atlantic Ocean, and the Strait of Magellan, this nearly treeless tableland reaches an elevation of 5,000 feet. Only a few streams run continuously through the region from the Andes to the Atlantic Ocean, including the Colorado, Chubut, and Santa Cruz. Others run when there is rainfall, a rare occurrence: Patagonia receives less than ten inches of rain a year. The average monthly temperature in the city of Comodoro Rivadavia, on the eastern coast of Argentina and in the middle of the Patagonian region, is 66°F in January and about 44°F in July. At the extreme southern tip of the continent of South America stretches the archipelago of rocky islands known as Tierra del Fuego, the land of fire: one large island and numerous smaller ones.

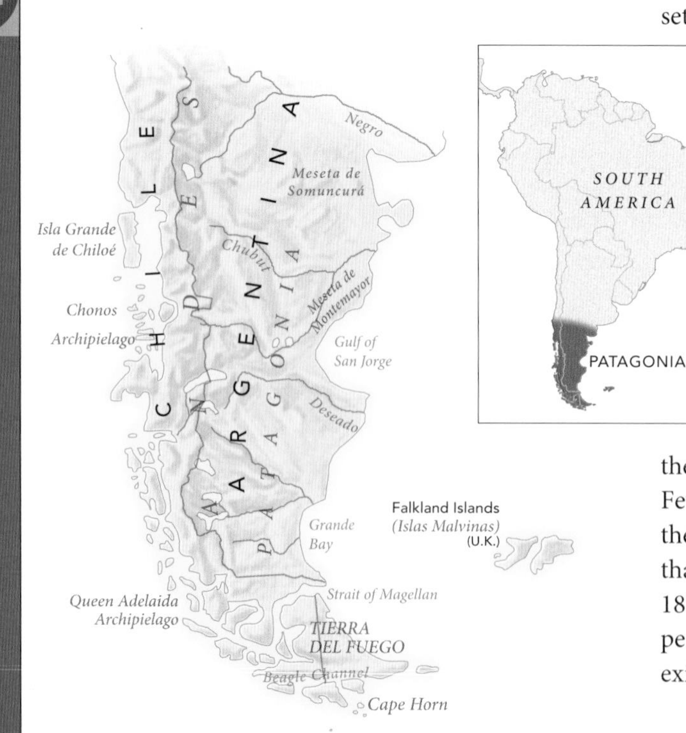

Argentines

Population: 43 million
Location: Argentina
Language family: Italic

From its northern reaches in the subtropics to its southern tip in Tierra del Fuego, Argentina stretches some 2,100 miles. The Andes form its western border and a vast, flat grassland, known as the pampas, covers its central area. With fertile soil for growing grain and nutritious graze for cattle, the pampas became the lifeblood of the country's settlement.

The Spanish colonized Argentina in the 16th century and gave the country its name, derived from the Latin word for the silver they hoped to find there. Although parts of the country were in touch with other colonial centers in Latin America, most early contact and influence came directly from Spain—giving Argentina its European focus from the beginning. Spanish and Italians formed the largest waves of initial settlers; the roster would grow to include Germans, French, English, Welsh, Scots, Lebanese, Polish, and Eastern Europeans and what would become the largest settlement of Jews in South America.

Each arrival reinforced the cosmopolitan nature of Argentina. Castilian Spanish became the national language, which the Argentines have made their own with some distinct pronunciations. Italian also is widely understood, and many Italian words have entered Argentine Spanish. About 70 percent of the people in Argentina are Roman Catholic, and about 10 percent are Protestant, a number that continues to grow.

The country's indigenous populations, now greatly diminished, were dominated by the Spanish and ultimately pressed into forced labor. Few descendants remain of African slaves imported by the Spanish in the colonial era. One theory suggested that most succumbed in the yellow fever epidemic of 1871. The gaucho, or mestizo cowboy, who was indispensable to the success of cattle ranching, now mostly exists as an enduring cultural icon.

Despite the size of the country, Argentine life and much of its livelihood centers on the pampas and the capital of Buenos Aires. Some 70 percent of Argentines

ARGENTINES A sultry couple dances the tango, Argentina's national dance, on a street in Buenos Aires.

live in this area, more than 30 percent in metropolitan Buenos Aires. Porteños, as citizens of the port city are called, remain aware of life in the countryside. Each year country and city merge as people flock to the Exposición Rural, a national agricultural exhibit held in Buenos Aires featuring champion livestock, equestrian competitions, and the latest in farm machinery and other necessities of rural life.

The gregarious Argentines love to talk and make many opportunities to do so. European-style café life is vibrant, and late nights are the norm. In general, Argentines have a strong sense of themselves as individuals, which may often make collaboration difficult, as people seek to have their contributions recognized—and admired. This focus on oneself and one's thoughts and feelings routinely takes an unusually large number of Argentines into psychotherapy, a widely accepted practice.

When it comes to Argentine cuisine, it's almost all about the meat, especially beef. The *asado,* or barbecue, is an Argentine culinary and social tradition that involves grilling large quantities of various cuts of meats. Men typically do both the shopping and the grilling for an asado. Various Italian pasta dishes also are eaten regularly. For dessert, Argentines opt for confections containing *dulce de leche,* a spread made by boiling down sweetened milk. It appears at breakfast as well, as a topping on toast or pancakes.

Maté drinking is a widespread social ritual that has its roots in Argentina's Indian cultures and was adopted by the gauchos. Maté is an infusion of the leaves of the yerba maté plant, a variety of holly. It usually is sipped from a decorated gourd or silver cup through a special metal straw, and it is often passed around a group of friends. A mild stimulant, maté enabled the gauchos to keep going in the saddle without stopping for rest or food.

Few traditions are as evocative of a place as the tango is of Argentina. Tango represents both dance and song. It originated in the back streets of Buenos Aires and was at first eschewed by the upper classes. Tango partners hold each other closely and glide in long, languorous steps, pausing often for dramatic effect. Tango music is plaintive and melancholy, and its best interpreters have become national celebrities.

Argentina currently is recovering from an economic downturn that thrust many of the middle class into poverty. Years of political conflict, government corruption, and inflation have held Argentina back from the economic rank it should hold due to its abundance of natural and human resources.

Chileans

Population: 17.5 million
Location: Chile
Language family: Italic

Natural features isolate the long and narrow country of Chile, covering more than 2,600 miles and 38 degrees of latitude. To the north is the bone-dry Atacama Desert; to the east the lofty Andes; to the south, the frigid waters separating Chile from Antarctica; and to the west, the vast Pacific.

Chile's native Araucanian Indians resisted Spanish encroachment for several centuries. Spain did prevail early in northern Chile, but its control also ended early with Chilean independence in 1810.

Most Chileans represent a mixture of indigenous and European descent. From the late 19th century, colonial Castilian and Andalusian settlers were joined by Basques, Germans, British, Italians, Croatians, Jews—and one of the largest groups of Palestinians outside Palestine. Chile's current indigenous populations, the Aymara in the Andes and the Mapuche, an Araucanian people of the south, are not well integrated into Chilean society. Mapuche tend to hold service positions, such as construction worker or nanny. Chile's territory includes Easter Island, 2,000 miles off its western coast and home to some 5,000 people, many of Polynesian origin.

More than 99 percent of Chile's population speak Spanish. Chilean Spanish is distinctive, known for dropped terminal consonants and frequent use of diminutive endings. It also displays few regional differences. In general, Chileans of non-Spanish immigrant descent tend to identify more with their Chilean affiliation than their ultimate country of origin.

The one who loves you will make you weep.

ARGENTINIAN PROVERB

Roman Catholics make up 66 percent of the populations. Protestants, mostly Evangelicals, count for 16 percent. The country celebrates many holidays based on the religious calendar.

The highly urbanized Chileans maintain strong family ties. Adult children tend to live at home until marriage, which occurs fairly early, and maintain frequent—often daily—contact with their families. Holidays are celebrated with family, including September 18, commemorating independence from Spain.

Foreign trade helps propel Chile's economy, and mining—particularly of copper—provides significant revenue. Agricultural products include orchard fruits, wine, and beef. The country glorifies the culture of the *huasos*, or cowboys, who work the cattle on its ranches. Chile's stable democracy and strong economy attract growing numbers of immigrants.

CHILEANS Chilean cowboys prepare a calf for a rodeo event at the annual Fiesta a la Chilena in Chilean Patagonia.

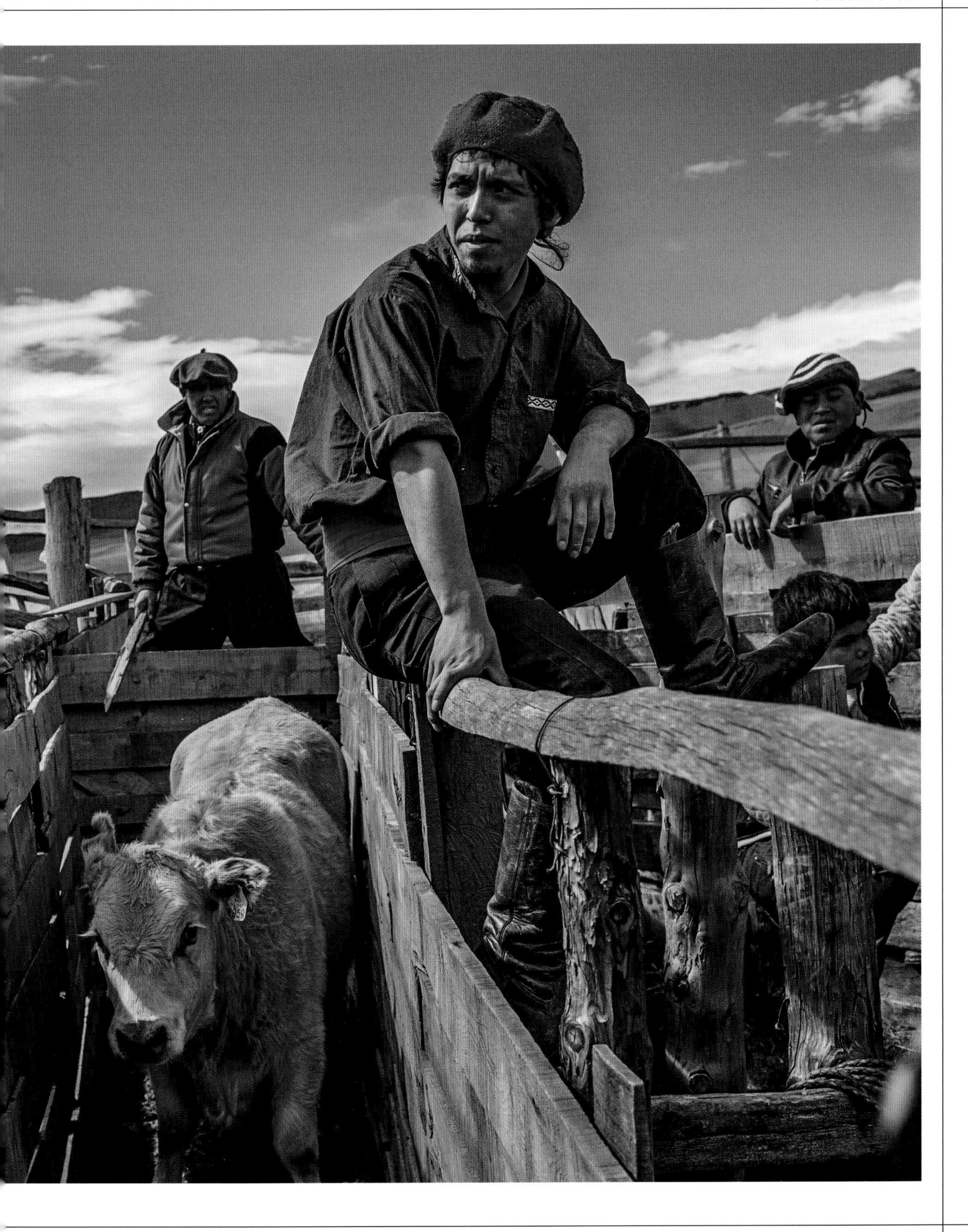

ACROSS CULTURES

MAKING A SPACE TO LIVE IN

Not only the natural landscape but also the shape, materials, and character of the homes that people build define their worldview.

What do you need to build a house? Bricks, mortar, glass, and wood are commonly used in modern architecture. But looking around the world, we find people who have built or are still building traditional houses made of clay, straw, reeds, felt, animal hides, snow, and pits dug into the ground. Shelter is a basic need, but humans thrive in vastly different climates, ecozones, and altitudes. Our species has always applied ingenuity to design shelters, including some of the more unusual and imaginative houses we explore here.

A house does much more than provide shelter. Its shape reflects deep cultural values of social hierarchy, aesthetics, and the proper organization of space, a notion perhaps best known by the Chinese term *feng shui.* Long before "green buildings" were fashionable and architecture became an art form, indigenous peoples used invention and artistry to build abodes of all types from many materials. Modernity has not yet swept aside this architectural legacy. Dung dwellings may still be found in India, felt houses in Mongolia, reed cottages on the floating islands of Peru's Lake Titicaca, tree houses in Papua New Guinea, goatskin tents in the Sahara, tipis in Siberia, and igloos in the Arctic North.

Traditional dwellings, along with some modern updates to them, showcase the genius of indigenous architecture. Houses may not be the pinnacle of human building arts. Societies of the past have also erected monumental tombs, public buildings, roads, cities, and pyramids for the gods. Other societies left behind nothing of the sort because they used impermanent materials to build, perhaps, or they directed their creative energies elsewhere. Regardless of their architectural prowess or grandeur, all human societies constructed dwellings for people. Traditional housing styles are now in decline worldwide as indigenous ways of life vanish and modern materials become more widely available. Unique building styles such as the Asmat longhouse, the Xingu *maloca* (a long oval house), the Evenki deerskin tipi, or the Inuit igloo vanish into the past. As they fade, we lose insight into human values and cultural traditions.

Mobile Homes

Those traditional housing styles that do survive today have much to teach us about culture. The nomads of Mongolia or Kyrgyzstan live in ergonomic round structures of felt overlaid on wood lattices. Called yurts in Central Asia and gers in Mongolia, these ingenious houses manage to keep out the elements, keep in warmth, and let smoke escape through an opening in the roof.

Constructing a new yurt requires intensive labor, usually by a prospective groom in preparation for his marriage. Great bundles of wool must first be collected by shearing sheep. The wool is then beaten by hand with metal rods to form thick felt. Roof poles and collapsible wall lattices are painstakingly planed by hand and painted bright reds and yellows. Ropes of horsehair are plaited by hand to gird the entire structure. No nails or glues are used; each joint is bored through and secured by a leather thong. The doorway is then decorated with intricate geometric patterns.

When the nomads need to move, these yurts can be rapidly disassembled, packed into a truck or onto a camel's back, and transported to a new place. Because nomads must migrate to find new grass for their animals, the yurt architecture perfectly fits their lifestyle. But the yurt is also a complex space that reflects social values. Stooping to enter through the low yurt door, you find yourself in a dim interior space with light coming

Children paddle a dugout past their community's houses on stilts in the shallows off the island of Borneo.

through the round roof opening. Immediately in front of you is a stove. The women's space is to the right and includes the stove front, cooking area, flour storage, utensils, space for processing milk and sheep entrails, and the main bed. The men's space, taking up the left half, is where food is served and guests seated. It is used for saddles and lassos, cuts of meat, a family shrine, clothing chests, and a bed for guests or children. Though the line separating the men's and women's spaces is invisible, it is keenly perceived and governs how people comport themselves inside the yurt. A guest who sits in the wrong place may be gently scolded or even asked to move. A blind person who walked into any one of many thousands of nomadic yurts in Mongolia could immediately locate most objects and furnishings, so regular is the interior layout. The door of the yurt must face south, and the proximity to neighboring yurts may indicate how closely the family is related.

Far away in the Sahara, the Tuareg people of Niger also move frequently to herd camels. Espousing mobile architecture, they spend their lives living in goatskin tents in the desert. Tents are the focus of their social interaction, beginning with ritual greetings outside the tent. Within a tent, the Tuareg are able to enjoy a protected space to conceal food, wealth, or information from prying eyes.

The sturdy tents, supported on curved sticks, reveal Tuareg ideas about gender roles. Built and owned by a woman, a tent is brought by her to the marriage as dowry and retained if the couple divorces. Nuptial tents are constructed for a young bride by old women, who may ritually rebuild and take down the tent for seven successive days during her wedding ceremony. The wedding tent is greeted by ceremonial songs and drumming, and there are special protocols for how the bride and groom must enter it. During a wedding, young people play a game of casting sticks across the divide between the left (female) and right (male) sides of the tent. A Tuareg woman maintains control over her tent and the use of space within it; for example, she may orient the tent doors and sleeping spaces differently depending on whether her husband is present or away traveling.

Tent life also extends to some of the northernmost peoples on Earth. The Evenki reindeer herders continue to live for much of the year in canvas- or deerskin-covered tipis, as did their ancestors, deep in the Siberian forest. They construct a conical frame of birch poles and cover it with deerskins, lining the interior floor with deerskins for sleeping. Mobile architecture in the form of conical lodges made of animal skins was also practiced by the Sami people of Norway but is now a vanishing art. Tipis built by the Lakota, Crow, Nez Percé, and other Native American tribes once graced America's northwest plains by the thousands. Highly practical, tipis in both Siberia and native North America provided warmth in the cold, cool in the heat, a sheltered space to cook and sleep, and mobility. American tipis may now be seen in modern forms at tribal powwows, where they help reconnect Plains peoples to their past.

Housing that is immobile can still be light, ecological, and ergonomic. The Kombai and Korowai people, hunter-gatherers of Papua New Guinea, may live as high as a hundred feet off the ground in treehouses made of wood lashed together with vines. The houses are so solid that their residents can haul small domestic animals up narrow ladders, light fires, cook, sleep, and eat—all high in the sky. Separate living spaces are maintained for men and women, another example of gender roles expressed in architecture. Primarily for protection from neighboring hostile tribes, the tree houses also shield from mosquitoes and provide a lookout point. When one deteriorates, a new one is quickly built from locally plentiful plant material.

Adaptive Architecture

Perhaps the most famed example of adaptive architecture is the igloo, historically built by Inuit peoples across a large swath of Canada and Greenland. Though it begins as a house of snow blocks, melting and condensation cause snow to become ice after a few days of residence. The igloo is thus perhaps the only house that beneficially changes its substance over time. Neighboring igloos can be joined by tunnels, creating a complex of domed structures, with sunken entryways and raised sleeping platforms inside. The ingenious

Although igloos are seldom used now by the Inuit, many still have the skills to construct one.

helix method for stacking the snow blocks produces a structure of enough strength to support a grown man standing on top. Though the Inuit no longer live in igloos, some still know how to build them and can demonstrate the technology. The Inuit, now settled in modern houses, preserve the art of the snow house as part of their repertoire of survival techniques. There is considerable nostalgia for this icon of survival in harsh climes, and there is even an igloo-inspired hotel in Greenland with aluminum igloos for guests.

Going from Arctic snows to high-altitude lakes, we find the Uros people of Peru's Lake Titicaca living on floating artificial islands made of tortora reeds. Once a way of life based on birding and fishing, artificial-island dwelling has now become a magnet for tourists. The Uros build elegant boats and simple houses from this humble material. The abundant reeds are collected and processed to obtain bundles of uniform thickness and length. They are dried, then bound into bundles to form the walls and sewn into mats for the roofs. Reed houses are both disposable and renewable. They can be updated with new technologies. Tin roofs have become common, and some reed houses now have solar panels, electricity, and televisions.

Local Materials

Most local building materials can be obtained for free, and the Dogon of Mali use mud as their main material. Dogon houses, towers of dried mud, are tightly clustered into geometrically complex arrangements. They may be round or quadratic, multistoried, and sometimes topped with peaked straw cones that resemble hats. Mud as a building material is constantly renewable, so any cracks can be repaired with a few handfuls of earth. As layers are added over the years, the structures take on fanciful, sand-castle shapes. Entire clusters of Dogon houses may be nestled in a rock crevice or on a cliff face, where they provide a natural fortress.

Elaborate and symbolic mahogany doors may protect a home, granary, or women's quarters. Bearing bas-relief images of crocodiles, snakes, turtles, and mythical beings, these portals serve as symbolic shields to ward off evil influences and protect people and possessions. Dogon doors have also become a valued commodity on the global art market.

Houses in traditional indigenous cultures may teach us something. Often fully biodegradable and made of local materials, their value lies more in the design and labor than the stuff of which they are made. Often mobile, lightweight, and readily replaceable, their worth lies more in their proximity to kin, grazing lands, water, and protective landscapes than in any ownership of fixed real estate. Often porous or easily accessed by outsiders, their true protective power lies in their ability to strengthen the social network, enfolding kin and cotribals into a private and intimate space.

And finally, with their infinite variety of design, material, geometry, and functionality, many indigenous housing styles represent the best ways people have yet devised to live in a low-impact manner in challenging environments.

—K. David Harrison

A Dogon boy rolls a bicycle tire past a row of mud houses in the town of Mopti in Mali.

GENEALOGY 101

A QUICK GUIDE TO UNDERTAKING YOUR PERSONAL ANCESTRAL QUEST

If you've always wanted to learn more about your ancestors but didn't know where to begin, take heart. Starting out on a personal genealogical quest has never been easier. Everybody's doing it—and there is a wealth of information and technology out there to aid your search. Here's a top ten list of tips to get you started.

1. Get Organized

Start with a plan and round up your tools. Before you download an app on your cell phone or computer, acquire some old-school items. You'll want a substantial notebook with serrated, hole-punched paper that can go in a binder later. (Even if digital is your go-to preference, you might want to assemble both digital and hard copy records.) Print some blank family tree sheets, available online. Add a digital audio or video recorder, or both, and a camera of good enough quality to photograph documents clearly. These lower-tech items will come in handy when you can't get a signal or Internet access or your phone needs a charge.

2. Treasure Hunt at Home

Resist the urge to go online right away. Play genealogical detective in your own home first. Search out photos and documents that will help you create a paper trail: birth certificates; baptismal, marriage, and military records; diplomas and report cards; letters and postcards; and, of course, diaries and journals. When you've combed through your own home, ask your parents and grandparents if you can do some sleuthing at their homes. Always ask permission before you take, copy, or photograph anything.

3. Talk to Your Elders

Even if you've heard family stories all you life, take time to systematically interview your elders: parents, grandparents, aunts, and uncles—and don't forget family friends of long standing. Bring your digital recorder, and prepare specific questions. Show photos you've found, and have them identify unfamiliar people in the images and tell their stories. Get as much detail as you can about each individual you learn about, including full name and names of siblings, birthplace and date, location (and ideally address) of family homes, nationality, ethnic background, occupation, hobbies, military service, and burial place. Don't rule out talking to younger relatives. They may have heard stories that you haven't or have access to information new to you. Cast a wide net.

4. Go Online

Here's where you take the information you've gathered and begin your search online. New resources, services, and options are added regularly on popular genealogy sites including FamilySearch.org, Ancestry.com, and its offshoots, RootsWeb.com and archives.com. You might want to start with the Church of Jesus Christ of Latter-Day Saints' free, nonprofit Family Search, the world's largest genealogical organization. Each week tens of millions of records are added to their digital database. You can browse the Family Search catalog of genealogical materials and request a free loan to a Family History Center (typically at a public library) where you can view the items in person. If you want a total-immersion experience, you can even visit the Family History Library in Salt Lake City. There, you can take classes, conduct searches, and use the library's collection of print, film, and digital resources. Ancestry.com offers everything you need for researching and writing your family history, including webinars and tutorials, as well as sharing your findings with others. Your local library may offer the Ancestry's Library Edition, which puts six specialized databases from their immigration and travel collection at your disposal.

5. Hit the Library

Libraries have amazing online resources, but they also have a lot of books that you can peruse. From basic "how-to-do-genealogy" books to more specialized works, you'll want to familiarize yourself with your library's collection. Be sure to check both books that circulate as well as those available only on reference shelves. For some suggestions, see page 375.

6. Be Social

Now that you're an online genealogical researcher, don't forget to enlist your favorite social networking and social media sites in your quest, such as Facebook, Twitter, Instagram, Pinterest, and Reddit. In addition to searching for and connecting with people who share your ancestral surname, look for local organizations, libraries, archives, tour guides, and genealogy-related services in your ancestor's hometown or region. But don't forget about sitting down and writing a snail mail letter if you get a lead with a postal address. Older people especially might relate better to mailed letters, and if language is a barrier, you might be able to find a relative or friend to help with translation.

Old photographs provide invaluable—and compelling—links to the genealogical past.

7. Get a DNA Test

Cutting-edge DNA testing kits such as Geno 2.0, available from National Geographic's Genographic Project, Family Tree DNA, and AncestryDNA can lead you to places and people you may never have found simply by following a paper trail. You'll want to read in more detail about ancestry testing on page 370 before you proceed.

8. Plan a Trip

Freelancing your own ancestral tour is a great option, especially if your destination is fairly accessible and language is not an issue. A self-planned trip can yield great results if you plan ahead. Contact known or suspected relatives; bring along printed copies of your family tree; make appointments at records offices, churches, and cemeteries; and in general read up on the area you plan to visit. If you have the time and the financial resources, you can outsource your travel planning to a genealogical travel specialist such as Hager's Journeys, European Focus, Ancestral Attic, and P.A.T.H. Finders, to name a few. Fees and services vary, and the process often begins with customized genealogical research that will pinpoint productive destinations.

9. Manage Your Expectations

Celebrity family history search stories on TV produce some blockbuster findings. For example, Sarah Jessica Parker, who played a resurrected Salem witch in the 1993 film *Hocus Pocus,* learned during her *Who Do You Think You Are?* reveal that her tenth great-grandmother actually escaped prosecution at the Salem witch trials. But you—and most celebrities—will likely uncover more mundane connections to ordinary people who nevertheless made a difference in your life and the lives of other family members.

10. Keep Going

Ancestry searches and travels sometimes present detours, roadblocks, and dead ends. Being open to all discoveries (even unsettling ones) and taking pleasure and reward from the process as well as the results will help you keep things in perspective. Once bitten by the genealogy bug, you might be hooked for life!

Note: See Further Resources on page 375 for Web addresses of many of the organizations listed here.

ANCESTRY TESTING

EXPLORING THE GENE IN GENEALOGY

The same kind of technology that allows us to discover genetic mutations related to disease and the degree of our connection to other members of the animal kingdom also allows us to learn about our own ancestry. Today a number of organizations make it possible for the average person to collect a DNA sample, send it in, and have it analyzed for ancestry markers that are autosomal (from the nonsex chromosomes across an individual's genome), Y-chromosomal (from the direct male line, which only males have), or mitochondrial (from the DNA of the maternal line of males or females.) The results come with interpretive information in varying degrees of detail.

The various DNA testing organizations don't offer the same kinds of results or supplemental information. Some reveal general ancestry at the deep ancestry level that can go back hundreds of generations, while others will link you up with other contemporary individuals in their database of testers who are a match and have a good probability of being related to you.

DNA tests are not free; they usually cost from $69 to more than $500. The testing companies run frequent discounted specials. All the main companies collect DNA through saliva collection or a cheek swab (cheek swab tests generally work better for babies and the elderly), and most agree to store the DNA for at least 25 years. Some share results with the pharmaceutical industry or other third-party research organizations. Some allow interorganization transfers of DNA results so that they can be compared against different databases of test takers. Some include estimates of Neanderthal ancestry. Not all provide ancestry matching with other database members, but many give general information such as maternal and paternal haplogroups, associations based on genetic markers that occurred during early human migrations and today have regional associations.

A caution: Ancestry testing is not invasive or painful, but it's not something to be taken lightly. Every so often, the results from a test yield information that substantially contradicts what an individual has been told about his or her background. Sometimes a family just has a wrong or incomplete idea of its origins, but in other cases, the results are at odds because the individual has a different—or no—genetic connection to the family, when adoption or issues of paternity or, more rarely, maternity come into play. If this hasn't been disclosed, it can be an upsetting revelation.

Nevertheless, some people can't seem to get enough of DNA testing and opt to test with a number of vendors in order to increase their chances of being matched with others who test with only one company and appear in only one database. The "all-in" testers literally want to cover all their bases and take advantage of the different levels of specificity in the results provided.

Here is some basic information about four of the main DNA ancestry testing organizations and what they offer. More details and updates are available, of course, from their websites. You are encouraged to learn as much as you can before you purchase a testing kit.

National Geographic Genographic Project Geno 2.0

The Geno 2.0 ancestry testing kit *(www.nationalgeographic/DNA.com)* is an integral part of the National Geographic Genographic Project that has conducted cutting-edge research on deep ancestry and early human migrations for more than a decade, while creating a worldwide database of genetic samples from members of indigenous cultures as well as from the general public. Its current database includes more than 700,000 individuals. The project and the test results focus on deep ancestry, but the test also offers the basics of personal ancestry, analyzing some 700,000 distinct markers. It also provides percentages of Neanderthal and Denisovan ancestry if present. Geno 2.0 does not offer ancestry matching, but the Genographic Project does have a partnership with

Family Tree DNA through which you can transfer your results and take advantage of the Family Tree DNA databases. Geno 2.0 participants can also post their stories online, providing contact information if they so wish. And by participating, they are contributing to globally significant research.

Family Tree DNA

Family Tree DNA *(www.familytreedna.com)* has a partnership with the National Geographic Genographic Project. It also independently provides ancestry DNA testing. This company offers testers a wide, comprehensive, and sophisticated range of test options and allows them to opt in to different groups, including a Surname Project, or its entire database. Those who opt in receive names and e-mail address of matches. Family Tree DNA members can add their results to the National Geographic Genographic Project database and contribute to its research on worldwide human ancestry for a nominal fee.

AncestryDNA

AncestryDNA *(www.ancestry.com/DNA)* is affiliated with Ancestry.com, the world's largest family history resource. Test takers join more than a million individuals in its database and receive results that predict genetic ethnicity across 26 regions and ethnicities. They are provided lists of predicted relatives and a means of messaging them. By additionally joining Ancestry.com, testers have access to associated family trees (and millions of others), as well as more than 15 billion historical records. Testers can also be placed in DNA circles.

23+Me

The testing protocol through 23+Me *(www.23andme.com)* offers a range of results, including general ancestry composition and results for maternal and paternal lineages and Neanderthal percentage. It has a database of more than a million test takers. The company offers an ancestry-sharing component called DNA Relatives, an opt-in service with different levels of sharing according to privacy preferences. 23+Me also offered a health assessment based on genetic factors that was suspended for U.S. subscribers in 2013, due to concerns from the U.S. Food and Drug Administration. It is available to residents of Canada and the United Kingdom.

The many faces of County Mayo capture the richness of their Irish heritage.

WHAT'S MY LINE?

HOW WE CONNECT THROUGH TIME AND SPACE

The human story is one of continuous pulses in cultural development, expansion, and discovery. As we became human, our species grew like a tree with roots in Africa and branches reaching in every direction of the globe. The following summaries are the stories of 12 ancestral branches, or haplogroups (six maternal and six paternal). These 12 haplogroups represent our ancestors and account for more than 80 percent of lineages in North America and Europe, and the stories detail the migratory paths, environmental circumstances, and cultural developments associated with them.

Maternal Haplogroups: North American Genographic Participants

Haplogroup H

One of the earliest waves of migration into western Europe was marked by the spread of the Aurignacian culture—a culture distinguished by advanced tools and tool types, such as end-scrapers for animal skins and tools for woodworking. But between 15,000 and 20,000 years ago, an ice age hit, and with that, settlers retreated to the warmer climates of Iberia, Italy, and the Balkans. There, population size and genetic diversity were greatly reduced. By 15,000 years ago, haplogroup H women had moved north to recolonize Europe, and because of the population growth that quickly followed this expansion, this haplogroup, and its two numerous daughters (H1 and H3), now dominate the European female landscape. Today, this haplogroup comprises more than 40 percent of the western and central European gene pools, while moving eastward the frequencies decreases, illustrating a west-to-east migratory path. Today, haplogroup H is found at 25 percent in Turkey, 15 percent in Central Asia, and 5 percent in northern Asia.

Haplogroup H1

From West Asia, this line spread north into Europe and west into West Africa. Descendants of some of those who traveled to Europe expanded out of southern Europe *refugia* (isolated habitable areas) after the last glacial maximum to recolonize the continent. Today, this large branch of haplogroup H makes up about 10 percent of maternal lineages in Denmark and around 8 percent of maternal lineages in Norway and Sweden. It accounts for around 9 percent of maternal lineages throughout Great Britain and Ireland, but as much as 12 percent of maternal lineages in Northern Ireland, suggesting a slight founder effect in that regional population. Elsewhere, it accounts for 8 percent of the population in Portugal and between 4 and 5 percent of maternal lineages in Croatia.

Haplogroup T1a

This branch arose in West Asia, and from there, members of this lineage spread across Anatolia (Turkey) and into most of eastern Europe. Today, it occurs in highest frequency in the Middle East, Turkey, the Caucasus, and southeastern Europe, specifically among people with ancestry from Greece, Turkey, Rumania, Ukraine, and southern Russia. It is also found in regions that border the Mediterranean Sea, including Italy, Spain, the Balkans, Morocco, and Egypt. This branch is among the most common branches within haplogroup T.

Haplogroup H3

In preagricultural Europe, the last glacial maximum forced members of this lineage into the few habitable places that remained in the southern parts of the continent. When the glaciers receded, this line expanded from the Franco-Cantabrian refugium (in southwestern France and Spain), and from there grew to become one of the most common maternal branches in the continent. Today, this lineage is common in Portugal, where it is about 10 percent of maternal lineages; it is between 4 and 8 percent of maternal lineages in the British Isles and about 5 percent of the populations of France and Croatia. It is also part of some Jewish Diaspora groups throughout Europe.

Haplogroup T2b

Born at the beginning of the Neolithic Revolution 10,000 years ago, this lineage likely originated in West Asia and migrated into Europe with the spread of agriculture a few thousand years later. Today, it is present at the highest frequencies in Croatia (12 percent), Tunisia (9 percent), and Greece (5 percent), but it is also part of some of the Jewish Diaspora population throughout Europe. Unlike its sister branch, T1, this one migrated farther north, and it accounts for about 7 percent of the population in Bulgaria, while elsewhere in Europe it occurs at a slightly lower frequency—around 6 percent of the population in Germany and about 5 percent in the British Isles, Italy, France, and the Netherlands.

Haplogroup V

This branch of the human tree is both Northern European and North African. Some of the first haplogroup V people weathered out the last ice age in isolated habitable areas, or refugia, one of which was in modern-day Iberia. As the Ice Age ended, those living in refugia moved north to recolonize Europe. Today, this lineage is part of west, central, and north European populations. It occurs in 12 percent of Basques, but less so in other Spanish groups. It is common in Norway (20 percent) and also found throughout central Europe between 5 and 10 percent, and in about 4 percent of people of British ancestry. It also occurs in Algeria and Morocco, suggesting this group also may have migrated south from the Iberian Peninsula.

Paternal Haplogroups: North American Genographic Participants

Haplogroup R1b

The earliest members of this lineage may have lived as hunter-gatherers on the savannas that stretched from East Asia to Europe. Yet the range and numbers likely decreased during the last ice age, rebounding in the subsequent warming periods. Since then, R1b lineages have moved west across Europe, while others moved south to their ancestors' homelands in the Levant and Africa. With the population boom following the Neolithic Revolution 10,000 years ago, this lineage came to dominate Europe. Today it accounts for more than 60 percent of western European male lineages and about 40 percent of central European males. Farther east, it represents 21 percent of eastern European males, 13 percent of Balkan males, and 7 percent of Russian men.

Haplogroup I1

After the last ice age, members of haplogroup I1 expanded from the Black Sea region across Northern Europe and back east toward Asia. Today, the highest frequency of I1 is found in Scandinavia, since it was likely first carried by small founder groups and hunting bands. It accounts for 40 percent of men in Norway and 35 percent in Finland. In Great Britain, it's between 15 and 20 percent of male lineages, while it is 15 percent of German and 10 percent of French paternal lineages. Farther south, it's about 4 percent of males in Spain, 3 percent of Italians, and 2 percent of Greeks. It also occurs in the Balkans, eastern Europe, and the Middle East at low frequencies.

Haplogroup R1a

The earliest members of haplogroup R1a lived on the grassy steppes between Central and South Asia, and were likely nomadic steppe dwellers. Over time, they spread as far east as East Asia, but also west to central Europe where its frequency is second to its brother lineage, R1b. Today, geneticists find R1a in 25 percent of the males in Iceland and between 10 and 20 percent of men in the Ukraine, Russia, and Caucasus countries. In Central Asia, it occurs in 14 percent of the population in Kyrgyzstan, while in South Asia the highest frequency is in Bangladesh (18 percent). Linguists have suggested that carrying R1a may have contributed to the spread of Indo-European languages, which include not only English, French, German, and Russian but also Bengali and Hindi.

Haplogroup E1b1b

This branch of E1b is the most common one among Mediterranean populations. It was born about 22,000 years ago in East Africa. While some members of this group remained in Africa, others spread north to the Levant region, becoming some of the earliest farmers. From there, E1b1b descendants spread by many roads to the Mediterranean and Anatolia in the north and west, as well as south to Sudan. Today, this lineage is common in southeastern European populations such as Albanians (46 percent), Greeks (25–35 percent), and southern Italians (10 percent)—as well as North Africans like Moroccan Arabs (40 percent) and Egyptians (30–40 percent). Smaller branches of E1b1b are found at low frequencies throughout Europe.

Haplogroup J2

The earliest members of haplogroup J2 were nomadic hunter-gatherers who weathered the last ice age in the mild climates of Western Asia. As the Earth warmed, this haplogroup led the expansion from the Middle East spurred by the Neolithic Revolution. It likely thrived in early agricultural societies and expanded into new lands carrying farming with it. Today, people from this lineage live as far east as China (2 percent) and as far west as Italy (10 percent) and Spain (10 percent). Researchers find it in highest frequencies in Bahrain and Iran, where it accounts for more than 33 percent of male lineages. Descendants of this line appear in highest frequencies in the Middle East, North Africa, and East Africa and at a lower frequency in Mediterranean Europe.

Haplogroup G

This branch was born in western Asia some 40,000 years ago. A few branches of haplogroup G survived there in small numbers, weathering out the Ice Age. As the ice melted, members of this lineage who lived in the Fertile Crescent quickly adopted farming, and this transition away from the hunter-gatherer lifestyle led to population booms and rapid cultural changes. The population boom led to the spread of this branch and its subtypes (G1 and G2) north into Turkey, the Caucasus Mountains, the Balkans, and across the Mediterranean. Today, it remains at low frequencies (less than 15 percent) throughout these regions.

ABOUT THE AUTHORS

K. David Harrison is a linguist and anthropologist specializing in endangered languages and co-founder of the Living Tongues Institute for Endangered Languages. He has conducted field research in Siberia, Mongolia, India, Bolivia, Micronesia, the United States, and other places where cultures are threatened by globalization. His book *The Last Speakers: The Quest to Save the World's Most Endangered Languages* explores the consequences of language loss and efforts at revitalization. He lives in Philadelphia and teaches at Swarthmore College.

Catherine Herbert Howell has conducted field research among urban women in India and among Indian immigrants in New York City. A former National Geographic staff member, she has authored a dozen Society publications and has contributed to dozens more, including previous editions of *Peoples of the World, Wonders of the Ancient World,* and the *Expeditions Atlas.* She was also the editor of *Out of Ireland,* a companion volume to the PBS documentary. She lives in Arlington, Virginia.

Spencer Wells is a geneticist, anthropologist, author, and entrepreneur. For more than a decade, he was an explorer-in-residence at the National Geographic Society and director of the Genographic Project, which analyzed DNA samples from hundreds of thousands of people to decipher how our ancestors populated the planet. Wells has appeared in numerous documentary films. His fieldwork has taken him to more than 90 countries, and he is the author of three books: *The Journey of Man, Deep Ancestry,* and *Pandora's Seed.* He lives in Austin, Texas.

Other Contributors

Wade Davis, an explorer-in-residence at the National Geographic Society, is an anthropologist and ethnobotanist whose several books include the best-seller *The Serpent and the Rainbow* and *The Lost Amazon.*

Robert E. Hart trained in linguistics at Swarthmore College and has done fieldwork in China and Papua New Guinea; he holds a Ph.D. in ethnobotany from the University of Missouri, St. Louis.

Elizabeth Kapu'uwailani Lindsey is a filmmaker and anthropologist who was raised by native Hawaiian elders. Her documentary film, *Then There Were None,* chronicles Hawaiian history.

Miranda Weinberg, currently a Ph.D. student in educational linguistics and anthropology at the University of Pennsylvania, has done fieldwork in India and Nepal.

FURTHER RESOURCES

Print

Davis, Wade. *Light at the Edge of the World: A Journey Through the Realm of Vanishing Cultures.* National Geographic Society, 2002.

Davis, Wade. *The Wayfinders: Why Ancient Wisdom Matters in the Modern World.* House of Anansi Press, 2009.

Dolan, Allison, and the editors of *Family Tree Magazine. The Family Tree Guidebook to Europe.* Family Tree Books, 2013.

Gruber, Ruth Ellen. *Jewish Heritage Travel.* National Geographic Society, 2007.

Harrison, K. David. *When Languages Die: The Extinction of the World's Languages and the Erosion of Human Knowledge.* Oxford University Press, 2007.

Harrison, K. David. *The Last Speakers: The Quest to Save the World's Most Endangered Languages.* National Geographic Society, 2010.

Helm, Matthew L., and April Leigh Helm. *Genealogy Online for Dummies,* 7th ed. Wiley, 2014.

Hendrickson, Nancy. *Discover Your Family History Online.* Family Tree Books, 2012.

Indian Nations of North America. National Geographic Society, 2010.

Journeys Home: Inspiring Stories, Plus Tips and Strategies to Find Your Family History. National Geographic Society, 2015.

National Geographic Family Reference Atlas of the World, 4th ed. National Geographic Society, 2015.

Price, David H. *Atlas of World Cultures: A Geographical Guide to Ethnographic Literature.* Blackburn Press, 2004.

Rising, Marsha Hoffman. *The Family Tree Problem Solver: Tried-and-True Tactics for Tracing Elusive Ancestors.* F+W Media, 2011.

Rose, Christine, and Kay Germain Ingalls. *The Complete Idiot's Guide to Genealogy.* Alpha Books, 2012.

Smolenyak, Megan. *Hey, America, Your Roots Are Showing.* Citadel, 2012.

Smolenyak, Megan. *Who Do You Think You Are?* Penguin, 2010.

Sturtevant, William C., ed. *Handbook of North American Indians.* 17 vols. Smithsonian Institution Press, various dates.

Webster, Donovan. *Meeting the Family: One Man's Journey Through His Human Ancestry.* National Geographic Society, 2010.

Wells, Spencer. *Deep Ancestry: Inside the Genographic Project.* National Geographic Society, 2007.

Williams, Heather Andrea. *Help Me to Find My People: The African American Search for Family Lost in Slavery.* University of North Carolina Press, 2012.

Online

Ancestral Attic, www.ancestralattic.com

Ancestry.com, www. ancestry.com

The World Factbook, www.cia.gov

The Ethnologue, www.ethnologue.org

European Focus, www.europeanfocus.com

Family History Library, Salt Lake City, Utah, www.familysearch.org/locations/saltlakecity-library

Family Search, www.familysearch.org

Family Tree Magazine, www.familytreemagazine.com

Global Language Hotspots, www.swarthmore.edu/SocSci/langhotspots/

Hager's Journeys, www.hagersjourneys.com

International Genealogical Society, www.issog.org

Living Tongues, www.livingtongues.org

National Geographic Genographic Project, genographic.nationalgeographic.com

P.A.T.H. Finders, www.pathfinders.cz

U.S. National Archives and Records Administration, www.archives.gov/research/genealogy

ILLUSTRATIONS CREDITS

Cover (top row, L to R), Simone Becchetti/Cultura Images RM/F1online; iStock.com/andresr; Jürgen Müller/imageBROKER/Corbis; Philippe Renault/hemis.fr/Getty Images; Jason Edwards/National Geographic Creative; (bottom row, L to R), William Albert Allard/National Geographic Creative; Patryce Bak/Getty Images; Alex Treadway/National Geographic Creative; Tancredi J. Bavosi/Getty Images; Alison Wright/National Geographic Creative; Back Cover (L to R), Alison Wright/National Geographic Creative; xPacifica/National Geographic Creative; Stephen Zeigler/Getty Images; Joe Petersburger/National Geographic Creative; Cristina Mittermeier/National Geographic Creative; Back Flap (top to bottom), Courtesy Catherine Herbert Howell; David Evans/National Geographic Creative; Jeremy Fahringer; 2-3, Finbarr O'Reilly/Reuters/Corbis; 4, John Dambik/Alamy; 7, Rena Effendi; 9, AFP Photo/Karim Sahib/Getty Images; 10, Tino Soriano/National Geographic Creative; 11, iStock.com/Vichly44; 14, Matthias Oesterle/ZUMA Press/Corbis; 15, iStock.com/cristaltran; 21, Kirsz Marcin/Shutterstock; 22, Evan D. Howell; 24, Per-Anders Pettersson/Getty Images; 25, Chris Johns/National Geographic Creative; 26, Robert Fried/Alamy; 27, Ed Kashi; 28, Sergi Reboredo/Alamy; 30-31, Ed Kashi/National Geographic Creative; 33, blickwinkel/Alamy; 34, Giuseppe Di Rocco/Getty Images; 35, Quick Shot/Shutterstock.com; 36, Carlos Mora/Alamy; 37, Finbarr O'Reilly/Reuters/Corbis; 38-9, Gordon Gahan/National Geographic Creative; 40, Ian Langsdon/epa/Corbis; 41, Charles O. Cecil/Alamy; 42, Paul Almasy/Corbis; 43, Courtesy of Charles Dorsey; 45, Jose Azel/National Geographic Creative; 46 (UP), Per-Anders Pettersson/Corbis; 46 (LO), Michael Nichols/National Geographic Creative; 47, George Mobley/National Geographic Creative; 49, George Steinmetz/National Geographic Creative; 50, iStock.com/FrankvandenBergh; 51, RnDmS/Shutterstock.com; 52-3, George Steinmetz/National Geographic Creative; 55, Courtesy of Xenia Fuller; 56, Matthieu Paley/National Geographic Creative; 57, iStock.com/Marja_Schwartz; 58, Pascal Maitre; 59, Sergio Pitamitz/National Geographic Creative; 60, Egill Bjarnason/Alamy; 61, blickwinkel/Alamy; 63, Kenneth Garrett/National Geographic Creative; 65, Steve McCurry; 66, iStock.com/Bartosz Hadyniak; 68, Maggie Steber/National Geographic Creative; 69, Jan Wlodarczyk/Alamy; 70-71, Kenneth Garrett/National Geographic Creative; 73, Novarc Images/Alamy; 74, Thomas J. Abercrombie/National Geographic Creative; 75, Westend61 GmbH/Alamy; 77, iStock.com/Joel Carillet; 78, iStock.com/fulyaatalay; 79, Courtesy of Fabrizio Scarselli; 80, dinosmichail/Shutterstock.com; 81, ZUMA Press, Inc./Alamy; 82, Morteza Nikoubazl/X01327/Reuters/Corbis; 83, Ian M Butterfield (Syria)/Alamy; 84, Alexandra Avakian; 85, Raul Touzon/National Geographic Creative; 87, Lynn Johnson/National Geographic Creative; 89, K. David Harrison; 91, Stephen Shaver/Getty Images; 92, iStock.com/paulprescott72; 94, Sadik Gulec/Shutterstock.com; 95, George Mobley/National Geographic Creative; 96, Dmitry Kostyukov/AFP/Getty Images; 97, Michael Coyne/National Geographic Creative; 99, Alexandra Avakian; 100, Jonathan Blair/National Geographic Creative; 101, Matthieu Paley/Corbis; 102, Matthieu Paley/National Geographic Creative; 103, timsimages/Alamy; 104, S. Sabawoon/epa/Corbis; 105, Steve McCurry; 107 (UP), Dai Kurokawa/epa/Corbis; 107 (LO), iStock.com/tupianlingang; 108, iStock.com/Grigorev_Vladimir; 109, Li Ga/Xinhua Press/Corbis; 111, iStock.com/Peeter Viisimaa; 112, Jodi Cobb/National Geographic Creative; 113, Courtesy of Aleena Torres; 114-15, iStock.com/Bartosz Hadyniak; 116, Ira Block/National Geographic Creative; 117, Keren Su/China Span/Alamy; 119, Manfred Bortoli/Alamy; 120, Pacific Press/Alamy; 121, Hemis/Alamy; 122, OlegD/Shutterstock.com; 123, Steve McCurry/National Geographic Creative; 124 (UP), Giulio Ercolani/Alamy; 124 (LO), Tjetjep Rustandi/Alamy; 125, Pamela Chen; 126, Tooykrub/Shutterstock.com; 127, Karen Kasmauski/National Geographic Creative; 129, iStock.com/Tarzan9280; 130, AFP/Getty Images; 131, epa european pressphoto agency b.v./Alamy; 132, iStock.com/Paul_Cooper; 133, Courtesy of Junaid Syed; 134, Jim Zuckerman/Alamy; 135, Swetapadma07/www.commons.wikimedia.org/wiki/File:Saura_Lady.JPG/www.creativecommons.org/licenses/by-sa/3.0/legalcode; 136, Kat Kallou/Alamy; 137, iStock.com/vicspacewalker; 139, Robert Harding Picture Library Ltd/Alamy; 140-41, Earl & Nazima Kowall/Corbis; 142, Arne Hodalic/Corbis; 144, Lynn Johnson/National Geographic Creative; 143, ITAR-TASS Photo Agency/Alamy; 145, Dean Conger/National Geographic Creative; 146, David Evans/National Geographic Creative; 147, Karen Kasmauski/National Geographic Creative; 148 (UP), Behrouz Mehri/AFP/Getty Images; 148 (LO), Kelly James Richardson; 149, Karen Kasmauski/National Geographic Creative; 151, Randy Olson/National Geographic Creative; 152, Jodi Cobb/National Geographic Creative; 154, Medford Taylor/National Geographic Creative; 155, Medford Taylor/National Geographic Creative; 156, Mark Kolbe/Getty Images; 157, Annie Griffiths/National Geographic Creative; 158, The AGE/Getty Images; 159, dave stamboulis/Alamy; 160, Ludo Kuipers/Corbis; 161, Philip Game/Alamy; 162-3, The Sydney Morning Herald/Getty Images; 164, John Van Hasselt/Sygma/Corbis; 165 (UP), Simon Fergusson/Getty Images; 165 (LO), The Sydney Morning Herald/Getty Images; 167, Danny Lehman/Corbis; 168, Michael Runkel/Robert Harding World Imagery/Corbis; 169, iStock.com/ArtPhaneuf; 170, Peter Hendrie/Lonely Planet Images/Getty Images; 171, Marc Dozier/Corbis; 172, Torsten Blackwood/AFP/Getty Images; 173, paulista/Shutterstock; 174, Hemis/Alamy; 175, Eric Lafforgue/Alamy; 177, Steve Raymer/National Geographic Creative; 178, Stanislav Fosenbauer/Shutterstock.com; 179, ZUMA Press, Inc./Alamy; 180, Neil Farrin/JAI/Corbis; 181, iStock.com/joeygil; 182, AFP Photo/Greg Wood/Getty Images; 183, Jodi Cobb/National Geographic Creative; 185, WaterFrame/Alamy; 186, Greg Vaughn;

187, Jon Arnold Images Ltd/Alamy; 188, Bob Krist/Corbis; 189, Robert Harding Picture Library Ltd/Alamy; 190, Nicholas DeVore III/National Geographic Creative; 191, David Boyer/National Geographic Creative; 192, Anders Ryman/Corbis; 193, Alvis Upitis/Getty Images; 195, iStock.com/Madzia71; 196, Atlantide Phototravel/Corbis; 198, RIA Novosti/Alamy; 199, Steve Raymer/National Geographic Creative; 200, Italianvideophotoagency/Shutterstock.com; 201, Joe McNally; 202, iStock.com/salajean; 204-205, iStock.com/UygarGeographic; 207, Anthony Suau; 209, Dado Ruvic/Reuters/Corbis; 210, iStock.com/paulprescott72; 211, Lukas Gojda/Shutterstock.com; 212, iStock.com/ProjectB; 213, iStock.com/Madzia71; 215, Courtesy of Roberto Pasqualetti; 216, Koca Sulejmanovic/epa/Corbis; 217, Peter Josek/Reuters/Corbis; 219, Srdjan Zivulovic/Reuters/Corbis; 221, iStock.com/DavorLovincic; 222, De Visu/Shutterstock.com; 224-5, Tino Soriano/National Geographic Creative; 227, Basque Country—Mark Baynes/Alamy; 228, WorldPictures/Shutterstock.com; 229, iStock.com/jarcosa; 230, iStock.com/TonyTaylorStock; 231, Cloud Mine Amsterdam/Shutterstock.com; 232-3, Catherine Karnow/National Geographic Creative; 235, Gerd Ludwig/National Geographic Creative; 236, Sean Pavone/Shutterstock.com; 237, David Evans/National Geographic Creative; 239, Jim Richardson/National Geographic Creative; 241, geogphoto/Alamy; 242, iStock.com/EdStock; 243, Catherine Karnow; 244, Morris Mac Matzen/Reuters/Corbis; 245, dpa picture alliance/Alamy; 246, Tierfotoagentur/Alamy; 247, Atlantide Phototravel/Corbis; 248, Chris Rainier; 249, Alison Wright/National Geographic Creative; 250, Ton Koene/VWPics/Alamy; 251, Courtesy of Holly Morse; 253, Jim Richardson/National Geographic Creative; 254, Raymond Patrick/National Geographic Creative; 255, Jim Richardson/National Geographic Creative; 257, Christian Bertrand/Shutterstock.com; 258, Anders Ryman/Corbis; 259, Hamid Sardar/Corbis; 261, Richard Ellis/Corbis; 262, Mark Skalny/Shutterstock.com; 264-5, Michael Melford/National Geographic Creative; 266, iStock.com/alazor; 267, Natalie Fobes; 269, David Coventry; 271 (UP), Lynn Johnson/National Geographic Creative; 271 (LO), Design Pics Inc/Alamy; 273 (UP), Christian Heeb/JAI/Corbis/Corbis; 273 (LO), NPS Photo/Alamy; 274, Maggie Steber/National Geographic Creative; 275, Marc Dozier/Corbis; 276, inga spence/Alamy; 277, PK Weis/SouthWestPhotoBank.com; 279, Lex van Doorn/Alamy; 280, Maggie Steber/National Geographic Creative; 281, allen russell/Alamy; 282, Aaron Huey/National Geographic Creative; 283, Phil Schermeister/National Geographic Creative; 285, David McLain; 286, Peter Turnley/Corbis; 287, Nativestock.com/Angel Wynn; 289, Skip Brown/National Geographic Creative; 290, glenda/Shutterstock.com; 291, Andy Clark/Reuters/Corbis; 292, Maggie Steber; 293, Courtesy of Ricardo Hernandez-Bonifacio; 294, Xinhua/Alamy; 295, Philip Scalia/Alamy; 297, Philip Gould/Corbis; 298, Thomas Nebbia/National Geographic Creative; 299, Education Images/Getty Images; 300, Jeff Greenberg 6 of 6/Alamy; 301, Tribune Content Agency LLC/Alamy; 303, AP Photo/David Goldman; 305, Prakash Singh/AFP/Getty Images; 306, Viviane Moos/Corbis; 307, Jeffrey L. Rotman/Corbis; 309, Marka/Alamy Stock Photo; 310, Tricia Daniel/Shutterstock.com; 312, Hector Retamal/Getty Images; 313, age fotostock/Alamy; 314, Andrea De Silva/Reuters/Corbis; 315, Thomas Coex/AFP/Getty Images; 316, iStock.com/Claudiad; 317, Ian Cumming/Design Pics/Corbis; 319, Thomas Koehler/Getty Images; 321, iStock.com/Wildroze; 322, iStock.com/folewu; 323, Feije Riemersma/Alamy; 324, XPacifica/National Geographic Creative; 325, Kenneth Garrett/National Geographic Creative; 326, Drake Sprague; 327, Maunus//www.commons.wikimedia.org/wiki/File:Nahua_man_of_Morelos.jpg//creative
commons.org/licenses/by-sa/3.0/legalcode; 328 (UP), Rodrigo Oropeza/Xinhua Press/Corbis; 328 (LO), Peter Langer/Design Pics/Corbis; 329, Richard Ellis/ZUMA Press/Corbis; 331, Katherine Needles; 332, Valter Campanato/ABr//commons.wikimedia.org/wiki/File:Bororo004.jpg//creativecommons.org/licenses/by/3.0/br/legalcode; 333, David Alan Harvey; 334, Jad Davenport/National Geographic Creative; 335, Courtesy Aisye Miraj Elevli; 336, David Lazar; 337, Paulo Whitaker/Reuters/Corbis; 338, Gerd Ludwig/National Geographic Creative; 339, Danny Lehman/Corbis; 340, Guillermo Legaria/Getty Images; 341, Robert Caputo/Aurora Photos; 342, Rodrigo Buendia/AFP/Getty Images; 343, Paulo Whitaker/Reuters/Corbis; 344, Robert Caputo/Aurora Photos; 345, ZUMA Press, Inc./Alamy; 347, Michael Nichols/National Geographic Creative; 349, iStock.com/urosr; 350, Hemis/Alamy; 351, Stuart Franklin; 352, iStock.com/Tituz; 353, Susie Post Rust; 354, Charton Franck/Hemis/Corbis; 355, iStock.com/MarkSkalny; 357, Fernando Bizerra Jr./EFE/Corbis; 358 (UP), Robert Harding Picture Library Ltd/Alamy; 358 (LO), imageBROKER/Alamy; 359, Mario Tama/Getty Images; 361, Kobby Dagan/Shutterstock.com; 362-3, Tomas Munita/National Geographic Creative; 365, Mohd Khairil Majid/Shutterstock.com; 366, Yvette Cardozo/Getty Images; 367, Aldo Pavan/SIME; 369, Steve Starr/Corbis; 371, Courtesy County Mayo and the Genographic Project.

INDEX

Boldface indicates illustrations.

D

E

F

G

PEOPLE OF THE WORLD

Catherine Herbert Howell
With K. David Harrison

PREPARED BY THE BOOK DIVISION

Hector Sierra, *Senior Vice President and General Manager*
Lisa Thomas, *Senior Vice President and Editorial Director*
Melissa Farris, *Creative Director*
Susan Tyler Hitchcock, *Senior Editor*
R. Gary Colbert, *Production Director*
Jennifer A. Thornton, *Director of Managing Editorial*
Susan S. Blair, *Director of Photography*
Meredith C. Wilcox, *Director, Administration and Rights Clearance*

STAFF FOR THIS BOOK

Barbara Payne, *Editor*
Catherine Herbert Howell, *Developmental Editor*
Elisa Gibson, *Art Director*
Charles Kogod, *Photo Editor*
Linda Makarov, *Designer*
Patrick Bagley, *Assistant Illustrations Editor*
Gregory Ugiansky, *Map Production*
Miguel Vilar, *Scientific Manager, National Geographic Genographic Project*
Jason Blue-Smith, *Contributing Writer*
Marshall Kiker, *Associate Managing Editor*
Judith Klein, *Senior Production Editor*
Lisa A. Walker, *Production Manager*
Rock Wheeler, *Rights Clearance Specialist*
Katie Olsen, *Design Production Specialist*
Nicole Miller, *Design Production Assistant*
Darrick McRae, *Manager, Production Services*

Since 1888, the National Geographic Society has funded more than 12,000 research, exploration, and preservation projects around the world. National Geographic Partners distributes a portion of the funds it receives from your purchase to National Geographic Society to support programs including the conservation of animals and their habitats.

National Geographic Partners
1145 17th Street NW
Washington, DC 20036-4688 USA

ISBN: 978-1-4262-1708-1

When humans first ventured out of Africa some 60,000 years ago, they left genetic footprints still visible today. By mapping the appearance and frequency of genetic markers in modern peoples, we create a picture of when and where ancient humans moved around the world. Please visit genographic.com to take part in learning about your own migratory story.

Printed in Hong Kong

16/THK/1